Design of Breast Walls

Design of Breast Walls

A Practical Solution Approach

Rajendra Chalisgaonkar

CRC Press
Taylor & Francis Group
Boca Raton   London   New York

CRC Press is an imprint of the
Taylor & Francis Group, an **informa** business

First edition published 2022
by CRC Press
6000 Broken Sound Parkway NW, Suite 300
Boca Raton, FL 33487–2742

and by CRC Press
2 Park Square, Milton Park, Abingdon, Oxon, OX14 4RN

© 2022 Taylor & Francis Group, LLC

CRC Press is an imprint of Taylor & Francis Group, LLC

ISBN: 978-0-367-75471-6 (hbk)
ISBN: 978-0-367-75568-3 (pbk)
ISBN: 978-1-003-16299-5 (ebk)

DOI: 10.1201/9781003162995

Typeset in Times
by Apex CoVantage, LLC

To
My Family

Contents

Preface

The design of breast walls is not an everyday design task, although they are commonly built in every nook and corner of the country for various earth-retaining purposes. During my 40 years of experience, I observed that many problems are encountered in the field as a result of the improper design of breast walls. These problems arise mainly due to the fact that engineers in the field are bogged down with numerous responsibilities and often do not find the time to carry out proper design. In my opinion, field and practicing engineers need some brushing up, particularly on some parameters that influence the technoeconomic designs.

Because a breast wall is a very simple structure, not much technical information about the design is available. This book is aimed at providing guidance to designers/field engineers, and standard designs are also provided that can be followed for normal soil conditions. However, for all important projects, site-specific designs should be carried out, and the standard designs should only be used as a reference.

An attempt has been made to condense, simplify and compile information from many sources, including my own experience, into this book. The book is divided into eight chapters dealing with the basics of earth pressure theories, parameters influencing earth pressures, gravity vis-à-vis breast walls and tables and charts for designing stone masonry and concrete breast walls. The analysis details are given in the tables so that the engineer or designer knows how, where and when to use them in their professional work.

Hopefully, it will ease the comfort level when designing breast walls and will give a good overview of the process and will become a helpful guideline that will benefit engineers/architects engaged in highway, geotechnical and water resources sectors and infrastructure development.

The author owes an enormous debt of gratitude to his friend Mr. Mukesh Mohan for reviewing the manuscript and providing constructive comments. The author is also thankful to other friends for their suggestions and encouragement. Last but not the least, the author is thankful to his wife, Deepa; children, Rahul-Madhurima and Rohan-Nandita; grandchild, Reyansh for their tolerance during the lockdown period.

Efforts have been made to keep errors to a minimum. However, they are inevitable. Suggestions are welcome at chalisgaonkar@gmail.com from readers to enhance the usefulness of the publication.

Rajendra Chalisgaonkar
Roorkee, India

Acknowledgements

I acknowledge the many engineers and colleagues who constantly challenged me for answers during services, and the book has evolved from such discussions. In the busy times we live in, there are many good intentions but there was not enough time to fulfill those intentions. Several competent friends were asked to review this book and had such good intentions, but the constraints of ongoing work commitments and balancing family life are understood. I acknowledge their contributions.

The author wishes to express his grateful thanks and acknowledgements to the following:

1. The Bureau of Indian Standards for according permission to include extracts from a number of relevant Indian Standard Codes of Practice.
2. The Indian Road Congress for extracts from a number of relevant IRC Codes (Copy Right with Indian Road Congress).
3. The Research Designs & Standards Organisation for according permission to include para 5.7.1 of IRS Code of Practice for Design of Substructure and Foundation of Bridges.
4. The authors and publishers of various books and technical papers, referred to in the appropriate places.
5. Website owners and others for giving permission for the use of illustrations.

The author thanks the publishers for bringing the book to fruition.

Author

Rajendra Chalisgaonkar, former Engineer-in-Chief, Irrigation Department, Govt. of Uttarakhand, India, graduated from Madhav Institute of Technology & Science, Gwalior, India, in 1978 in Civil Engineering and then obtained an M.E. (Structural Engineering) in 1982 from the University of Roorkee, now IIT Roorkee. He was also Managing Director, Uttarakhand Project Development and Construction Corporation Ltd., a government of Uttarakhand enterprise. He was responsible for the development of irrigation, policy making at the state level, planning, management and construction of water resources and infrastructure projects. He has more than 40 years of varied experience, which includes investigation, planning, design, research and CAD applications, construction and has also worked on prestigious projects of Uttar Pradesh and Uttarakhand in India. He has authored a book on *Computer-Aided Concrete Mix Design* and more than 125 research papers in various journals and symposia. After superannuation, he was engaged as Adjunct Professor at MITS, Gwalior, Water Resources & Hydropower Consultant to IIT Roorkee, DMR Hydroengineering & Infrastructures Ltd., DHI, India, Water & Environment Pvt. Ltd. and other organizations. He is the recipient of the following prestigious awards: CBIP-Young Engineers Award, IEI-HYDROSEM Award, IEI-Eminent Engineers Award, IEI-Engineering Excellence in Field Award, CBIP-D.B. Anand Medal, Gwalior Ratna, among other honours.

Symbols

The notations have been explained, wherever they appear. The following notations are more commonly used in the book:

α   Angle which the earth face of the wall makes with the vertical

δ   Angle of friction between the masonry/concrete wall and the backfill

φ   Angle of repose of the backfill

ι   Angle of the surcharge of the backfill

μ   Coefficient of friction between the masonry/concrete and the soil/rock mass

θ   Horizontal angle of the earth face of the wall

γ_{sat} Unit weight of saturated earthfill

γ_{sub} Unit weight of submerged earthfill

γ' Unit weight of the earthfill in front of wall

$\gamma_{concrete}$ Unit weight of concrete

$\gamma_{masonry}$ Unit weight of random rubble masonry

γ_{water} Unit weight of water

B   Base width of a breast wall

D_{sk} Depth of the shear key below the base of the wall

K_a Coefficient of the active earth pressure

K_{ah} Horizontal component of the coefficient of the active earth pressure

K_p Coefficient of the passive earth pressure

F_{SLD} Factor of safety against sliding

F_{OVT} Factor of safety against overturning

H   Height of a breast wall

H'/h_1 Height of a fill in front of the breast wall

h_2 Depth of the shear key below the natural ground level

P_a Active earth pressure along the length of the wall

P_p Passive earth pressure along the length of the wall

T_w Top width of a breast wall

X_{sk} Location of the shear key from the toe

1 Introduction

1.1 EVOLUTION OF RETAINING WALLS

Retaining walls are structures used to retain soil and to resist the lateral pressure of the soil against the wall. The retaining walls are normally not intended to stabilize slope failures. They are mainly meant to support the active or passive earth pressure from the assumed failure wedge above the base of the wall.

In 1 million BC, or thereabouts, an anonymous man, or woman, laid a row of stones atop another row to keep soil from sliding into their camp. Thus was constructed an early retaining wall, and we've been keeping soil in place ever since . . . with increasingly better methods and understanding.

The early engineers in the ancient cultures of Egypt, Greece, Rome and the Mayans were masters at invention and experimentation, learning primarily through intuition and trial and error. Even the most casual observer looks in wonder at the magnificent structures they created and that have stood for thousands of years— including countless retaining walls. With great skill, they cut, shaped and set stone with such precision that the joints were paper thin. Reinforced concrete would not be developed for 1,000 years, but they used what they had and learned how to do it better with each succeeding structure. Consider the Great Wall of China, for example, where transverse bamboo poles were used to tie the walls together—a forerunner of today's *mechanically stabilized earth*. Those early engineers also discovered that by battering a wall so that it leaned slightly backwards, the lateral pressure was relieved and the height could be extended—an intuitive understanding of the soil wedge theory.

Major advances in understanding how retaining walls work and how soil generates forces against walls appeared in the 18th and 19th centuries with the work of French engineer Charles Coulomb (1776), who is better remembered for his work on electricity, and later by William Rankine in 1857. Today, their equations are familiar to most civil engineers. A significant body of work was the introduction of soil mechanics as a science through the pioneering work of Karl Terzaghi in the 1920s.

Indeed, soil mechanics and the design of retaining structures have advanced dramatically in recent decades, giving us new design concepts, a better understanding of soil behaviour and hopefully safer and more economical designs.

1.2 RETAINING WALLS

A retaining wall is a structure constructed and designed to resist the lateral pressure of the soil. Apart from the soil, the lateral pressure is caused by the pressure of liquid, earthfilling, sand or other granular material filled behind the wall after it's constructed. These walls are key components of highway and water resources,

DOI: 10.1201/9781003162995-1

infrastructure and other projects and are commonly employed in the construction of hill roads. The type of material used for the construction of the wall depends on the site conditions, the type of material to be retained and the height of the wall to be constructed.

A retaining wall is defined as a structure whose primary purpose is to provide lateral support for the soil or rock. Granular soils (sand or gravels) are the standard recommendation for backfill material, and there are several reasons for this recommendation.

1.2.1 PREDICTABLE BEHAVIOUR

Granular backfill generally has a more predictable behaviour in terms of earth pressure exerted on the wall. If silt or clay is used as backfill material, expansive soil-related forces can be generated by these soil types.

1.2.2 DRAINAGE SYSTEM

To prevent the build-up of hydrostatic water pressure on the retaining wall, a drainage system is often constructed at the heel of the wall. This system will be more effective if highly permeable granular soil is used as backfill.

1.2.3 FROST ACTION

In cold climates, the formation of ice lenses in the backfill soil can cause so much lateral movement that the retaining wall will become unstable. Backfill soil consisting of granular soil and the installation of a drainage system at the heel of the wall will help protect the wall from frost action.

1.3 TYPES OF RETAINING WALLS

Basically, retaining walls are classified into four types, as shown in Figure 1.1.

1.3.1 GRAVITY RETAINING WALLS

Gravity retaining wall depends on its own weight only to resist lateral earth pressure. Generally, a gravity retaining wall is massive because it requires a significant gravity load to counteract the horizontal soil pressure. When designing a gravity retaining wall, sliding, overturning and bearing forces should be taken into consideration. Chalisgaonkar (1987) carried out a computer-aided analysis of gravity retaining walls. It can be constructed from different materials such as stone, concrete and masonry units (Figure 1.2). Crib retaining walls and gabion retaining walls are the types of gravity retaining walls being used more frequently nowadays.

Breast walls are also gravity-type walls, generally used to protect the slopes for highway and riverbank toe-erosion projects, especially in the hilly areas (Figure 1.3).

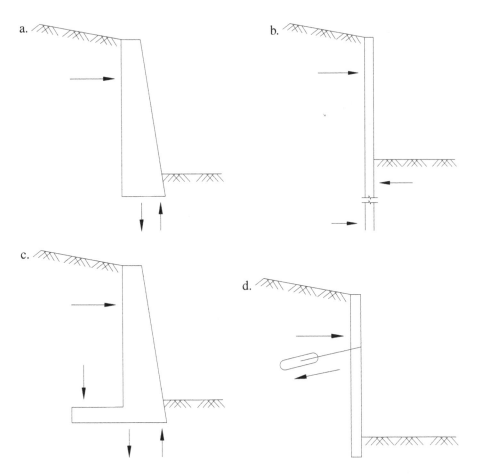

FIGURE 1.1   Types of Retaining Walls: (a) Gravity Wall, (b) Sheet Pile Wall, (c) Cantilever Wall, (d) Anchored Wall

Source: www.aboutcivil.org/retaining-wall-definition-types-uses-retaining-walls.html.

1.3.1.1 Semi-Gravity Retaining Walls

Semi-gravity retaining walls are a specialized form of gravity walls. These types of retaining walls have some tension-reinforcing steel included to minimize the thickness of the wall without requiring extensive reinforcement. They are a blend of the cantilever wall and gravity wall designs. These walls are often constructed of reinforced concrete, unreinforced concrete or stone masonry. Rigid gravity walls develop their soil-retaining capacity from their deadweights. Semi-gravity walls, such as cast-in-place concrete cantilever walls, develop resistance to overturning and sliding from self-weight and the weight of the soil above the wall footing.

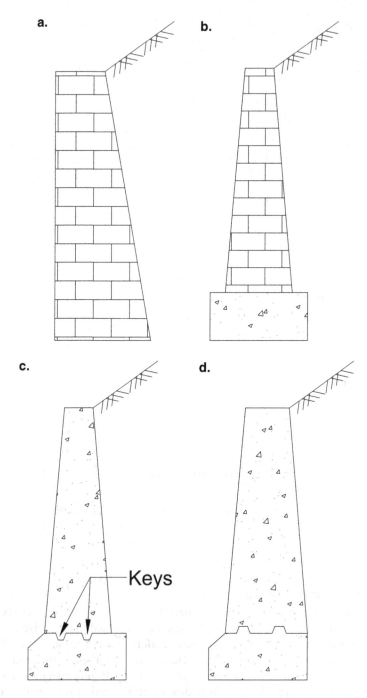

FIGURE 1.2 Types of Gravity Retaining Walls: (a) Stone Wall, (b) Brick Wall, (c) Plain Concrete Wall with Shear Key, (d) Concrete Wall with Rough Surface

Source: www.globalspec.com/reference/61891/203279/chapter-11-retaining-walls.

FIGURE 1.3 Breast Wall in Hilly Areas

1.3.1.2 Prefabricated Modular Gravity Walls

These walls include crib walls, bin walls, and gabion walls (Figure 1.4). A crib wall, concrete or timber, is a gravity retaining structure constructed by interlocking individual boxes made from precast concrete or timber. The boxes are filled with crushed stone or other coarse granular materials to form a free-draining structure. A bin wall, concrete or

a. Crib Wall
Source: www.vectorex.co.uk/.

b. Bin Wall
Source: https://precast.org/wp-content/uploads/2015/02/
Precast-concrete-gravity-wall-retaining-wall-install-3.1.jpg.

c. Gabion Wall

FIGURE 1.4   Prefabricated Modular Gravity Walls

metal, is constructed of adjoining closed-face or open-face bins. Each bin unit is filled with compacted granular soil. Gabion walls consist of baskets made of galvanized steel mesh or PVC-coated wire mesh. Gabion walls are multi-celled rectangular wire-mesh boxes which are filled with rocks or other suitable materials. It is constructed for the construction of erosion-control structures. Gabion retaining walls are also used to stabilize steep slopes.

1.3.2 PILE RETAINING WALLS

Sheet pile walls (Figure 1.5a) are built using steel sheets into a slope or excavations up to a required depth, but they cannot resist very high pressure. Sheet pile retaining walls are economical up to a height of 6 m. Reinforced concrete pile retaining walls are constructed by driving reinforced cement concrete (R.C.C.) piles near each other (Figure 1.5b). Piles are forced into a depth satisfactory for countering the force that tries to push over the wall.

a. Sheet Pile Wall
Source: https://3.bp.blogspot.com/-Qzxvm4J-MC4/UK1AEj9v0HI/AAAAAA-AABSc/GfpglwDVWgM/s1600/design-1.jpg.

b. R.C.C. Pile Wall
Source: https://www.underpin.com/wp-content/uploads/2015/12/beech-House.jpg.

FIGURE 1.5 Pile Retaining Walls

1.3.3 CANTILEVER RETAINING WALLS

These walls (Figure 1.6) are constructed of reinforced concrete. This wall consists of a thin stem and base slab. The base of this retaining wall is divided into two parts, namely the heel and the toe. The heel is a part of the base under the backfill. This wall uses much less concrete than gravity retaining walls, but it needs careful construction and design. Chalisgaonkar (1993) had developed the design charts for the design of R.C.C. cantilever walls. These walls can either be precast in a factory or formed on-site.

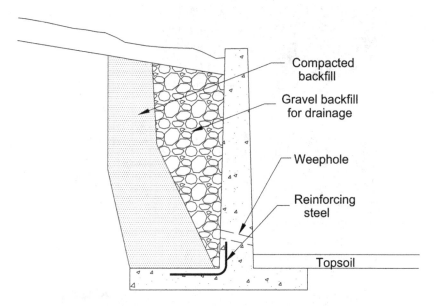

FIGURE 1.6 Cantilever Retaining Wall

1.3.3.1 Non-Gravity Cantilevered Walls

These types of retaining walls develop lateral resistance through the embedment of vertical wall elements and support retained soil with wall-facing elements. Vertical wall elements are normally extended deep into the ground to provide lateral and vertical support. The vertical wall elements can be piles, drilled shafts, and steel sheet piles, among others. Wall faces can be reinforced concrete, metal, or timber.

1.3.4 ANCHORED RETAINING WALLS

For high retaining walls, deep cable rods or wires are driven deep sideways into the earth; then the ends are filled with concrete to provide an *anchor*. These are also known as tiebacks. These types of retaining walls (Figure 1.7) typically consist of the same elements as the non-gravity cantilevered walls but derive additional lateral resistance from one or more tiers of anchors. They are constructed when a thinner retaining wall is needed or space is limited to install other types of retaining walls.

FIGURE 1.7 Anchored Retaining Wall

Source: https://civildigital.com/wp-content/uploads/2018/01/Anchored-retaining-wall.jpg.

They are very effective for loose soils over solid rocks. Anchored retaining walls are usually employed in lots of highways construction projects, where they are used to keep rocks from falling on the roads by accidents. Other types of retaining walls are explained in the following sections.

1.3.5 COUNTERFORT RETAINING WALLS

Counterfort retaining walls (Figure 1.8) are the same as cantilever walls except they have thin vertical concrete webs at regular intervals along the backside of the wall, and these webs are known as counterforts. The counterforts tie the slab and base together, and the purpose of them is to reduce the bending moments and shear forces imposed on the wall by the soil. The secondary effect of these webs is to increase the weight of the wall from the added concrete. It can be precast or formed on site.

1.3.6 MECHANICALLY STABILIZED EARTH (MSE) WALLS

These types of retaining walls normally include a facing element and a reinforcement element embedded in the backfill behind the facing. The facing element can be concrete, segmental block or panel, or steel wire mesh. The reinforcement element can be either geo-synthetic (geotextile, geo-grid) or metallic (strip, grid, wire mesh). These types of retaining walls (Figure 1.9) are often used in constructing the approach roads to bridges in India.

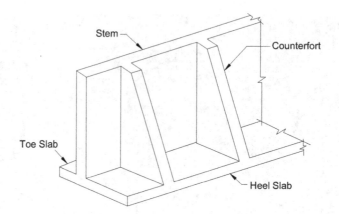

FIGURE 1.8 Counterfort Retaining Wall

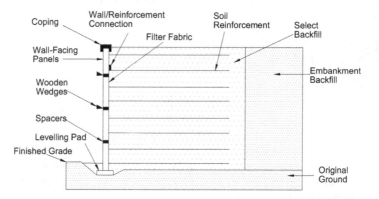

FIGURE 1.9 MSE Walls

Source: www.garwarefibres.com/.

1.3.7 RETAINING WALL WITH A PRESSURE RELIEF SHELF

A retaining wall with a pressure relief shelves is one of the special types of retaining wall. High reinforced concrete retaining walls may be used economically by providing relief shelves on the backfill side of the wall. Such walls may be termed as retaining walls with pressure relief shelf (Figure 1.10) (Donkada and Menon, 2012).

According to Jumikis (1964), the provision of one or more relief shelves and extending them to the rupture surface can considerably increase the stability of the retaining wall. The relief shelves have the advantage of decreasing the overall lateral earth pressure on the wall and increasing the stability of the structure. This results in an economical design because less material goes into the wall compared to the massive structure of cantilever or even counterfort retaining walls without shelves (Padhye and Ulligaddi, 2011).

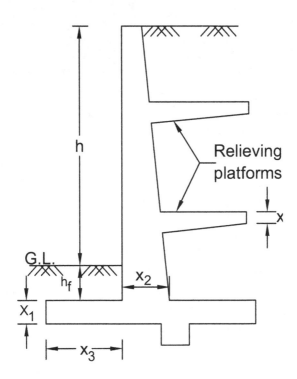

FIGURE 1.10   Retaining Wall with a Pressure Relief Shelf

1.3.8 HYBRID SYSTEMS

Retaining walls that use both mass and reinforcement for stability are termed composite retaining walls or hybrid systems (Figure 1.11).

FIGURE 1.11 Hybrid Walls

Source: http://www.constructionenquirer.com/wp-content/uploads/JP-Concrete%E2%80%99s-new-gabion-baskets-will-hold-their-shape-longer-600x414.jpg.

2 Revisiting Earth Pressure Theories

2.1 LATERAL EARTH PRESSURES

The soil pressure is the lateral pressure exerted on the soil retaining structure by the soil behind the retaining structure. In the retaining wall design, the soil behind the wall is also known as the backfill. There are three types of lateral earth pressures based on the direction of the retaining wall movement:

1. *At-rest earth pressure*, when the retaining wall stands vertically and does not move either away from or into the backfill
2. *Active earth pressure*, when the retaining wall moves away from the backfill, that is the backfill actively pushes the wall
3. *Passive earth pressure*, when the retaining wall moves into the backfill, that is the backfill is passively pushed. The rotation of the wall into the backfill can be caused by external lateral forces on the wall.

The three types of lateral soil pressures are illustrated in Figure 2.1. The determination of lateral soil pressures is based on the limit equilibrium principle: the backfill is at the onset of failure. So the active and passive earth pressures are defined as the pressures when a failure shear surface is developed in the backfill (Figures 2.1b and 2.1c), and the failure occurs only when there is a sufficient wall movement. Definitions of lateral earth pressures, based on the maximum wall movement until failure, are depicted in Figure 2.2.

2.2 EARTH PRESSURE THEORIES

The main force acting on the retaining wall is constituted by lateral earth pressure which tends to bend, slide and overturn it. The classical earth pressure theories, which are followed in estimating the earth pressures, were published by Coulomb in 1776 and Rankine in 1857. Developments since 1920, largely due to the influence of Dr. Karl Terzaghi, have led to a better understanding of how retaining walls work and how soil generates forces against walls. Most lateral pressure theories are based on the sliding soil wedge theory. This, in simple terms, is based upon the assumption that if the wall is suddenly removed, a triangular wedge of soil will slide down along a rupture plane, and it is this wedge of soil that the wall must retain. The basis for determining the magnitude and direction of the earth pressure is the principles of soil mechanics. The behaviour of lateral earth pressure is similar to that of a fluid, with its

DOI: 10.1201/9781003162995-2

a.

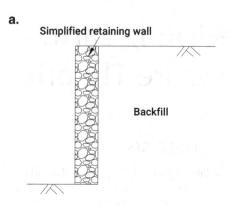

b.

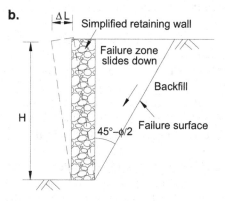

c.

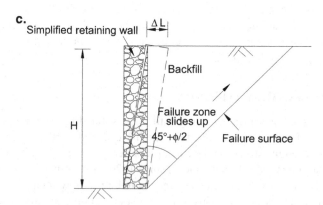

FIGURE 2.1 Lateral Earth Pressures: (a) Earth Pressure at Rest, (b) Active Earth Pressure and (c) Passive Earth Pressure

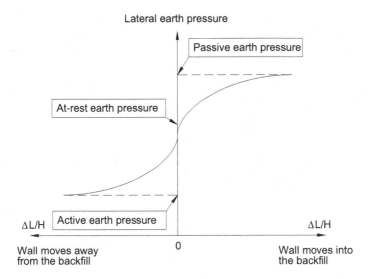

FIGURE 2.2 Definitions of Lateral Earth Pressures

magnitude pressure increasing nearly linearly with increasing depth, h, for moderate depths below the surface:

$$P = K\gamma h \qquad (2.1)$$

where
   γ is the unit weight of the backfill;
   K is the coefficient that depends on its physical properties and whether the pressure is active or passive:
      Active Earth Pressure Coefficient K_a, in the case of active pressure, and
      Passive Earth Pressure Coefficient K_p, in the case of passive pressure; and
   H is the height of the wall.

2.2.1 COULOMB'S THEORY

Coulomb (1776) developed a method for the determination of the earth pressure considering the equilibrium of the sliding wedge formed due to movement of the retaining wall. In the active earth pressure case, the sliding wedge moves downwards and outwards on a slip surface relative to the intact backfill, and in the case of passive earth pressure, the sliding wedge moves upwards and inwards. The pressure on the wall is, in fact, a force of reaction which it has to exert to keep the sliding wedge in equilibrium. The lateral pressure on the wall is equal and opposite to the reactive force exerted by the wall in order to keep the sliding wedge in equilibrium. Assumptions made in Coulomb's theory include the following:

1. The backfill is cohesionless, dry, homogeneous, isotropic and ideally plastic material.

2. The slip surface is a plane surface that passes through the heel of the wall.
3. The sliding wedge behaves like a rigid body, and the magnitude of the earth pressure is obtained by considering the equilibrium of the wedge as a whole.
4. The back of the wall is rough.
5. The position and direction of the resultant earth pressure are known. It acts at a distance one-third the height of the wall above the base and is inclined at an angle, δ, to the normal to the back of wall, where δ is the angle of friction between the wall and the backfill.

The general conditions for the development of active and passive earth pressures encountered in the design of retaining walls are illustrated in Figures 2.3 and 2.4.

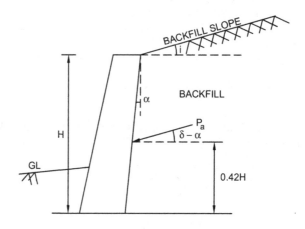

FIGURE 2.3   Active Earth Pressure

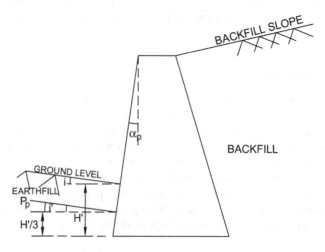

FIGURE 2.4   Passive Earth Pressure

2.2.1.1 Active Pressure Due to the Backfill

The active pressure exerted against the wall should be

$$P_a = \frac{1}{2}\gamma H^2 K_a \qquad (2.2)$$

where

P_a is the active earth pressure acting at an angle $(\alpha + \delta)$ with horizontal
γ is the unit weight of the backfill,
H is the height of the wall,
K_a is the coefficient of active pressure, which takes into account backfill slope, the friction angle at the wall face, the angle of repose of the backfill and the angle of the wall face with the vertical:

$$K_a = \frac{cos^2(\varphi - \alpha)}{cos^2\alpha . cos(\delta + \alpha)\left[1 + \sqrt{\dfrac{sin(\varphi + \delta)sin(\varphi - i)}{cos(\alpha - i)cos(\delta + \alpha)}}\right]^2} \qquad (2.3)$$

where

ϕ is the angle of internal friction of the soil;
δ is the angle of friction between the wall and the backfill. Where the value of δ is not determined by actual tests, the following values may be assumed:

1. $\delta = 1/3\phi$ for concrete structures
2. $\delta = 2/3\phi$ for masonry structures

i is the angle that the earth surface makes with the horizontal behind the earth retaining structure; and
α is the angle the earth face of the wall makes with the vertical.

If the backfill is levelled, the earth face of the wall is vertical, and if zero friction is assumed between the soil and the wall, then the Coulomb equation reduces to the familiar Rankine's equation:

$$K_a = \frac{1 - sin\varphi}{1 + sin\varphi}. \qquad (2.4)$$

IRC:6 (2017) specifies that Coulomb's theory is acceptable, subject to the modification that the centre of pressure exerted by the backfill, when considered dry, is located at an elevation of 0.42 of the height of the wall above the base instead of 0.33 of that height, mentioned in IS:1893.

2.2.1.2 Passive Pressure Due to the Earthfill

The passive pressure exerted against the wall should be

$$P_p = \frac{1}{2}\gamma' H'^2 K_p \qquad (2.5)$$

where

P_p is the passive earth pressure acting at an angle ($\alpha + \delta$) with the horizontal,

γ' is the unit weight of the earthfill,

H' is the height of the earthfill in front of the wall,

K_p is the coefficient of passive pressure, which takes into account the earthfill slope in front of the wall, the friction angle at the wall face, the angle of repose of the earthfill, and the angle of wall face with the vertical:

$$K_p = \frac{\cos^2\left(\varphi + \alpha\right)}{\cos^2\alpha.\cos(\delta - \alpha)\left[1 - \sqrt{\dfrac{\sin(\varphi + \delta)\sin(\varphi + i)}{\cos(\alpha - i)\cos(\delta - \alpha)}}\right]^2} . \qquad (2.6)$$

If the earthfill in front of the wall is levelled, the outer face of the wall is vertical, and zero friction is assumed between the soil and the wall, then the Coulomb equation reduces to the familiar Rankine's equation:

$$K_p = \frac{1 + \sin\varphi}{1 - \sin\varphi} . \qquad (2.7)$$

2.2.2 RANKINE'S THEORY

Rankine (1857) considered the equilibrium of a soil element at any depth in the backfill behind a retaining wall and determined the active earth pressure. The Rankine equation is a simplified version of the Coulomb equation that does not take into account wall batter or friction at the wall–soil interface. As such, it is a conservative approach to the design of retaining walls. In the case of vertical walls with a levelled backfill and zero wall friction, the lateral pressure factor, K_a, will be the same by either approach. Following assumptions were made in the originally proposed Rankine's theory:

1. The soil mass is homogeneous and semi-infinite.
2. The soil mass is cohesionless and dry.
3. The surface of the soil is a plane which may be horizontal or inclined.
4. The back of the wall is vertical.
5. The back of the wall is smooth so that there will be no shearing stresses between the wall and soil. Because of this assumption, the stress relationship for any element adjacent to the wall is the same as that for any other element far away from the wall.
6. The wall yields about the base and thus satisfies the deformation condition for plastic equilibrium.

2.2.2.1 Active Pressure Due to the Backfill

The active pressure exerted against the wall should be

$$P_a = \frac{1}{2}\gamma H^2 K_a \qquad (2.8)$$

where
  K_a is the Rankine's coefficient of active pressure acting at an angle, ι, with the horizontal,

$$K_a = cos i. \frac{cos i - \sqrt{cos^2 i - cos^2 \phi}}{cos i + \sqrt{cos^2 i - cos^2 \phi}} \qquad (2.9)$$

where
  ϕ is the angle of repose of the backfill and
  ι is the surcharge angle of the backfill slope.

If the backfill is levelled, the Rankine equation can be written as

$$K_a = \frac{1 - \sin \varphi}{1 + \sin \varphi}. \qquad (2.10)$$

2.2.2.2 Passive Pressure Due to the Earthfill

The passive pressure exerted against the wall should be

$$P_p = \frac{1}{2} \gamma' H'^2 K_p \qquad (2.11)$$

where
  K_p is the Rankine's coefficient of the passive pressure acting at an angle ι with the horizontal,

$$K_p = cos i. \frac{cos i + \sqrt{cos^2 i - cos^2 \phi}}{cos i - \sqrt{cos^2 i - cos^2 \phi}}. \qquad (2.12)$$

If the earthfill in front of the wall is levelled, the Rankine equation can be written as

$$K_p = \frac{1 + \sin \phi}{1 - \sin \phi}. \qquad (2.13)$$

2.3 RELIABILITY OF LATERAL EARTH PRESSURES

Several sets of wall tests were performed to check the validity of the Coulomb and Rankine active and passive earth pressure methods by Terzaghi (1934, 1941), Peck and Ireland (1961), Rowe and Peaker (1965), Mackey and Kirk (1967), James and Bransby (1970), Rehnman and Broms (1972) and Coyle et al. (1972), among others.

Rankine's theory assumes the back of the wall to be smooth. No frictional forces are assumed to exist between the soil and the wall. Hence, the lateral pressure is assumed to act parallel to the surface of the backfill. But in practice, considerable friction will develop between the soil and the wall due to the movement of the wall.

As a consequence, the earth pressure will be inclined at a certain angle to the normal to the wall. The assumption of a smooth wall surface results in an overestimation of the active earth pressure and an underestimation of the passive earth pressure.

It is obvious that the assumptions made in Rankine's theory are not realistic. The wall will never be perfectly smooth and will have some degree of roughness. Hence, there will invariably be friction/adhesion developed between the wall and the soil. Hence, the assumption that no shear forces develop on the back of the wall is not true.

Coulomb's earth pressure theory is a practical method because it takes into consideration friction between the wall and the soil, the inclination of the wall face and the inclination of the ground surface behind the wall. It is generally known that earth pressure values derived by Coulomb's method agree with the values derived by Rankine's method when these parameters take certain values. Experience has shown that, if conventional procedures are followed with a full understanding of the assumptions on which they are based, the resulting designs will be safe and as economical as present knowledge permits. Nowadays, it is a general consensus amongst soil engineers that for the solution of most earth pressure problems, the Coulomb theory should be used.

2.4 REDUCING ACTIVE EARTH PRESSURE

The lateral movement of a retaining wall primarily depends on the size of the failure wedge, which is influenced by many parameters. One particular approach for reducing lateral earth pressures is to minimize the size of the failure wedge developed behind a wall. This can be simply accomplished by modifying the shape of the earth face of a wall. Figure 2.5 (Chalisgaonkar, 2020) shows the typical failure wedges for

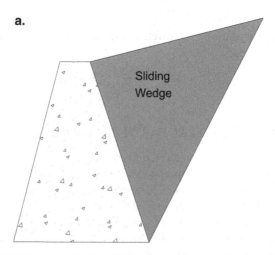

FIGURE 2.5 Wall Shape and Failure Wedges for the Active Earth Pressure: (a) Wall Leaning Outward (Positive Batter)

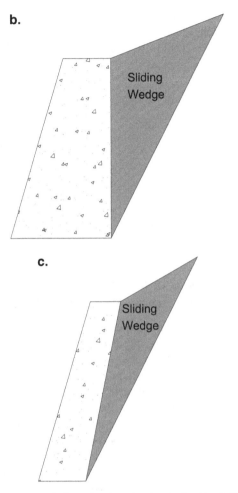

FIGURE 2.5 Wall Shape and Failure Wedges for the Active Earth Pressure: (b) Vertical Earth Face, (c) Wall Face Leaning towards the Backfill (Negative Batter)

the earth face of a wall leaning outwards (positive batter), the vertical earth face of a wall and the earth face of a wall leaning towards backfill (negative batter). It can be easily observed that the earth face of a wall leaning towards backfill (negative batter) is subject to a smaller backfill failure wedge and therefore a smaller lateral thrust. On the other hand, a larger failure wedge and lateral thrust develop behind the earth face of a wall leaning outwards (positive batter).

2.5 INFLUENCE OF THE PARAMETERS ON EARTH PRESSURE

The magnitude and direction of the active and passive pressures, as well as their distribution, depend on many variables, such as the angle of repose of the backfill, the angle of friction between the wall and the backfill, the slope of the backfill, the angle of the face of the wall, the level of the water table, the surcharge loads and the degree of soil compaction, amongst others. The following design parameters were considered (Chalisgaonkar, 2018) to study the influence on active and passive earth pressures using Coulomb's equations:

Angle of repose of the backfill, ϕ	:	$20°, 25°, 30°, 35°, 40°, 45°$
Angle of friction between the wall and the backfill, δ	:	$0°, 1/3\phi, 2/3\phi$
Angle of surcharge of backfill, ι	:	$-1/4\phi, 0°, 1/4\phi, 1/2\phi, 3/4\phi, \phi$
Angle the earth face of the wall makes with the vertical, α	:	$14.03°, 11.31°, 5.71°, 0°,$ $-5.71°, -11.31°, -14.03°$
Angle the outer face of the wall makes with the vertical, α_p	:	$0°, 5.71°, 11.31°, 21.79°, 30.95°$

2.5.1 INFLUENCE OF THE ANGLE OF REPOSE OF THE BACKFILL, Φ AND THE ANGLE OF FRICTION BETWEEN THE WALL AND THE BACKFILL, δ

Figures 2.6a and 2.6b show the variation of the coefficients of the active earth pressure, K_a, and passive earth pressure, K_p, for the angle of repose of the backfill, ϕ, varying from 20° to 45° when the angle of wall friction between the wall and the backfill is zero, $1/3\phi$ and $2/3\phi$; the angle of the surcharge of the backfill, ι, and the angles of the face of the walls with the vertical, α and α_p, are zero.

A perusal of Figure 2.6a indicates that the coefficient of the passive earth pressure, K_p, increases substantially with the angle of repose of the earthfill, ϕ. It is also observed that when the angle of friction between the wall and earthfill, δ, is also considered, the increase in the coefficient of the passive earth pressure, K_p, is phenomenal when the angle of repose of the earthfill, ϕ, is more than 30°.

A perusal of Figure 2.6b indicates that the coefficient of active earth pressure, K_a, decreases substantially with the increase in the angle of repose of the backfill, ϕ. It is also observed that the coefficient of active earth pressure, K_a, decreases when the angle of friction between the wall and backfill δ is also considered.

2.5.2 INFLUENCE OF THE ANGLE OF THE SURCHARGE, ι, ON THE ACTIVE EARTH PRESSURE

Figure 2.7 shows the variation of the coefficient of active earth pressure, K_a, for the angle of repose of the backfill, ϕ, varying from 20° to 45° when the angle of the surcharge of the backfill ι is $-1/4\phi$, $0°$, $1/4\phi$, $1/2\phi$, $3/4\phi$, and ϕ and the angle of wall friction between the wall and the backfill, δ, is neglected and the earth face of wall is vertical.

It can be seen from Figure 2.7 that the coefficient of the active earth pressure, K_a, increases with the angle of the surcharge of the backfill, ι, and the coefficient of the

a.

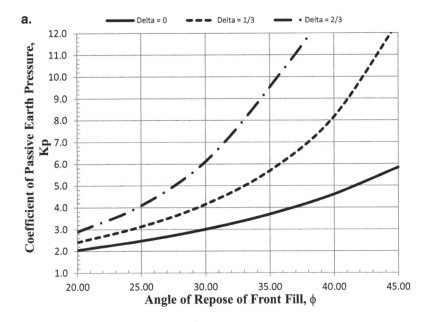

b.

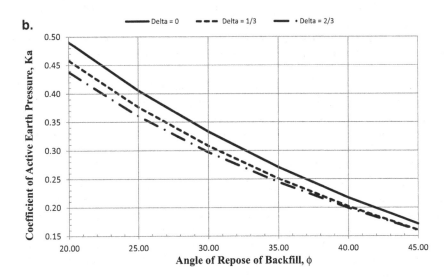

FIGURE 2.6 Influence of the Angle of Repose, ϕ, and the Angle of Friction between the Wall and the Backfill, δ: (a) Passive Earth Pressure, (b) Active Earth Pressure

active earth pressure, K_a, is very high when the angle of the surcharge of the backfill, ι, is equal to the angle of repose of the backfill, ϕ.

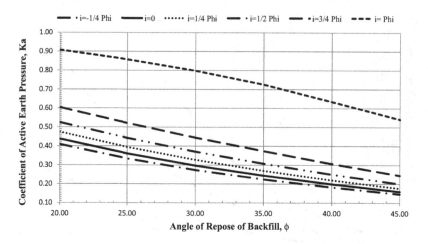

FIGURE 2.7   Influence of the Angle of the Surcharge, ι

2.5.3 Influence of the Vertical Angle of the Earth Face of Wall, α, on the Active Earth Pressure

Figure 2.8a shows the variation of the coefficient of the passive earth pressure, K_p, for the angle of repose of the backfill, ϕ, varying from 20° to 45° when the vertical angle of the outer face of the wall, α_p, is varied from 0° to 30.95°; the angle of friction between the wall and the backfill, δ, is neglected; and the angle of the surcharge of the earthfill, ι′, is zero.

Figure 2.8b shows the variation of the coefficient of the active earth pressure, K_a, for the angle of repose of the backfill, ϕ, varying from 20° to 45° when the vertical angle of the earth face of the wall, α, is −14.03°, −11.31°, −5.71°, 0°, 5.71°, 11.31°, and 14.03°; the angle of friction between the wall and the backfill, δ, is neglected; and the angle of the surcharge of backfill, ι, is zero.

Figure 2.8c shows the variation of the coefficient of the active earth pressure, K_a, for the angle of repose of the backfill, ϕ, varying from 20° to 45° when the vertical angle of the earth face of wall, α, is either 0° or $\phi - 90$°; the angle of friction between the wall and the backfill, δ, is neglected; and the angle of the surcharge of backfill, ι, is zero.

A perusal of Figure 2.8a indicates that the coefficient of the passive earth pressure, K_p, decreases with an increase of the vertical angle of the outer face of wall, α_p. Thus, it can be said that it is better to keep the outer face of the wall as vertical to the extent possible.

A perusal of Figure 2.8b indicates that the coefficient of the active earth pressure, K_a, decreases when the vertical angle of the earth face of wall, α, is given a negative batter, and it substantially reduces the active earth pressure, K_a. A study conducted on inclined retaining walls by Chalisgaonkar (1988) has also indicated that when a wall leans towards the backfill, the reduction in the active pressure coefficient is considerable.

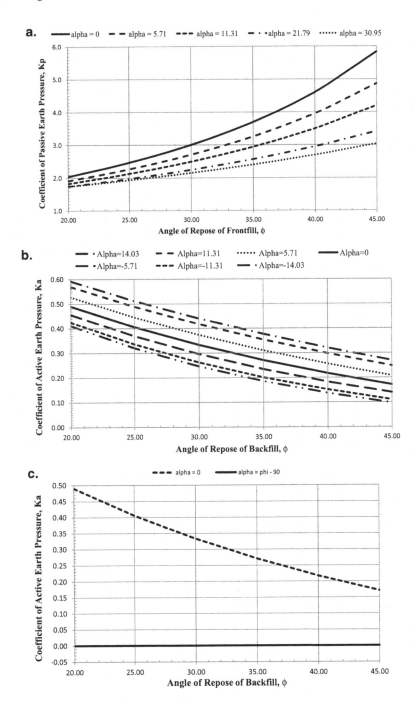

FIGURE 2.8 Influence of the Vertical Angle of the Face of Wall, α: (a) Passive Earth Pressure, (b) Active Earth Pressure, (c) Active Earth Pressure when the Vertical Angle α is $\phi - 90°$

A perusal of Figure 2.8c indicates that when the vertical angle of the earth face of the wall, α, is kept as $\phi - 90°$, the coefficient of the active earth pressure, K_a, becomes zero. In other words, theoretically, when the horizontal angle of the wall towards the backfill leans by an angle of repose of the backfill, ϕ, no active earth pressures are exerted on the wall. However, the construction of retaining walls with negative batter in the field will depend on the purpose and loading on the retaining walls.

2.6 INFERENCE FROM THE PARAMETRIC STUDY

The following inferences can be drawn from the study:

1. The coefficient of the active earth pressure, K_a, decreases with the angle of repose of the backfill, ϕ, and the coefficient of the active earth pressure, K_a, is much less for backfills with a higher value of ϕ. Therefore, the angle of repose of the backfill, ϕ, can be estimated on-site for cost-effective design of retaining wall.
2. The angle of friction between the wall and the backfill, δ, should be considered for estimating the coefficient of the active earth pressure, K_a, because K_a is maximum for a smooth wall (i.e. $\delta = 0$) and it decreases with δ. Guidelines issued by Indian Railway Standard Code of Practice for the Design of Sub Structure and Foundation of Bridges be followed for adopting the value of angle of friction between the wall and backfill, δ, if its value is not known.
3. The vertical angle of the earth face of the wall, α, will greatly influence the coefficient of the active earth pressure, K_a, and even a small tilt towards the backfill can reduce the active earth pressure on the wall substantially.
4. The coefficient of the passive earth pressure also depends on ϕ, δ, ι, and δ_p, but the passive pressure is generally used in checking the stability of the wall against sliding; therefore, it is the active earth pressure which governs the design of the wall.

2.7 COEFFICIENTS OF THE ACTIVE AND PASSIVE EARTH PRESSURES

The coefficients of the active and passive earth pressures for different values of the angle of repose of the backfill, ϕ; the surcharge angle, ι; the friction angle, δ; and the vertical angle of the earth face of the wall, α, have been shown in Tables 2.1 and 2.2, respectively.

TABLE 2.1
Coefficient of Active Earth Pressure K_a for Different Values of the Angle of Repose of the Backfill, ϕ; the Surcharge Angle, ι; the Friction Angle, δ; and the Vertical Angle of the Earth Face of the Wall, α

S. No.	Φ	δ	ι	Slope of the Earth Face (H:V)	α	θ	K_a
1	20	0.00	0.00	0:1	0.00	90.00	0.490
2	25						0.406
3	30						0.333
4	35						0.271
5	40						0.217
6	20	6.67	0.00	0:1	0.00	90.00	0.458
7	25	8.33					0.377
8	30	10.00					0.308
9	35	11.67					0.251
10	40	13.33					0.202
11	20	13.33	0.00	0:1	0.00	90.00	0.438
12	25	16.67					0.361
13	30	20.00					0.297
14	35	22.50					0.244
15	40	22.50					0.199
16	20	0.00	10.00	0:1	0.00	90.00	0.569
17	25		12.50				0.481
18	30		15.00				0.402
19	35		17.50				0.330
20	40		20.00				0.266
21	20	6.67	10.00	0:1	0.00	90.00	0.541
22	25	8.33	12.50				0.455
23	30	10.00	15.00				0.378
24	35	11.67	17.50				0.311
25	40	13.33	20.00				0.252
26	20	13.33	10.00	0:1	0.00	90.00	0.524
27	25	16.67	12.50				0.442
28	30	20.00	15.00				0.371
29	35	22.50	17.50				0.307
30	40	22.50	20.00				0.251
31	20	0.00	20.00	0:1	0.00	90.00	0.883
32	25		25.00				0.821
33	30		30.00				0.750
34	35		35.00				0.671
35	40		40.00				0.587

(Continued)

TABLE 2.1 (Continued)

Coefficient of Active Earth Pressure K_a for Different Values of the Angle of Repose of the Backfill, ϕ; the Surcharge Angle, ι; the Friction Angle, δ; and the Vertical Angle of the Earth Face of the Wall, α

S. No.	Φ	δ	ι	Slope of the Earth Face (H:V)	α	θ	K_a
36	20	6.67	20.00	0:1	0.00	90.00	0.889
37	25	8.33	25.00				0.830
38	30	10.00	30.00				0.761
39	35	11.67	35.00				0.685
40	40	13.33	40.00				0.603
41	20	13.33	20.00	0:1	0.00	90.00	0.907
42	25	16.67	25.00				0.857
43	30	20.00	30.00				0.798
44	35	22.50	35.00				0.726
45	40	22.50	40.00				0.635
46	20	0.00	0.00	1:10	−5.71	84.29	0.456
47	25						0.369
48	30						0.296
49	35						0.235
50	40						0.183
51	20	6.67	0.00	1:10	−5.71	84.29	0.423
52	25	8.33					0.339
53	30	10.00					0.271
54	35	11.67					0.214
55	40	13.33					0.167
56	20	13.33	0.00	1:10	−5.71	84.29	0.401
57	25	16.67					0.322
58	30	20.00					0.258
59	35	22.50					0.207
60	40	22.50					0.163
61	20	0.00	10.00	1:10	−5.71	84.29	0.528
62	25		12.50				0.436
63	30		15.00				0.355
64	35		17.50				0.283
65	40		20.00				0.221

S. No.	Φ	δ	ι	Slope of the Earth Face (H:V)	α	θ	K_a
66	20	6.67	10.00	1:10	−5.71	84.29	0.497
67	25	8.33	12.50				0.407
68	30	10.00	15.00				0.330
69	35	11.67	17.50				0.263
70	40	13.33	20.00				0.205
71	20	13.33	10.00	1:10	−5.71	84.29	0.477
72	25	16.67	12.50				0.392
73	30	20.00	15.00				0.319
74	35	22.50	17.50				0.256
75	40	22.50	20.00				0.203
76	20	0.00	20.00	1:10	−5.71	84.29	0.824
77	25		25.00				0.750
78	30		30.00				0.669
79	35		35.00				0.583
80	40		40.00				0.495
81	20	6.67	20.00	1:10	−5.71	84.29	0.820
82	25	8.33	25.00				0.747
83	30	10.00	30.00				0.668
84	35	11.67	35.00				0.583
85	40	13.33	40.00				0.497
86	20	13.33	20.00	1:10	−5.71	84.29	0.827
87	25	16.67	25.00				0.760
88	30	20.00	30.00				0.687
89	35	22.50	35.00				0.606
90	40	22.50	40.00				0.514
91	20	0.00	0.00	1:5	−11.31	78.69	0.425
92	25						0.336
93	30						0.262
94	35						0.201
95	40						0.151
96	20	6.67	0.00	1:5	−11.31	78.69	0.390
97	25	8.33					0.306
98	30	10.00					0.237
99	35	11.67					0.182
100	40	13.33					0.136

(Continued)

TABLE 2.1 (Continued)

Coefficient of Active Earth Pressure K_a for Different Values of the Angle of Repose of the Backfill, ϕ; the Surcharge Angle, ι; the Friction Angle, δ; and the Vertical Angle of the Earth Face of the Wall, α

S. No.	Φ	δ	ι	Slope of the Earth Face (H:V)	α	θ	K_a
101	20	13.33	0.00	1:5	−11.31	78.69	0.367
102	25	16.67					0.287
103	30	20.00					0.224
104	35	22.50					0.173
105	40	22.50					0.132
106	20	0.00	10.00	1:5	−11.31	78.69	0.491
107	25		12.50				0.395
108	30		15.00				0.312
109	35		17.50				0.241
110	40		20.00				0.181
111	20	6.67	10.00	1:5	−11.31	78.69	0.458
112	25	8.33	12.50				0.365
113	30	10.00	15.00				0.286
114	35	11.67	17.50				0.220
115	40	13.33	20.00				0.165
116	20	13.33	10.00	1:5	−11.31	78.69	0.436
117	25	16.67	12.50				0.347
118	30	20.00	15.00				0.273
119	35	22.50	17.50				0.212
120	40	22.50	20.00				0.161
121	20	0.00	20.00	1:5	−11.31	78.69	0.774
122	25		25.00				0.688
123	30		30.00				0.598
124	35		35.00				0.506
125	40		40.00				0.414
126	20	6.67	20.00	1:5	−11.31	78.69	0.761
127	25	8.33	25.00				0.676
128	30	10.00	30.00				0.587
129	35	11.67	35.00				0.496
130	40	13.33	40.00				0.406
131	20	13.33	20.00	1:5	−11.31	78.69	0.759
132	25	16.67	25.00				0.678
133	30	20.00	30.00				0.593
134	35	22.50	35.00				0.506
135	40	22.50	40.00				0.414

S. No.	Φ	δ	ι	Slope of the Earth Face (H:V)	α	θ	K_a
136	20	0.00	0.00	1:4	−14.03	75.97	0.411
137	25						0.320
138	30						0.246
139	35						0.186
140	40						0.137
141	20	6.67	0.00	1:4	−14.03	75.97	0.376
142	25	8.33					0.290
143	30	10.00					0.221
144	35	11.67					0.167
145	40	13.33					0.122
146	20	13.33	0.00	1:4	−14.03	75.97	0.352
147	25	16.67					0.271
148	30	20.00					0.208
149	35	22.50					0.158
150	40	22.50					0.118
151	20	0.00	10.00	1:4	−14.03	75.97	0.474
152	25	0.00	12.50				0.376
153	30	0.00	15.00				0.293
154	35	0.00	17.50				0.222
155	40	0.00	20.00				0.163
156	20	6.67	10.00	1:4	−14.03	75.97	0.440
157	25	8.33	12.50				0.345
158	30	10.00	15.00				0.267
159	35	11.67	17.50				0.201
160	40	13.33	20.00				0.148
161	20	13.33	10.00	1:4	−14.03	75.97	0.417
162	25	16.67	12.50				0.327
163	30	20.00	15.00				0.253
164	35	22.50	17.50				0.192
165	40	22.50	20.00				0.143
166	20	0.00	20.00	1:4	−14.03	75.97	0.752
167	25		25.00				0.661
168	30		30.00				0.566
169	35		35.00				0.470
170	40		40.00				0.377

(Continued)

TABLE 2.1 (Continued)

Coefficient of Active Earth Pressure K_a for Different Values of the Angle of Repose of the Backfill, ϕ; the Surcharge Angle, ι; the Friction Angle, δ; and the Vertical Angle of the Earth Face of the Wall, α

S. No.	Φ	δ	ι	Slope of the Earth Face (H:V)	α	θ	K_a
171	20	6.67	20.00	1:4	−14.03	75.97	0.736
172	25	8.33	25.00				0.644
173	30	10.00	30.00				0.550
174	35	11.67	35.00				0.457
175	40	13.33	40.00				0.366
176	20	13.33	20.00	1:4	−14.03	75.97	0.730
177	25	16.67	25.00				0.642
178	30	20.00	30.00				0.552
179	35	22.50	35.00				0.461
180	40	22.50	40.00				0.370

TABLE 2.2

Coefficient of the Passive Earth Pressure, K_p, for Different Values of the Angle of Repose of the Earthfill, ϕ, and the Friction Angle, δ

S. No.	Φ	δ	K_p
1	20	0.00	2.040
2	25	0.00	2.465
3	30	0.00	3.001
4	35	0.00	3.692
5	40	0.00	4.602
6	20	6.67	2.415
7	25	8.33	3.125
8	30	10.00	4.146
9	35	11.67	5.686
10	40	13.33	8.157
11	20	13.33	2.890
12	25	16.67	4.083
13	30	20.00	6.112
14	35	22.50	9.521
15	40	22.50	13.842

3 Design Methodology of Breast Walls

3.1 DESIGN METHODOLOGY

The design of a breast wall consists of two principal parts, the evaluation of loads and pressures that may act on the structure and the design of the structure to withstand these loads and pressures.

3.1.1 DESIGN LOADS

In general, the following forces are considered in the design of breast walls:

1. Self-weight of the wall
2. Earth pressure acting on the wall
3. Surcharge
4. Water pressure
5. Allowable soil-bearing pressure
6. Frictional forces on the base against the sliding

Forces acting on the breast walls have been depicted in Figure 3.1.

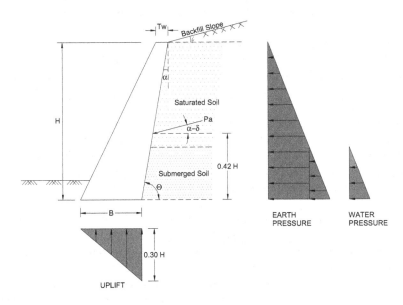

FIGURE 3.1 Forces Acting on the Breast Wall

DOI: 10.1201/9781003162995-3

1. *Self-Weight of the Wall*

The self-weight of the breast wall should be taken into consideration on the basis of the unit weight of materials, as per the Indian Standard Code of Practice for Design Loads (Other than Earthquake) for Buildings and Structures IS-875 (Part 1).

2. *Earth Pressure Acting on the Wall*

The active and passive earth pressures behind the walls have been calculated in accordance with paragraphs 22.1.1 and 22.1.2 of the Indian Standard Code of Practice: Criteria for Earthquake Resistant Design of Structures Part 3—Bridges and Retaining Walls, IS:1893:Part 3 (2014) in the present study. However, paragraph 214.1 of IRC:6–2017 specifies that *Coulomb's theory shall be acceptable, subject to the modification that the centre of pressure exerted by the backfill, when considered dry, is located at an elevation of 0.42 of the height of the wall above the base instead of 0.33 of that height,* mentioned in IS:1893 Part 3 (2014) and have been used accordingly in the study.

Paragraph 5.7.1 of the Indian Railway Standard Code of Practice for the Design of Sub Structure and Foundation of Bridges (2003) states that *while calculating earth pressure, the angle of friction between the wall and the earth fill, where value of δ is not determined by actual tests, the following values may be assumed:*

(i) $\delta = 1/3\phi$ *for concrete structures.*
(ii) $\delta = 2/3\phi$ *for masonry structures.*

3. *Surcharge*

No surcharge has been considered in the study.

4. *Water Pressure*

Paragraph 5.1 of the Indian Standard Retaining Wall for Hill Area—Guidelines, Part 2 Design of Retaining/Breast Walls, IS:14458 (Part 2) (1997) states that *at least 30 percent water pressure shall always be considered even in the case of provision of good efficient pressure release system.* When the water pressure is considered, the uplift pressure at the base is also considered in computing the stability of the wall.

5. *Earthquake Forces*

Paragraph 3.1.1 of the Indian Standard Retaining Wall for Hill Area—Guidelines, Part 1 Selection of Type of Wall, IS:14458 (Part 1) 1998 states that *for hilly roads, being of low volume, walls may not be designed for earthquake forces. It is economical to repair failed walls after an earthquake.* Therefore, earthquake forces have been ignored in developing the design charts.

6. *Allowable Soil-Bearing Pressure*

The foundations should be designed to take care of the shear capacity of the soil/rock formation and permissible settlements which the structure can tolerate. It is generally the settlement criterion that governs the value of the allowable bearing pressure and is used for designing foundations. The foundations of breast walls are normally so proportioned that no tension is created at the foundation plane under static loading.

7. *Frictional Forces on the Base against the Sliding*

Paragraph 706.3.4 of the Standard Specifications and Code of Practice for Road Bridges Section: VII Foundations and Substructure IRC:78 (2014) states that the *frictional coefficient between concrete and soil/rock will be Tan ϕ, ϕ being the angle of friction. Founding soil in the foundation of a bridge being generally properly consolidated, the following values may be adopted:*

Friction coefficient between the soil and the concrete, μ = 0.50
Friction coefficient between the rock and the concrete, μ = 0.80 for good rock
  = 0.70 for fissured rock

3.2 STABILITY ANALYSIS OF BREAST WALLS

As a minimum requirement, all breast walls must satisfy *global stability* and vertical settlement limits. Total global stability should be checked during the stage of design. However, it is common that the *global stability* and the expected settlement are verified by geotechnical engineers and that the structural engineer verifies that the walls have sufficient resistance against overturning, sliding and sufficient bearing capacity to resist the loads (Huntington, 1957).

3.2.1 STABILITY AGAINST OVERTURNING

If the breast wall structure were to overturn, it would do so with the toe acting as the centre of rotation. Thus, the safety factor of wall stability against overturning is defined as the ratio between the sum of resisting moments and the sum of overturning moments about the toe. In evaluating these moments, the vertical component of the active thrust on the wall may be considered in two different ways: decreasing the overturning moment or increasing the resisting one.

$$F.S._{overturning} = \frac{\Sigma M_{resisting}}{\Sigma M_{overturning}} \qquad (3.1)$$

Paragraph 5.2 of the Indian Standard Retaining Wall for Hill Area—Guidelines, Part 2 Design of Retaining/Breast Walls, IS:14458 (Part 2) (1997) states that *for static loads, a factor of safety as 2.0 against overturning shall be ensured for stability.*

3.2.2 STABILITY AGAINST SLIDING

The resistance against sliding is essentially provided by the friction between the base and the supporting soil μ. The calculated factor of safety for sliding along the wall base is

$$FS_{sliding} = \frac{0.90 * \mu * \Sigma W}{\Sigma H} \geq 1.5 \qquad (3.2)$$

where
  μ is the coefficient of friction between the base and the supporting soil,
  ΣW is the summation of vertical weights and
  ΣH is the summation of horizontal pressures.

Paragraph 5.2 of the Indian Standard Retaining Wall for Hill Area—Guidelines, Part 2 Design of Retaining/Breast Walls, IS:14458 (Part 2) (1997) states that *for static loads, a factor of safety as 1.5 against sliding shall be ensured for stability.*

3.2.3 SHEAR KEY

When lateral pressures are relatively high, it is generally difficult to mobilize the required factor of safety against sliding by frictional resistance below the footing alone. In such a situation, it is advantageous to provide a shear key projecting below the base of the wall and extending throughout the length of the wall.

Several procedures have been proposed for estimating the passive resistance, P_p (Pillai and Menon, 2005). A simple and conservative is obtained by considering the pressure developed over a region, $h_2 - h_1$, below the toe:

$$P_p = K_p \cdot \gamma (h_2^2 - h_1^2)/2 \tag{3.3}$$

where h_1 and h_2 are the depths below the natural ground level, as indicated in Figure 3.2. It may be noted that the overburden due to the top 300 mm of earth below ground level is usually ignored in the calculation.

The shear key is best positioned to be located near the heel to get the maximum advantage.

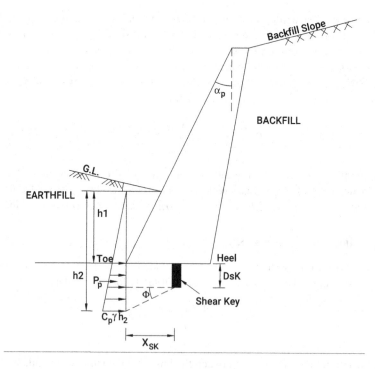

FIGURE 3.2 Passive Resistance Due to the Shear Key

3.2.4 BEARING PRESSURE

Similar to any footing design, the width of the base of a breast wall must be adequate to distribute the vertical reaction to the foundation soil without causing excessive settlement or rotation. When the base pressure exerted exceeds the safe bearing capacity of the soil, the wall fails due to bearing pressure.

4 Comparison of Gravity Retaining Walls and Breast Walls

4.1 DESIGN PARAMETERS

It has been largely observed from the literature review that generally, the retaining walls with the earth face leaning away from backfill (positive batter) and a vertical face have also been constructed at the project sites to retain soil and resist the lateral pressure of the soil against the wall. However, from the figures presented in Chapter 2, it can be easily said that the retaining walls with the earth face leaning towards the backfill (negative batter) give an economical section, when the walls are to be constructed to retain soil only.

In this chapter, an attempt is made to compare the designs of generally adopted shapes of gravity walls with breast walls (Chalisgaonkar, 2019).

The shapes of the walls considered for comparing the design of 4.0-m-high random rubble (R. R.) masonry walls are shown in Figure 4.1, and the parameters varied for the design of gravity retaining walls are as follows:

Angle of repose of the backfill, ϕ	: 25°, 30°, 35°, 40°
Angle of the surcharge of the backfill, ι	: 0°
Height of the gravity retaining wall, H	: 4.0 m
Angle of friction between a masonry wall and backfill, δ	: $\frac{2}{3}\phi$ or 22.50°, whichever is less

Angle which the earth face of the wall makes with the vertical, α / Slope 1.0(H):x(V)	:		
		5.71°	1.0(H):10.0(V)
		11.31°	1.0(H):5.0(V)
		14.03°	1.0(H):4.0(V)
		0°	Vertical Face
		−5.71°	1.0(H):10.0(V)
		−11.31°	1.0(H):5.0(V)
		−14.03°	1.0(H):4.0(V)

Coefficient of friction between the masonry and the soil/rock mass, μ	: 0.80

DOI: 10.1201/9781003162995-4

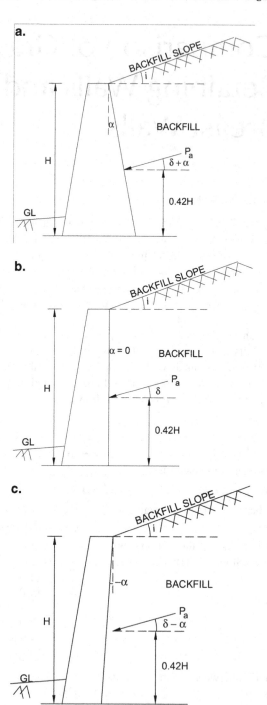

FIGURE 4.1 Typical Section of R.R. Masonry Retaining Walls: (a) Earth Face of a Wall with a Positive Batter, (b) Vertical Earth Face of a Wall and (c) Earth Face of a Wall with a Negative Batter

Other parameters assumed for carrying out the design of a wall include the following:

Unit weight of saturated earthfill, γ_{sat}	:	1.76 t/m³ or 17.27 kN/m³
Unit weight of submerged earthfill, γ_{sub}	:	0.76 t/m³ or 7.46 kN/m³
Unit weight of water, γ_{water}	:	1.00 t/m³ or 9.81 kN/m³
Unit weight of masonry, $\gamma_{masonry}$	:	2.30 t/m³ or 22.56 kN/m³
Effect of water pressure	:	30% of total height
Top width of the wall	:	200 mm
Width of the shear key at the bottom	:	150 mm
Location of the shear key	:	At the centre of the base

4.1.1 Design Loads

The following forces have been considered in the design of a 4.0-m-high gravity retaining wall:

1. Self-weight of the wall
2. Active and passive earth pressures acting on the wall
3. Water pressure
4. Uplift pressure
5. Frictional forces on the base

The base of the gravity retaining walls has been so proportioned in the analysis that there is nearly zero stress near the heel under static loading, the factor of safety against overturning is equal to or more than 1.50, and the maximum permissible bearing pressure is 100 kN/m².

The results obtained after carrying out the analysis of a 4.0-m-high gravity retaining wall for different parametric variations are as follows.

4.1.2 Coefficient of the Active Earth Pressure

The coefficient of the active earth pressure, K_a, for the different angle of repose of backfill, ϕ, for different angles which the earthface of the wall makes with vertical, α, varying from −14.03° to 14.03°, have been calculated using Coulomb's theory and are shown in Figure 4.2.

A perusal of Figure 4.2 clearly indicates that the coefficient of the active earth pressure, K_a, reduces with an increase in the angle of repose of the backfill, ϕ. It also indicates that the coefficient of the active earth pressure, K_a, reduces substantially when the angle of the earth face of the wall is given a negative batter instead of a vertical face or a positive batter. The effect of a negative batter can further be understood from the basic sliding wedges formed in each case.

For a clear understanding of the impact of the slope of the earth face of the wall, the superimposed sliding wedge failure for a positive batter, a vertical face and a negative batter is shown in Figure 4.3.

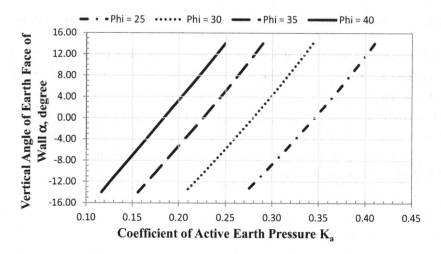

FIGURE 4.2   Coefficient of the Active Earth Pressure, K_a

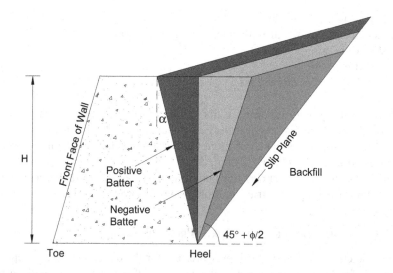

FIGURE 4.3   Failure Wedges for Active Earth Pressure When the Earth Face of the Wall Has a Positive Batter, a Vertical Face and a Negative Batter

4.1.3 BASE WIDTH

The base width of the gravity retaining wall, obtained for economic section, with nearly zero stress at heel, for a 4.0-m-high gravity retaining wall for a different angle of the repose of backfill, ϕ, for different angles when the earth face of the wall, α, has a positive batter, a vertical face, and a negative batter are shown in Figure 4.4.

A perusal of Figure 4.4 clearly indicates that the base width of the gravity retaining wall is reduced substantially when the angle of the earth face of the wall is given

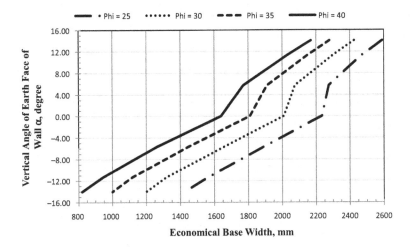

FIGURE 4.4 Economical Base Width of a Gravity Retaining Wall

a negative batter instead of vertical face or positive batter for cohesionless backfill having any angle of repose, ϕ.

4.1.4 VOLUME OF MASONRY IN A GRAVITY RETAINING WALL

The total volume of masonry wall per m length for the construction of a 4.0-m-high gravity retaining wall for different angles of repose of the backfill, ϕ, and for different angles, α, when the earth face of the wall has a positive batter, a vertical face, and a negative batter are shown in Figure 4.5.

A perusal of Figure 4.5 clearly indicates that the volume of masonry in a gravity retaining wall is reduced substantially when the angle of the earth face of the wall, α, is given a negative batter instead of a vertical face or a positive batter for cohesionless backfill having any angle of repose, ϕ.

4.1.5 FACTOR OF SAFETY AGAINST SLIDING

The factor of safety against sliding and the depth of the shear key required to attain a factor of safety of 1.5 for a 4.0-m-high gravity retaining wall for different angles of repose of the backfill, ϕ, and the coefficient of friction between the masonry and the soil/rock mass, $\mu = 0.80$, for different angles when the earth face of the wall, α, has a positive batter, a vertical face and a negative batter are shown in Figure 4.6.

A perusal of Figure 4.6 indicates that the factor of safety against sliding goes on reducing with the change in the earth face of the wall angle, α, from a positive batter to a negative batter for cohesionless backfill having any angle of repose of the backfill, ϕ. It can also be observed from Figure 4.6 that the factor of safety against sliding is below 1.50 when the angle of repose of the soil, ϕ, is less and the earth face

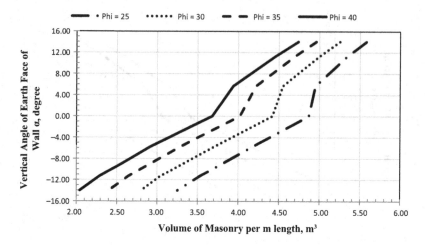

FIGURE 4.5   Volume of Masonry in a Gravity Retaining Wall

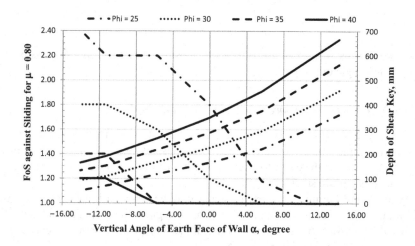

FIGURE 4.6   Factor of Safety against Sliding and Depth of the Shear Key

of the wall angle, α, is less than $0°$. The maximum depth of the shear key for achieving a factor of safety of 1.50 is 700 mm when the angle of repose of the soil, ϕ, is $25°$. However, the depth of the shear key will increase with a decrease in the coefficient of friction between the masonry and the soil/rock mass, μ.

4.2 REMARKS ON THE COMPARISON OF DESIGNS

From a perusal of Figures 4.2 to 4.6 for the designs of 4.0-m-high walls of different shapes and parameters, it can be clearly said that wherever gravity retaining walls are built to retain soil and resist the lateral pressure of the soil against the wall only, the earth face of the wall with a negative batter should be constructed to achieve economy without sacrificing safety.

4.3 INFLUENCE OF THE SURCHARGE AND THE DENSITY OF MATERIALS

In actual practice, on-site, the parameters assumed for designing a breast wall may not be exactly the same; therefore, the influence of some of the parameters on a wall with a vertical face and different vertical angle of the wall, α, have been carried out for 4.0-m-high walls.

4.3.1 INFLUENCE OF THE UNIT WEIGHT OF THE SOIL ON THE BASE WIDTH

In order to see the influence of the unit weight of the soil on the design of masonry breast walls, a 4.0-m-high wall with an angle of repose of the soil, ϕ, as 30°; the angle of surcharge, ι, as 0°, ½ϕ and ϕ; the angle of friction between the masonry wall and the backfill, δ, equal to ⅔ϕ; the unit weight of the masonry, $\gamma_{masonry}$, equal to 22.56 kN/m³; and different vertical angles, α, were considered with the saturated weight of the soil, γ_{sat}, equal to 15.70 kN/m³(1.6 t/m³), 17.27 kN/m³ (1.76 t/m³) and 19.62 kN/m³ (2.0 t/m³), respectively. The economical base widths obtained for different combinations of parameters are shown in Figure 4.7.

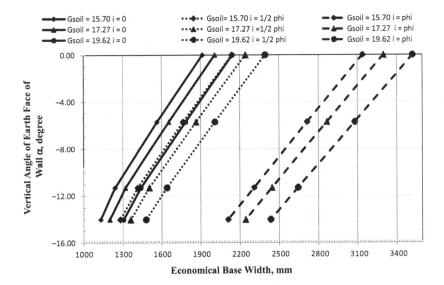

FIGURE 4.7 Variation of Economical Base Widths with the Unit Weight of the Soil

From a perusal of Figure 4.7, it can be clearly seen that the economical base width required for stability increases with an increase in the unit weight of saturated soil behind the wall. The effect is more prominent when the angle of the surcharge, ι, is equal to the angle of repose of the soil, ϕ.

4.3.2 Influence of the Unit Weight of Masonry on the Base Width

In order to see the influence of the unit weight of masonry on the design of masonry breast walls, a 4.0-m-high wall with an angle of repose of the soil, ϕ, as 30°; an angle of surcharge, ι, as 0°, $\frac{1}{2}\phi$ and ϕ; an angle of friction between the masonry wall and the backfill, δ, equal to $\frac{2}{3}\phi$; unit weight of saturated soil, γ_{sat}, equal to 17.27 kN/m³; and different vertical angles, α, were considered with unit weights of masonry, $\gamma_{masonry}$, equal to 21.58 kN/m³ (2.2 t/m³), 22.56 kN/m³ (2.3 t/m³) and 23.54 kN/m³ (2.4 t/m³), respectively. The economical base widths obtained for different combinations of parameters are shown in Figure 4.8.

From a perusal of Figure 4.8, it can be clearly seen that the base width decreases when the value of the unit weight of masonry is increased from 22.56 kN/m³ to 23.54 kN/m³ (2.3 t/m³ to 2.4 t/m³). It has been observed that the decrease in the base width with respect to (w.r.t.) 22.56 kN/m³ (2.3 t/m³) varies from 2.4% to 3.4% only. Therefore, it can be said that if the unit weight of masonry is more than 22.56 kN/m³ (2.3 t/m³), the base width computed for the unit weight of masonry as 22.56 kN/m³ (2.3 t/m³) may be assumed.

Similarly, from a perusal of Figure 4.8, it can also be clearly seen that the base width increases when the value of the unit weight of masonry is decreased from 22.56 kN/m³ to 21.58 kN/m³ (2.3 t/m³ to 2.2 t/m³). It has been observed that the

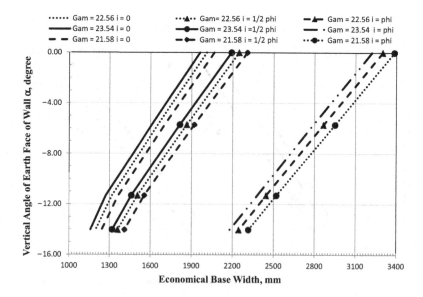

FIGURE 4.8 Variation of Economical Base Widths with the Unit Weight of Masonry

increase in the base width w.r.t. 22.56 kN/m³ (2.3 t/m³) varies from 2.6% to 3.75%. Therefore, it can be said that if the unit weight of masonry is less than 22.56 kN/m³ (2.3 t/m³), the base width computed may be increased by 4% to 5% for the unit weight of masonry as 21.58 kN/m³ (2.2 t/m³), obtained for the unit weight of masonry equal to 22.56 kN/m³ (2.3 t/m³).

4.3.3 Influence of the Angle of Surcharge on the Base Width

In order to see the influence of the angle of surcharge, ι, on the design of masonry breast walls, a 4.0-m-high wall with an angle of repose of soil, ϕ, as 25°, 30°, 35° and 40°; an angle of friction between the masonry wall and the backfill, δ, equal to ⅔ϕ; a unit weight of saturated of soil, γ_{sat} equal to 17.27 kN/m³, a unit weight of masonry, $\gamma_{masonry}$, equal to 22.56 kN/m³; and different vertical angles, α, were considered with angles of surcharge, ι, as 0°, ½ϕ and ϕ, respectively. The economical base widths obtained for different combinations of parameters are shown in Figure 4.9.

From a perusal of Figure 4.9, it can be clearly seen that the base width increases linearly with the increase in the angle of the surcharge from 0° to the angle of repose, ϕ. Therefore, if the angle of the surcharge is not equal to the assumed value, the base width can be obtained by linear interpolation for an intermediate value of the angle of the surcharge, ι.

4.3.4 Influence of Materials of Construction on the Base Width

In order to see the influence of construction materials on the design of masonry and concrete breast walls, a 4.0-m-high wall with an angle of repose of the soil, ϕ, as 30°; the angle of friction between the masonry wall and the backfill, δ, equal to

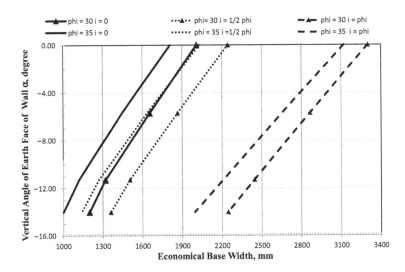

FIGURE 4.9 Variations of Economical Base Widths with the Angle of the Surcharge

⅔φ, angle of friction between the concrete wall and the backfill, δ, equal to ⅓φ; the unit weight of saturated soil, γ_{sat}, equal to 17.27 kN/m³; and different vertical angles, α, were considered with the unit weight of masonry $\gamma_{masonry}$, equal to 22.56 kN/m³ and the unit weight of concrete, $\gamma_{concrete}$, equal to 24.52 kN/m³, respectively. The economical base widths and volumes of concrete/masonry walls obtained for concrete and masonry breast walls for different combinations of parameters are depicted in Figures 4.10 and 4.11.

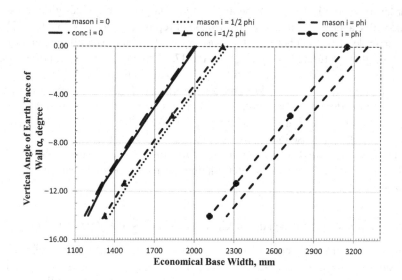

FIGURE 4.10 Economical Base Widths for Masonry and Concrete Walls

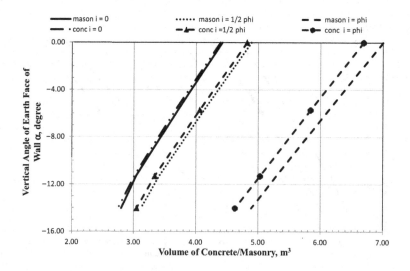

FIGURE 4.11 Volume of Concrete/Masonry in Gravity Breast Walls

From a perusal of Figure 4.10, it can be seen that the base width required for a concrete breast wall is slightly less as compared to a masonry breast wall, and the reduction in base width is more significant when the angle of the surcharge, ι, is nearly equal to the angle of repose of the soil, ϕ.

From a perusal of Figure 4.11, it can be seen that the volume of the concrete wall per running m is invariably less than the volume of the masonry wall. The reduction in the volume of concrete wall is not substantial for a low angle of the surcharge, ι, but the reduction in the volume of concrete wall is about 4.4% to 5.4% when the angle of the surcharge, ι, is nearly equal to the angle of repose of soil, ϕ. However, the decision of constructing a concrete wall in place of a masonry wall will depend on the length of the wall to be constructed along with serviceability and other considerations.

5 Charts for Stone Masonry Breast Walls

5.1 PARAMETERS FOR THE DEVELOPMENT OF THE CHARTS

The typical section adopted for the design of a breast wall is shown in Figure 5.1.

The following assumptions are made for carrying out the analysis of breast walls:

Top width of the wall	: 200 mm for the height of the wall <3.0 m
	300 mm for the height of the wall ≥3.0 m
Angle of friction between the masonry wall and the backfill, δ	: $\frac{2}{3}\phi$ or 22.50°, whichever is less
Effect of water for uplift pressure	: 30% of the total height
Unit weight of water, γ_{water}	: 1.00 t/m³ or 9.81 kN/m³
Maximum allowable soil-bearing pressure	: 100 kN/m²
Width of the shear key at the bottom	: 150 mm
Location of the shear key	: At the centre of the base

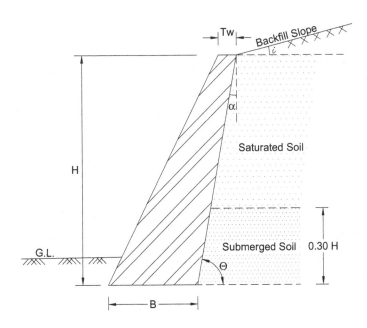

FIGURE 5.1 Typical Section of a Masonry Breast Wall

DOI: 10.1201/9781003162995-5

The following parameters have been varied for developing the design charts for breast walls:

Height of the breast wall, H	:	1.0 m
Angle of the surcharge of the backfill, ι	:	0°, ½ϕ, ϕ
Unit weight of masonry, $\gamma_{masonry}$	:	2.30 t/m³ or 22.56 kN/m³
Unit weight of saturated earthfill, γ_{sat}	:	1.60 t/m³ or 15.70 kN/m³
		1.76 t/m³ or 17.27 kN/m³

For a 1.00-m-high masonry breast wall, the stability analysis has been carried out for the preceding 6 combinations for 16 combinations of the following parameters:

Angle of repose of the backfill, ϕ	:	25°, 30°, 35°, 40°
Angle the earth face of the wall makes with vertical, α / Slope 1.0(H):x(V)	:	0° vertical face
		−5.71°1.0(H):10.0(V)
		−11.31°1.0(H):5.0(V)
		−14.03°1.0(H):4.0(V)
Coefficient of the friction between the masonry and the soil/rock mass, μ	:	0.50, 0.70, 0.80

The base width of the stone masonry breast wall has been fixed to satisfy the following conditions:

1. Factor of safety against overturning is 1.50 or more.
2. There is no tension at the heel.
3. The maximum allowable soil-bearing pressure is 100 kN/m². In other words, pressure at the toe does not exceed 100 kN/m².

For all the previous combinations, the following wall heights have been considered for developing the charts/tables:

Height of breast wall, H   :   1.0 m, 1.5 m, 2.0 m, 2.5 m, 3.0 m, 3.5 m, 4.0 m, 4.5 m, 5.0 m

5.2 CHARTS FOR THE DESIGN OF A 1.0-M-HIGH MASONRY BREAST WALL

The charts/tables developed for a 1.0-m-high stone masonry breast wall for the previously described six combinations are presented in Tables 5.1 through 5.6 and Figures 5.2 through 5.7.

TABLE 5.1

H = 1.0 m, $\gamma_{masonry}$ = 22.56 kN/m³, γ_{sat} = 15.70 kN/m³ and ι = 0°

A. Results of Stability Analysis

				Pressure at		$\mu = 0.50$		$\mu = 0.70$		$\mu = 0.80$	
ϕ Degree	α Degree	K_{ah}	F_{OVT}	Heel, kN/m²	Toe, kN/m²	F_{SLD}	Net Force, kN	F_{SLD}	Net Force, kN	F_{SLD}	Net Force, kN
25	0.00	0.346	1.94	0.18	28.56	1.07	1.30	1.49	0.02	1.71	0.00
25	−5.71	0.316	1.73	0.17	31.02	0.98	1.44	1.38	0.34	1.57	0.00
25	−11.31	0.286	1.54	0.98	34.20	0.90	1.54	1.26	0.62	1.44	0.17
25	−14.03	0.271	1.51	4.57	33.10	0.87	1.55	1.21	0.70	1.39	0.28
30	0.00	0.279	1.96	0.07	29.69	1.19	0.78	1.66	0.00	1.90	0.00
30	−5.71	0.250	1.75	0.70	32.20	1.10	0.93	1.53	0.00	1.75	0.00
30	−11.31	0.221	1.52	2.05	36.77	0.99	1.06	1.38	0.24	1.58	0.00
30	−14.03	0.207	1.53	10.49	31.69	0.96	1.06	1.35	0.30	1.54	0.00
35	0.00	0.226	2.02	0.43	30.34	1.33	0.37	1.86	0.00	2.12	0.00
35	−5.71	0.198	1.73	0.15	35.03	1.20	0.56	1.69	0.00	1.93	0.00
35	−11.31	0.170	1.61	9.13	33.06	1.12	0.65	1.56	0.00	1.79	0.00
35	−14.03	0.156	1.94	31.79	10.39	1.19	0.50	1.66	0.00	1.90	0.00
40	0.00	0.184	2.06	0.62	31.19	1.47	0.06	2.06	0.00	2.35	0.00
40	−5.71	0.156	1.72	0.24	37.43	1.32	0.28	1.85	0.00	2.12	0.00
40	−11.31	0.130	2.02	26.05	16.13	1.35	0.20	1.90	0.00	2.17	0.00
40	−14.03	0.117	2.06	37.72	0.19	1.21	0.38	1.69	0.00	1.93	0.00

B. Physical Dimensnsions for a Technoeconomic Wall Section

					Depth of Shear Key D_{sk}, mm			
ϕ Degree	α Degree	Top Width, mm	Bottom Width, mm	Horizontal Angle θ Degree	$\mu = 0.50$	$\mu = 0.70$	$\mu = 0.80$	Volume of Wall m³
25	0.00	200	495	90.00	300	100	0	0.348
25	−5.71	200	390	84.29	300	200	0	0.295
25	−11.31	200	290	78.69	300	200	100	0.245
25	−14.03	200	250	75.97	300	200	100	0.225
30	0.00	200	445	90.00	200	0	0	0.323
30	−5.71	200	340	84.29	200	0	0	0.270
30	−11.31	200	235	78.69	200	100	0	0.218
30	−14.03	200	200	75.97	200	100	0	0.200
35	0.00	200	405	90.00	100	0	0	0.303
35	−5.71	200	290	84.29	100	0	0	0.245

(Continued)

TABLE 5.1 (Continued)

$H = 1.0$ m, $\gamma_{masonry} = 22.56$ kN/m³, $\gamma_{sat} = 15.70$ kN/m³ and $\iota = 0°$

ϕ Degree	α Degree	Top Width, mm	Bottom Width, mm	Horizontal Angle θ Degree	Depth of Shear Key D_{sk}, mm			Volume of Wall m³
					$\mu = 0.50$	$\mu = 0.70$	$\mu = 0.80$	
35	−11.31	200	200	78.69	100	0	0	0.200
35	−14.03	200	200	75.97	100	0	0	0.200
40	0.00	200	370	90.00	100	0	0	0.285
40	−5.71	200	250	84.29	100	0	0	0.225
40	−11.31	200	200	78.69	100	0	0	0.200
40	−14.03	150	185	75.97	100	0	0	0.168

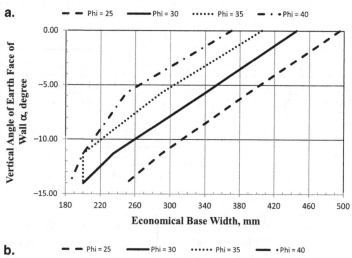

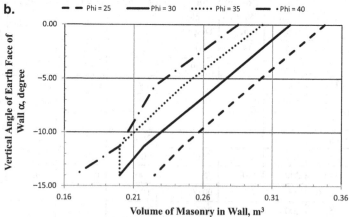

FIGURE 5.2 Design Charts for a Masonry Wall $H = 1.0$ m, $\gamma_{masonry} = 22.56$ kN/m³, $\gamma_{sat} = 15.70$ kN/m³ and $\iota = 0°$: (a) Economical Base Width for a Wall, (b) Volume of Masonry in Wall

TABLE 5.2

H = 1.0 m, $\gamma_{masonry}$ = 22.56 kN/m³, γ_{sat} = 17.27 kN/m³ and ι = 0°

A. Results of Stability Analysis

ϕ Degree	α Degree	K_{ah}	F_{OVT}	Pressure at		$\mu = 0.50$		$\mu = 0.70$		$\mu = 0.80$	
				Heel, kN/m²	Toe, kN/m²	F_{SLD}	Net Force, kN	F_{SLD}	Net Force, kN	F_{SLD}	Net Force, kN
25	0.00	0.346	1.93	0.22	28.07	1.01	1.59	1.42	0.27	1.62	0.00
25	−5.71	0.316	1.74	0.26	30.23	0.94	1.70	1.32	0.56	1.50	0.00
25	−11.31	0.286	1.54	0.07	34.11	0.86	1.79	1.20	0.84	1.37	0.36
25	−14.03	0.271	1.50	3.12	33.22	0.83	1.78	1.16	0.90	1.33	0.46
30	0.00	0.279	1.97	0.43	28.79	1.13	1.00	1.59	0.00	1.81	0.00
30	−5.71	0.250	1.74	0.46	31.70	1.04	1.13	1.46	0.09	1.67	0.00
30	−11.31	0.221	1.50	0.18	37.49	0.94	1.26	1.32	0.41	1.51	0.00
30	−14.03	0.207	1.51	7.33	33.28	0.92	1.23	1.29	0.45	1.47	0.06
35	0.00	0.226	2.01	0.54	29.70	1.26	0.54	1.77	0.00	2.02	0.00
35	−5.71	0.198	1.74	0.45	33.73	1.16	0.71	1.62	0.00	1.85	0.00
35	−11.31	0.170	1.52	3.76	37.87	1.05	0.83	1.47	0.06	1.68	0.00
35	−14.03	0.156	1.79	25.08	17.11	1.10	0.68	1.55	0.00	1.77	0.00
40	0.00	0.184	2.03	0.40	30.95	1.39	0.21	1.95	0.00	2.23	0.00
40	−5.71	0.156	1.76	1.17	35.16	1.28	0.38	1.79	0.00	2.05	0.00
40	−11.31	0.130	1.86	20.48	21.70	1.26	0.36	1.77	0.00	2.02	0.00
40	−14.03	0.117	2.05	33.47	3.07	1.18	0.45	1.65	0.00	1.88	0.00

B. Physical Dimensions for a Technoeconomic Wall Section

ϕ Degree	α Degree	Top Width, mm	Bottom Width, mm	Horizontal Angle θ Degree	Depth of Shear Key D_{sk}, mm			Volume of Wall, m³
					$\mu = 0.50$	$\mu = 0.70$	$\mu = 0.80$	
25	0.00	200	520	90.00	300	100	0	0.360
25	−5.71	200	415	84.29	300	200	0	0.308
25	−11.31	200	310	78.69	300	200	200	0.255
25	−14.03	200	270	75.97	300	200	200	0.235
30	0.00	200	470	90.00	200	0	0	0.335
30	−5.71	200	360	84.29	200	100	0	0.280
30	−11.31	200	250	78.69	200	100	0	0.225
30	−14.03	200	215	75.97	200	100	100	0.208
35	0.00	200	425	90.00	100	0	0	0.313
35	−5.71	200	310	84.29	100	0	0	0.255

(Continued)

TABLE 5.2 (Continued)

H = 1.0 m, $\gamma_{masonry}$ = 22.56 kN/m³, γ_{sat} = 17.27 kN/m³ and ι = 0°

ϕ Degree	α Degree	Top Width, mm	Bottom Width, mm	Horizontal Angle θ Degree	Depth of Shear Key D_{sk}, mm			Volume of Wall, m³
					μ = 0.50	μ = 0.70	μ = 0.80	
35	−11.31	200	205	78.69	200	100	0	0.203
35	−14.03	200	200	75.97	100	0	0	0.200
40	0.00	200	385	90.00	100	0	0	0.293
40	−5.71	200	270	84.29	100	0	0	0.235
40	−11.31	200	200	78.69	100	0	0	0.200
40	−14.03	150	200	75.97	100	0	0	0.175

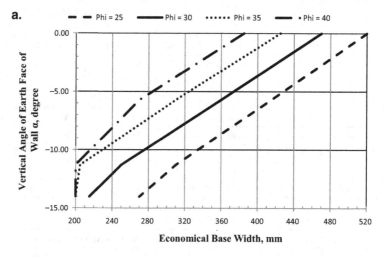

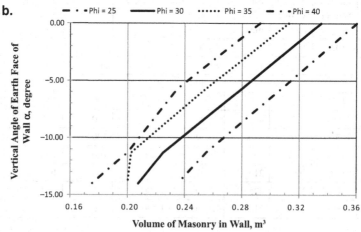

FIGURE 5.3 Design Charts for Masonry Wall H = 1.0 m, $\gamma_{masonry}$ = 22.56 kN/m³, γ_{sat} = 17.27 kN/m³ and ι = 0°: (a) Economical Base Width for a Wall, (b) Volume of Masonry in Wall

TABLE 5.3
$H = 1.0$ m, $\gamma_{masonry} = 22.56$ kN/m^3, $\gamma_{sat} = 15.70$ kN/m^3 and $\iota = \frac{1}{2}\phi$

A. Results of Stability Analysis

ϕ Degree	α Degree	K_{ah}	F_{OVT}	Pressure at Heel, kN/m^2	Toe, kN/m^2	$\mu = 0.50$ F_{SLD}	Net Force, kN	$\mu = 0.70$ F_{SLD}	Net Force, kN	$\mu = 0.80$ F_{SLD}	Net Force, kN
25	0.00	0.424	1.93	0.34	27.49	0.96	1.92	1.35	0.55	1.54	0.00
25	−5.71	0.385	1.74	0.11	29.76	0.90	1.98	1.26	0.79	1.44	0.20
25	−11.31	0.346	1.55	0.29	32.80	0.83	2.01	1.16	1.01	1.33	0.51
25	−14.03	0.326	1.50	2.11	33.07	0.80	1.99	1.12	1.07	1.28	0.61
30	0.00	0.348	1.96	0.48	28.17	1.07	1.31	1.49	0.02	1.71	0.00
30	−5.71	0.309	1.74	0.25	31.09	0.99	1.38	1.39	0.29	1.59	0.00
30	−11.31	0.270	1.53	0.60	35.43	0.91	1.43	1.28	0.54	1.46	0.10
30	−14.03	0.251	1.51	5.74	33.08	0.89	1.40	1.25	0.58	1.43	0.17
35	0.00	0.284	1.97	0.22	29.43	1.18	0.82	1.65	0.00	1.89	0.00
35	−5.71	0.246	1.75	0.56	32.53	1.10	0.89	1.55	0.00	1.77	0.00
35	−11.31	0.208	1.53	2.97	36.71	1.01	0.97	1.42	0.16	1.62	0.00
35	−14.03	0.190	1.64	17.34	24.84	1.03	0.88	1.44	0.12	1.64	0.00
40	0.00	0.232	2.01	0.43	30.19	1.31	0.41	1.83	0.00	2.10	0.00
40	−5.71	0.194	1.75	0.92	34.26	1.22	0.52	1.71	0.00	1.96	0.00
40	−11.31	0.158	1.71	13.97	28.21	1.18	0.52	1.65	0.00	1.88	0.00
40	−14.03	0.142	2.11	37.92	4.27	1.27	0.34	1.78	0.00	2.04	0.00

B. Physical Dimensions for a Technoeconomic Wall Section

ϕ Degree	α Degree	Top Width, mm	Bottom Width, mm	Horizontal Angle θ Degree	Depth of Shear Key D_{sk}, mm $\mu = 0.50$	$\mu = 0.70$	$\mu = 0.80$	Volume of Wall, m^3
25	0.00	200	550	90.00	300	200	0	0.375
25	−5.71	200	440	84.29	300	200	100	0.320
25	−11.31	200	335	78.69	300	200	200	0.268
25	−14.03	200	290	75.97	300	200	200	0.245
30	0.00	200	500	90.00	200	100	0	0.350
30	−5.71	200	385	84.29	200	100	0	0.293
30	−11.31	200	275	78.69	200	200	100	0.238
30	−14.03	200	235	75.97	200	200	100	0.218
35	0.00	200	450	90.00	200	0	0	0.325
35	−5.71	200	335	84.29	200	0	0	0.268

(Continued)

TABLE 5.3 (Continued)

H = 1.0 m, $\gamma_{masonry}$ = 22.56 kN/m³, γ_{sat} = 15.70 kN/m³ and ι = ½φ

φ Degree	α Degree	Top Width, mm	Bottom Width, mm	Horizontal Angle θ Degree	Depth of Shear Key D_{sk}, mm			Volume of Wall, m³
					μ = 0.50	μ = 0.70	μ = 0.80	
35	−11.31	200	225	78.69	200	100	0	0.213
35	−14.03	200	200	75.97	200	100	0	0.200
40	0.00	200	410	90.00	100	0	0	0.305
40	−5.71	200	290	84.29	100	0	0	0.245
40	−11.31	200	200	78.69	100	0	0	0.200
40	−14.03	200	200	75.97	100	0	0	0.200

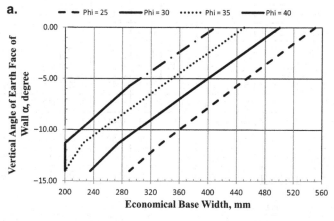

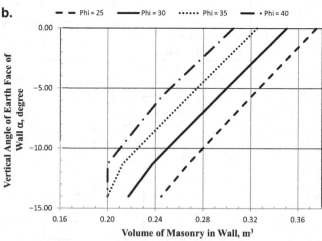

FIGURE 5.4 Design Charts for Masonry Wall H = 1.0 m, $\gamma_{masonry}$ = 22.56 kN/m³, γ_{sat} = 15.70 kN/m³ and ι = ½φ: (a) Economical Base Width for a Wall, (b) Volume of a Masonry in Wall

TABLE 5.4

$H = 1.0$ m, $\gamma_{masonry} = 22.56$ kN/m³, $\gamma_{sat} = 17.27$ kN/m³ and $\iota = \frac{1}{2}\phi$

A. Results of Stability Analysis

ϕ Degree	α Degree	K_{ah}	F_{OVT}	Pressure at		$\mu = 0.50$		$\mu = 0.70$		$\mu = 0.80$	
				Heel, kN/m²	Toe, kN/m²	F_{SLD}	Net Force, kN	F_{SLD}	Net Force, kN	F_{SLD}	Net Force, kN
25	0.00	0.424	1.90	0.12	27.35	0.91	2.31	1.27	0.89	1.45	0.18
25	−5.71	0.385	1.75	0.45	28.77	0.86	2.30	1.20	1.06	1.38	0.44
25	−11.31	0.346	1.56	0.04	32.12	0.80	2.30	1.11	1.26	1.27	0.74
25	−14.03	0.326	1.51	1.68	32.27	0.77	2.26	1.08	1.30	1.24	0.82
30	0.00	0.348	1.92	0.04	28.26	1.01	1.63	1.41	0.30	1.61	0.00
30	−5.71	0.309	1.74	0.39	30.24	0.95	1.63	1.33	0.50	1.52	0.00
30	−11.31	0.270	1.55	0.98	33.69	0.88	1.64	1.23	0.71	1.41	0.24
30	−14.03	0.251	1.51	4.30	33.01	0.86	1.61	1.20	0.75	1.37	0.32
35	0.00	0.284	1.94	0.03	29.19	1.12	1.06	1.56	0.00	1.79	0.00
35	−5.71	0.246	1.74	0.37	31.96	1.05	1.10	1.47	0.06	1.68	0.00
35	−11.31	0.208	1.51	1.12	37.31	0.97	1.14	1.35	0.32	1.55	0.00
35	−14.03	0.190	1.51	9.16	33.03	0.95	1.10	1.33	0.34	1.52	0.00
40	0.00	0.232	2.00	0.51	29.60	1.25	0.59	1.74	0.00	1.99	0.00
40	−5.71	0.194	1.73	0.22	34.20	1.16	0.68	1.63	0.00	1.86	0.00
40	−11.31	0.158	1.57	7.16	35.02	1.09	0.71	1.53	0.00	1.75	0.00
40	−14.03	0.142	1.95	31.83	10.35	1.19	0.50	1.66	0.00	1.90	0.00

B. Physical Dimensions for a Technoeconomic Wall Section

ϕ Degree	α Degree	Top Width, mm	Bottom Width, mm	Horizontal Angle θ Degree	Depth of Shear Key D_{sk}, mm			Volume of Wall, m³
					$\mu = 0.50$	$\mu = 0.70$	$\mu = 0.80$	
25	0.00	200	575	90.00	300	200	100	0.388
25	−5.71	200	470	84.29	300	200	200	0.335
25	−11.31	200	360	78.69	300	200	200	0.280
25	−14.03	200	315	75.97	300	200	200	0.258
30	0.00	200	520	90.00	200	100	0	0.360
30	−5.71	200	410	84.29	200	200	0	0.305
30	−11.31	200	300	78.69	200	200	100	0.250
30	−14.03	200	255	75.97	200	200	100	0.228
35	0.00	200	470	90.00	200	0	0	0.335
35	−5.71	200	355	84.29	200	100	0	0.278

(Continued)

TABLE 5.4 (Continued)

$H = 1.0$ m, $\gamma_{masonry} = 22.56$ kN/m^3, $\gamma_{sat} = 17.27$ kN/m^3 and $\iota = \frac{1}{2}\phi$

ϕ Degree	α Degree	Top Width, mm	Bottom Width, mm	Horizontal Angle θ Degree	Depth of Shear Key D_{sk}, mm			Volume of Wall, m^3
					$\mu = 0.50$	$\mu = 0.70$	$\mu = 0.80$	
35	−11.31	200	240	78.69	200	100	0	0.220
35	−14.03	200	200	75.97	200	100	0	0.200
40	0.00	200	430	90.00	100	0	0	0.315
40	−5.71	200	305	84.29	100	0	0	0.253
40	−11.31	200	200	78.69	100	0	0	0.200
40	−14.03	200	200	75.97	100	0	0	0.200

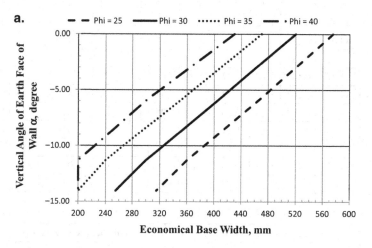

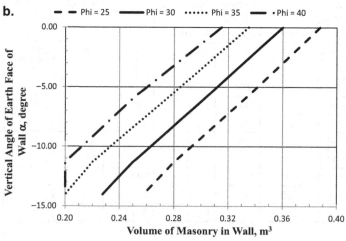

FIGURE 5.5 Design Charts for a Masonry Wall, $H = 1.0$ m, $\gamma_{masonry} = 22.56$ kN/m^3, $\gamma_{sat} = 17.27$ kN/m^3 and $\iota = \frac{1}{2}\phi$: (a) Economical Base Width for a Wall, (b) Volumen of a Masonry in Wall

TABLE 5.5
$H = 1.0$ m, $\gamma_{masonry} = 22.56$ kN/m³, $\gamma_{sat} = 15.70$ kN/m³ and $\iota = \phi$

A. Results of Stability Analysis

ϕ Degree	α Degree	K_{ah}	F_{OVT}	Pressure at		$\mu = 0.50$		$\mu = 0.70$		$\mu = 0.80$	
				Heel, kN/m²	Toe, kN/m²	F_{SLD}	Net Force, kN	F_{SLD}	Net Force, kN	F_{SLD}	Net Force, kN
25	0.00	0.821	1.86	0.14	25.34	0.68	5.37	0.95	3.60	1.08	2.72
25	−5.71	0.746	1.76	0.36	26.10	0.66	5.02	0.92	3.45	1.05	2.67
25	−11.31	0.675	1.64	0.49	27.34	0.63	4.71	0.89	3.34	1.01	2.65
25	−14.03	0.641	1.57	0.45	28.29	0.62	4.58	0.86	3.30	0.99	2.66
30	0.00	0.750	1.87	0.18	25.58	0.71	4.73	0.99	3.03	1.14	2.17
30	−5.71	0.666	1.75	0.23	26.72	0.69	4.32	0.97	2.83	1.11	2.09
30	−11.31	0.587	1.61	0.08	28.66	0.67	3.97	0.94	2.69	1.07	2.06
30	−14.03	0.549	1.55	0.48	29.39	0.66	3.80	0.92	2.62	1.05	2.03
35	0.00	0.671	1.90	0.59	25.48	0.76	4.00	1.06	2.36	1.21	1.54
35	−5.71	0.580	1.74	0.11	27.49	0.74	3.59	1.04	2.19	1.19	1.49
35	−11.31	0.496	1.60	0.22	29.66	0.72	3.21	1.01	2.03	1.15	1.44
35	−14.03	0.456	1.52	0.55	30.94	0.70	3.04	0.99	1.96	1.13	1.42
40	0.00	0.587	1.90	0.38	26.19	0.81	3.29	1.14	1.74	1.30	0.96
40	−5.71	0.492	1.76	0.46	27.92	0.80	2.84	1.13	1.52	1.29	0.87
40	−11.31	0.406	1.58	0.35	31.14	0.78	2.48	1.09	1.40	1.25	0.86
40	−14.03	0.366	1.51	1.65	32.07	0.77	2.30	1.08	1.33	1.23	0.84

B. Physical Dimensions for a Technoeconomic Wall Section

ϕ Degree	α Degree	Top Width, mm	Bottom Width, mm	Horizontal Angle θ Degree	Depth of Shear Key D_{sk}, mm			Volume of Wall, m³
					$\mu = 0.50$	$\mu = 0.70$	$\mu = 0.80$	
25	0.00	200	770	90.00	500	400	300	0.485
25	−5.71	200	660	84.29	500	400	300	0.430
25	−11.31	200	550	78.69	400	400	300	0.375
25	−14.03	200	495	75.97	400	400	300	0.348
30	0.00	200	735	90.00	400	300	300	0.468
30	−5.71	200	615	84.29	400	300	300	0.408
30	−11.31	200	495	78.69	300	300	300	0.348
30	−14.03	200	440	75.97	300	300	300	0.320
35	0.00	200	700	90.00	300	200	200	0.450
35	−5.71	200	565	84.29	300	200	200	0.383
35	−11.31	200	440	78.69	300	200	200	0.320
35	−14.03	200	380	75.97	300	200	200	0.290

(Continued)

TABLE 5.5 (Continued)

H = 1.0 m, $\gamma_{masonry}$ = 22.56 kN/m³, γ_{sat} = 15.70 kN/m³ and $\iota = \phi$

					Depth of Shear Key D_{sk}, mm			
ϕ Degree	α Degree	Top Width, mm	Bottom Width, mm	Horizontal Angle θ Degree	$\mu = 0.50$	$\mu = 0.70$	$\mu = 0.80$	Volume of Wall, m³
40	0.00	200	650	90.00	100	100	0	0.425
40	−5.71	200	515	84.29	200	200	100	0.358
40	−11.31	200	380	78.69	200	200	100	0.290
40	−14.03	200	320	75.97	200	200	100	0.260

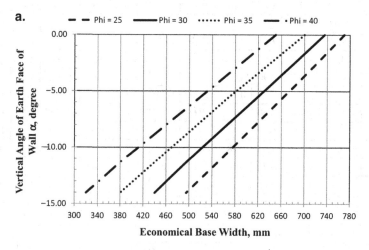

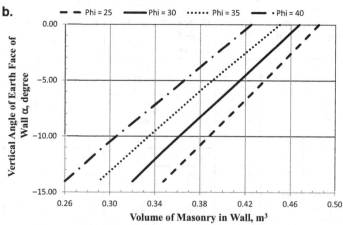

FIGURE 5.6 Design Charts for a Masonry Wall H = 1.0 m, $\gamma_{masonry}$ = 22.56 kN/m³, γ_{sat} = 15.70 kN/m³ and $\iota = \phi$: (a) Economical Base Width for a Wall, (b) Volume of a Masonry in Wall

TABLE 5.6

$H = 1.0$ m, $\gamma_{masonry} = 22.56$ kN/m^3, $\gamma_{sat} = 17.27$ kN/m^3 and $\iota = \phi$

A. Results of Stability Analysis

ϕ Degree	α Degree	K_{ah}	F_{OVT}	Pressure at Heel, kN/m^2	Toe, kN/m^2	$\mu = 0.50$ F_{SLD}	Net Force, kN	$\mu = 0.70$ F_{SLD}	Net Force, kN	$\mu = 0.80$ F_{SLD}	Net Force, kN
25	0.00	0.821	1.86	0.14	25.05	0.64	6.16	0.90	4.32	1.02	3.40
25	−5.71	0.746	1.76	0.43	25.64	0.63	5.72	0.88	4.08	1.00	3.26
25	−11.31	0.675	1.64	0.24	27.09	0.60	5.35	0.84	3.92	0.96	3.20
25	−14.03	0.641	1.58	0.23	27.90	0.59	5.18	0.83	3.84	0.94	3.17
30	0.00	0.750	1.87	0.29	25.15	0.67	5.43	0.94	3.66	1.08	2.77
30	−5.71	0.666	1.75	0.11	26.45	0.66	4.95	0.92	3.40	1.05	2.62
30	−11.31	0.587	1.62	0.10	28.03	0.64	4.51	0.90	3.17	1.02	2.50
30	−14.03	0.549	1.57	0.59	28.53	0.63	4.29	0.88	3.05	1.01	2.42
35	0.00	0.671	1.87	0.21	25.59	0.71	4.66	1.00	2.97	1.14	2.12
35	−5.71	0.580	1.75	0.21	26.93	0.71	4.12	0.99	2.66	1.13	1.92
35	−11.31	0.496	1.62	0.59	28.53	0.69	3.64	0.97	2.40	1.11	1.77
35	−14.03	0.456	1.54	0.38	30.25	0.68	3.44	0.95	2.31	1.08	1.75
40	0.00	0.587	1.88	0.15	26.11	0.77	3.85	1.07	2.24	1.23	1.44
40	−5.71	0.492	1.75	0.34	27.56	0.76	3.28	1.07	1.92	1.22	1.23
40	−11.31	0.406	1.59	0.53	30.10	0.75	2.82	1.05	1.69	1.20	1.13
40	−14.03	0.366	1.52	1.00	31.70	0.74	2.62	1.03	1.61	1.18	1.10

B. Physical Dimensions for a Technoeconomic Wall Section

ϕ Degree	α Degree	Top Width, mm	Bottom Width, mm	Horizontal Angle θ Degree	Depth of Shear Key D_{sk}, mm $\mu = 0.50$	$\mu = 0.70$	$\mu = 0.80$	Volume of Wall, m^3
25	0.00	200	810	90.00	500	400	400	0.505
25	−5.71	200	700	84.29	500	400	400	0.450
25	−11.31	200	585	78.69	400	400	400	0.393
25	−14.03	200	530	75.97	400	400	400	0.365
30	0.00	200	775	90.00	400	300	300	0.488
30	−5.71	200	650	84.29	400	300	300	0.425
30	−11.31	200	530	78.69	400	300	300	0.365
30	−14.03	200	475	75.97	300	300	300	0.338
35	0.00	200	730	90.00	200	100	100	0.465
35	−5.71	200	600	84.29	300	200	200	0.400
35	−11.31	200	475	78.69	300	200	200	0.338
35	−14.03	200	410	75.97	300	200	200	0.305

(Continued)

TABLE 5.6 (Continued)

$H = 1.0$ m, $\gamma_{masonry} = 22.56$ kN/m³, $\gamma_{sat} = 17.27$ kN/m³ and $\iota = \phi$

					Depth of Shear Key D_{sk}, mm			
ϕ Degree	α Degree	Top Width, mm	Bottom Width, mm	Horizontal Angle θ Degree	$\mu = 0.50$	$\mu = 0.70$	$\mu = 0.80$	Volume of Wall, m³
40	0.00	200	680	90.00	100	100	100	0.440
40	−5.71	200	545	84.29	200	200	200	0.373
40	−11.31	200	410	78.69	200	200	200	0.305
40	−14.03	200	345	75.97	200	200	100	0.273

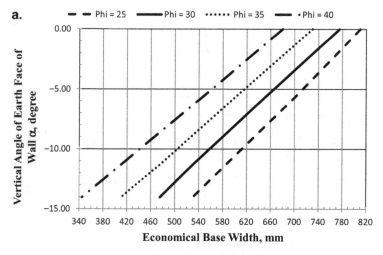

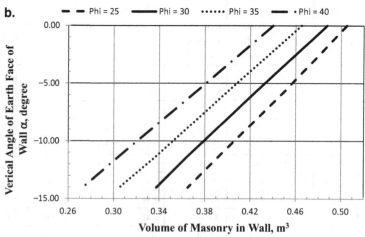

FIGURE 5.7 Design Charts for a Masonry Wall $H = 1.0$ m, $\gamma_{masonry} = 22.56$ kN/m³, $\gamma_{sat} = 17.27$ kN/m³ and $\iota = \phi$: (a) Economical Base Width for a Wall, (b) Volume of a Masonry in Wall

5.3 CHARTS FOR THE DESIGN OF A 1.5-M-HIGH MASONRY BREAST WALL

The charts/tables developed for a 1.5-m-high stone masonry breast wall for the previously discussed six combinations are presented in Tables 5.7 to 5.12 and Figures 5.8 to 5.13.

TABLE 5.7

$H = 1.5$ m, $\gamma_{masonry} = 22.56$ kN/m^3, $\gamma_{sat} = 15.70$ kN/m^3 and $\iota = 0°$

A. Results of Stability Analysis

				Pressure at		$\mu = 0.50$		$\mu = 0.70$		$\mu = 0.80$	
ϕ Degree	α Degree	K_{ah}	F_{OVT}	Heel, kN/ m^2	Toe, kN/ m^2	F_{SLD}	Net Force, kN	F_{SLD}	Net Force, kN	F_{SLD}	Net Force, kN
25	0.00	0.346	1.87	0.38	37.95	0.97	3.57	1.36	0.95	1.55	0.00
25	−5.71	0.316	1.70	0.52	39.92	0.89	3.79	1.25	1.56	1.43	0.44
25	−11.31	0.286	1.53	0.95	42.73	0.81	3.96	1.14	2.10	1.30	1.17
25	−14.03	0.271	1.51	5.53	39.82	0.79	3.92	1.10	2.18	1.26	1.32
30	0.00	0.279	1.88	0.14	39.25	1.07	2.44	1.49	0.03	1.71	0.00
30	−5.71	0.250	1.68	0.25	41.95	0.97	2.71	1.36	0.70	1.56	0.00
30	−11.31	0.221	1.50	2.00	44.79	0.88	2.91	1.23	1.27	1.40	0.45
30	−14.03	0.207	1.50	9.40	39.94	0.85	2.88	1.19	1.37	1.36	0.61
35	0.00	0.226	1.91	0.51	39.93	1.18	1.54	1.65	0.00	1.88	0.00
35	−5.71	0.198	1.66	0.04	44.27	1.06	1.89	1.48	0.08	1.69	0.00
35	−11.31	0.170	1.50	5.67	44.91	0.95	2.09	1.33	0.63	1.52	0.00
35	−14.03	0.156	1.51	18.42	36.08	0.92	2.08	1.29	0.75	1.47	0.09
40	0.00	0.184	1.92	0.33	41.30	1.28	0.89	1.79	0.00	2.05	0.00
40	−5.71	0.156	1.66	0.95	45.62	1.15	1.26	1.61	0.00	1.84	0.00
40	−11.31	0.130	1.52	12.23	43.24	1.03	1.48	1.44	0.19	1.65	0.00
40	−14.03	0.117	1.54	35.50	26.16	0.99	1.49	1.39	0.33	1.59	0.00

B. Physical Dimensions for a Technoeconomic Wall Section

					Depth of Shear Key D_{sk}, mm			
ϕ Degree	α Degree	Top Width, mm	Bottom Width, mm	Horizontal Angle θ Degree	$\mu = 0.50$	$\mu = 0.70$	$\mu = 0.80$	Volume of Wall, m^3
25	0.00	200	760	90.00	400	200	0	0.720
25	−5.71	200	615	84.29	400	300	200	0.611
25	−11.31	200	475	78.69	400	300	200	0.506
25	−14.03	200	425	75.97	400	300	300	0.469
30	0.00	200	680	90.00	300	100	0	0.660

(Continued)

TABLE 5.7 (Continued)

$H = 1.5$ m, $\gamma_{masonry} = 22.56$ kN/m³, $\gamma_{sat} = 15.70$ kN/m³ and $\iota = 0°$

ϕ Degree	α Degree	Top Width, mm	Bottom Width, mm	Horizontal Angle θ Degree	Depth of Shear Key D_{sk}, mm $\mu = 0.50$	$\mu = 0.70$	$\mu = 0.80$	Volume of Wall, m³
30	−5.71	200	530	84.29	300	200	0	0.548
30	−11.31	200	390	78.69	300	200	100	0.443
30	−14.03	200	340	75.97	300	200	200	0.405
35	0.00	200	615	90.00	200	0	0	0.611
35	−5.71	200	455	84.29	200	100	0	0.491
35	−11.31	200	320	78.69	200	100	0	0.390
35	−14.03	200	270	75.97	200	200	100	0.353
40	0.00	200	555	90.00	100	0	0	0.566
40	−5.71	200	395	84.29	200	0	0	0.446
40	−11.31	200	260	78.69	200	100	0	0.345
40	−14.03	200	210	75.97	200	100	0	0.308

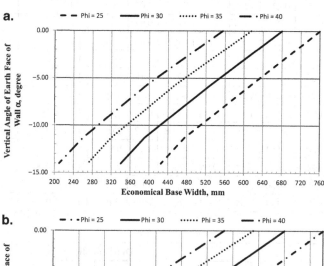

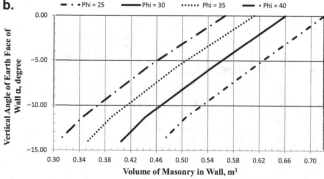

FIGURE 5.8 Design Charts for a Masonry Wall $H = 1.5$ m, $\gamma_{masonry} = 22.56$ kN/m³, $\gamma_{sat} = 15.70$ kN/m³ and $\iota = 0°$: (a) Economical Base Width for a Wall, (b) Volume of a Masonry in Wall

TABLE 5.8

H = 1.5 m, $\gamma_{masonry}$ = 22.56 kN/m³, γ_{sat} = 17.27 kN/m³ and ι = 0°

A. Results of Stability Analysis

ϕ Degree	α Degree	K_{ah}	F_{OVT}	Pressure at Heel, kN/m²	Toe, kN/m²	$\mu = 0.50$ F_{SLD}	Net Force, kN	$\mu = 0.70$ F_{SLD}	Net Force, kN	$\mu = 0.80$ F_{SLD}	Net Force, kN
25	0.00	0.346	1.85	0.08	37.86	0.92	4.25	1.29	1.54	1.48	0.18
25	−5.71	0.316	1.69	0.19	39.66	0.85	4.40	1.20	2.07	1.37	0.91
25	−11.31	0.286	1.53	0.62	42.08	0.78	4.49	1.10	2.53	1.25	1.55
25	−14.03	0.271	1.51	3.81	40.50	0.76	4.43	1.06	2.62	1.21	1.71
30	0.00	0.279	1.87	0.22	38.67	1.02	2.95	1.43	0.45	1.63	0.00
30	−5.71	0.250	1.69	0.50	40.91	0.94	3.14	1.31	1.04	1.50	0.00
30	−11.31	0.221	1.51	1.38	44.17	0.85	3.29	1.19	1.57	1.36	0.71
30	−14.03	0.207	1.51	8.02	39.70	0.83	3.22	1.16	1.64	1.32	0.84
35	0.00	0.226	1.90	0.47	39.45	1.12	1.93	1.57	0.00	1.80	0.00
35	−5.71	0.198	1.67	0.19	43.19	1.02	2.22	1.43	0.33	1.63	0.00
35	−11.31	0.170	1.50	4.27	44.78	0.92	2.37	1.29	0.85	1.48	0.09
35	−14.03	0.156	1.51	15.39	36.99	0.90	2.32	1.26	0.94	1.44	0.24
40	0.00	0.184	1.90	0.11	40.99	1.22	1.21	1.71	0.00	1.96	0.00
40	−5.71	0.156	1.67	0.77	44.78	1.11	1.51	1.55	0.00	1.78	0.00
40	−11.31	0.130	1.51	9.32	44.29	1.00	1.69	1.40	0.34	1.60	0.00
40	−14.03	0.117	1.51	27.09	32.42	0.96	1.70	1.34	0.50	1.53	0.00

B. Physical Dimensions for a Technoeconomic Wall Section

ϕ Degree	α Degree	Top Width, mm	Bottom Width, mm	Horizontal Angle θ Degree	Depth of Shear Key D_{sk}, mm $\mu = 0.50$	$\mu = 0.70$	$\mu = 0.80$	Volume of Wall, m³
25	0.00	200	795	90.00	400	300	100	0.746
25	−5.71	200	650	84.29	400	300	200	0.638
25	−11.31	200	510	78.69	400	300	300	0.533
25	−14.03	200	455	75.97	400	300	300	0.491
30	0.00	200	715	90.00	300	100	0	0.686
30	−5.71	200	565	84.29	300	200	0	0.574
30	−11.31	200	420	78.69	300	200	200	0.465
30	−14.03	200	370	75.97	300	200	200	0.428
35	0.00	200	645	90.00	200	0	0	0.634
35	−5.71	200	485	84.29	200	100	0	0.514
35	−11.31	200	345	78.69	200	200	100	0.409
35	−14.03	200	295	75.97	200	200	100	0.371

(Continued)

TABLE 5.8 (Continued)

H = 1.5 m, $\gamma_{masonry}$ = 22.56 kN/m³, γ_{sat} = 17.27 kN/m³ and ι = 0°

ϕ Degree	α Degree	Top Width, mm	Bottom Width, mm	Horizontal Angle θ Degree	Depth of Shear Key D_{sk}, mm			Volume of Wall, m³
					μ = 0.50	μ = 0.70	μ = 0.80	
40	0.00	200	580	90.00	200	0	0	0.585
40	−5.71	200	420	84.29	200	0	0	0.465
40	−11.31	200	280	78.69	200	100	0	0.360
40	−14.03	200	225	75.97	200	100	0	0.319

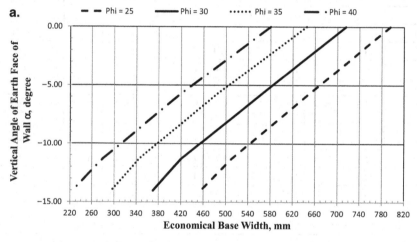

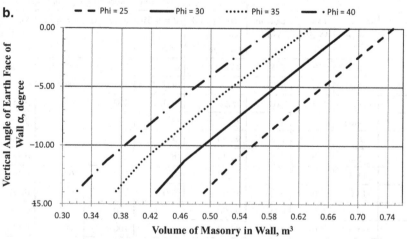

FIGURE 5.9 Design Charts for a Masonry Wall H = 1.5 m, $\gamma_{masonry}$ = 22.56 kN/m³, γ_{sat} = 17.27 kN/m³ and ι = 0°: (a) Economical Base Width for a Wall, (b) Volume of a Masonry in Wall

TABLE 5.9

$H = 1.5$ m, $\gamma_{masonry} = 22.56$ kN/m³, $\gamma_{sat} = 15.70$ kN/m³ and $\iota = \frac{1}{2}\phi$

A. Results of Stability Analysis

ϕ Degree	α Degree	K_{ah}	F_{OVT}	Pressure at		$\mu = 0.50$		$\mu = 0.70$		$\mu = 0.80$	
				Heel, kN/m²	Toe, kN/m²	F_{SLD}	Net Force, kN	F_{SLD}	Net Force, kN	F_{SLD}	Net Force, kN
25	0.00	0.424	1.85	0.12	37.37	0.88	4.99	1.23	2.16	1.41	0.74
25	−5.71	0.385	1.69	0.14	39.10	0.82	5.01	1.15	2.57	1.32	1.36
25	−11.31	0.346	1.54	0.34	41.51	0.76	4.99	1.06	2.94	1.22	1.92
25	−14.03	0.326	1.50	2.58	40.80	0.74	4.91	1.03	3.02	1.18	2.07
30	0.00	0.348	1.86	0.12	38.21	0.96	3.63	1.35	1.01	1.54	0.00
30	−5.71	0.309	1.69	0.36	40.26	0.90	3.68	1.26	1.47	1.44	0.37
30	−11.31	0.270	1.50	0.02	44.46	0.82	3.74	1.15	1.94	1.31	1.04
30	−14.03	0.251	1.51	6.29	40.06	0.80	3.60	1.13	1.93	1.29	1.10
35	0.00	0.284	1.87	0.05	39.26	1.06	2.53	1.48	0.11	1.69	0.00
35	−5.71	0.246	1.68	0.49	41.84	0.98	2.63	1.38	0.63	1.57	0.00
35	−11.31	0.208	1.51	3.18	44.30	0.90	2.68	1.26	1.08	1.44	0.28
35	−14.03	0.190	1.51	12.05	38.53	0.87	2.60	1.22	1.15	1.40	0.42
40	0.00	0.232	1.89	0.19	40.16	1.16	1.65	1.62	0.00	1.86	0.00
40	−5.71	0.194	1.66	0.17	44.30	1.07	1.83	1.49	0.03	1.71	0.00
40	−11.31	0.158	1.51	7.58	44.04	0.98	1.90	1.37	0.49	1.56	0.00
40	−14.03	0.142	1.50	21.74	35.32	0.94	1.88	1.31	0.62	1.50	0.00

B. Physical Dimensions for a Technoeconomic Wall Section

ϕ Degree	α Degree	Top Width, mm	Bottom Width, mm	Horizontal Angle θ Degree	Depth of Shear Key D_{sk}, mm			Volume of Wall, m³
					$\mu = 0.50$	$\mu = 0.70$	$\mu = 0.80$	
25	0.00	200	840	90.00	500	300	200	0.780
25	−5.71	200	690	84.29	500	300	300	0.668
25	−11.31	200	545	78.69	500	400	300	0.559
25	−14.03	200	485	75.97	500	400	300	0.514
30	0.00	200	760	90.00	300	200	0	0.720
30	−5.71	200	605	84.29	300	200	100	0.604
30	−11.31	200	450	78.69	300	300	200	0.488
30	−14.03	200	400	75.97	300	300	200	0.450
35	0.00	200	685	90.00	200	100	0	0.664
35	−5.71	200	525	84.29	200	100	0	0.544
35	−11.31	200	375	78.69	200	200	100	0.431
35	−14.03	200	320	75.97	200	200	100	0.390

(Continued)

TABLE 5.9 (Continued)

$H = 1.5$ m, $\gamma_{masonry} = 22.56$ kN/m^3, $\gamma_{sat} = 15.70$ kN/m^3 and $\iota = \frac{1}{2}\phi$

					Depth of Shear Key D_{sk}, mm			
ϕ Degree	α Degree	Top Width, mm	Bottom Width, mm	Horizontal Angle θ Degree	$\mu = 0.50$	$\mu = 0.70$	$\mu = 0.80$	Volume of Wall, m^3
40	0.00	200	620	90.00	100	0	0	0.615
40	−5.71	200	450	84.29	200	100	0	0.488
40	−11.31	200	305	78.69	200	100	0	0.379
40	−14.03	200	245	75.97	200	100	0	0.334

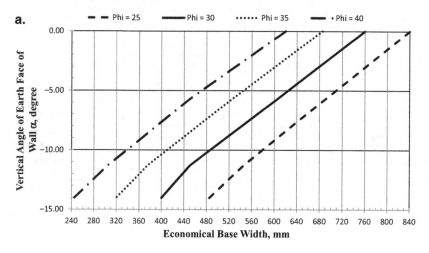

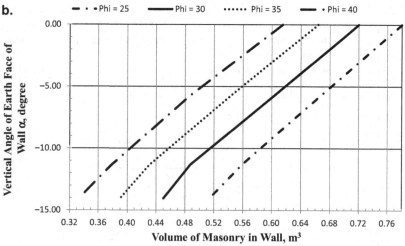

FIGURE 5.10 Design Charts for a Masonry Wall $H = 1.5$ m, $\gamma_{masonry} = 22.56$ kN/m^3, $\gamma_{sat} = 15.70$ kN/m^3 and $\iota = \frac{1}{2}\phi$: (a) Economical Base Width for a Wall, (b) Volume of a Masonry in Wall

TABLE 5.10

$H = 1.5$ m, $\gamma_{masonry} = 22.56$ kN/m^3, $\gamma_{sat} = 17.27$ kN/m^3 and $\iota = \frac{1}{2}\phi$

A. Results of Stability Analysis

ϕ Degree	α Degree	K_{ah}	F_{OVT}	Pressure at Heel, kN/m^2	Pressure at Toe, kN/m^2	$\mu = 0.50$ F_{SLD}	$\mu = 0.50$ Net Force, kN	$\mu = 0.70$ F_{SLD}	$\mu = 0.70$ Net Force, kN	$\mu = 0.80$ F_{SLD}	$\mu = 0.80$ Net Force, kN
25	0.00	0.424	1.85	0.27	36.81	0.84	5.81	1.17	2.86	1.34	1.39
25	−5.71	0.385	1.71	0.46	38.17	0.79	5.73	1.11	3.18	1.26	1.90
25	−11.31	0.346	1.55	0.30	40.70	0.73	5.64	1.03	3.49	1.17	2.41
25	−14.03	0.326	1.50	1.59	40.86	0.71	5.54	0.99	3.55	1.13	2.56
30	0.00	0.348	1.86	0.24	37.65	0.92	4.29	1.29	1.56	1.47	0.20
30	−5.71	0.309	1.69	0.08	39.92	0.86	4.27	1.21	1.96	1.38	0.81
30	−11.31	0.270	1.53	0.91	42.34	0.80	4.19	1.12	2.28	1.28	1.33
30	−14.03	0.251	1.50	4.67	40.50	0.78	4.07	1.09	2.32	1.24	1.45
35	0.00	0.284	1.87	0.11	38.72	1.01	3.05	1.41	0.54	1.62	0.00
35	−5.71	0.246	1.68	0.06	41.57	0.94	3.08	1.32	1.00	1.51	0.00
35	−11.31	0.208	1.50	1.44	44.91	0.86	3.07	1.21	1.40	1.38	0.57
35	−14.03	0.190	1.50	9.28	39.77	0.85	2.94	1.18	1.42	1.35	0.66
40	0.00	0.232	1.88	0.12	39.72	1.11	2.06	1.55	0.00	1.77	0.00
40	−5.71	0.194	1.67	0.37	43.16	1.03	2.15	1.44	0.27	1.65	0.00
40	−11.31	0.158	1.52	6.21	43.73	0.95	2.15	1.33	0.67	1.52	0.00
40	−14.03	0.142	1.51	18.50	36.00	0.92	2.09	1.29	0.77	1.47	0.10

B. Physical Dimensions for a Technoeconomic Wall Section

ϕ Degree	α Degree	Top Width, mm	Bottom Width, mm	Horizontal Angle θ Degree	Depth of Shear Key D_{sk}, mm $\mu = 0.50$	Depth of Shear Key D_{sk}, mm $\mu = 0.70$	Depth of Shear Key D_{sk}, mm $\mu = 0.80$	Volume of Wall, m^3
25	0.00	200	885	90.00	500	300	300	0.814
25	−5.71	200	735	84.29	500	400	300	0.701
25	−11.31	200	585	78.69	500	400	300	0.589
25	−14.03	200	520	75.97	500	400	300	0.540
30	0.00	200	800	90.00	300	200	100	0.750
30	−5.71	200	640	84.29	300	200	200	0.630
30	−11.31	200	490	78.69	300	300	200	0.518
30	−14.03	200	430	75.97	300	300	200	0.473
35	0.00	200	720	90.00	200	0	0	0.690
35	−5.71	200	555	84.29	300	200	0	0.566
35	−11.31	200	400	78.69	300	200	100	0.450
35	−14.03	200	345	75.97	200	200	100	0.409

(Continued)

TABLE 5.10 (Continued)

$H = 1.5$ m, $\gamma_{masonry} = 22.56$ kN/m³, $\gamma_{sat} = 17.27$ kN/m³ and $\iota = \frac{1}{2}\phi$

ϕ Degree	α Degree	Top Width, mm	Bottom Width, mm	Horizontal Angle θ Degree	Depth of Shear Key D_{sk}, mm			Volume of Wall, m³
					$\mu = 0.50$	$\mu = 0.70$	$\mu = 0.80$	
40	0.00	200	650	90.00	100	0	0	0.638
40	−5.71	200	480	84.29	200	100	0	0.510
40	−11.31	200	330	78.69	200	100	0	0.398
40	−14.03	200	270	75.97	200	100	100	0.353

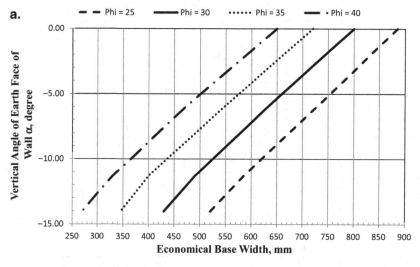

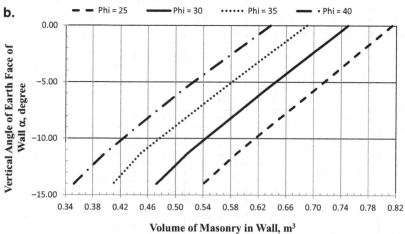

FIGURE 5.11 Design Charts for Masonry Wall $H = 1.5$ m, $\gamma_{masonry} = 22.56$ kN/m³, $\gamma_{sat} = 17.27$ kN/m³ and $\iota = \frac{1}{2}\phi$: (a) Economical Base Width for a Wall, (b) Volume of a Masonry in Wall

TABLE 5.11

$H = 1.5$ m, $\gamma_{masonry} = 22.56$ kN/m³, $\gamma_{sat} = 15.70$ kN/m³ and $\iota = \phi$

A. Results of Stability Analysis

ϕ Degree	α Degree	K_{ah}	F_{OVT}	Pressure at Heel, kN/m²	Toe, kN/m²	$\mu = 0.50$ F_{SLD}	Net Force, kN	$\mu = 0.70$ F_{SLD}	Net Force, kN	$\mu = 0.80$ F_{SLD}	Net Force, kN
25	0.00	0.821	1.83	0.24	34.90	0.64	12.64	0.89	8.89	1.02	7.02
25	−5.71	0.746	1.73	0.33	35.71	0.62	11.83	0.87	8.50	0.99	6.84
25	−11.31	0.675	1.61	0.05	37.21	0.59	11.10	0.83	8.21	0.95	6.76
25	−14.03	0.641	1.55	0.12	37.88	0.58	10.74	0.81	8.05	0.93	6.70
30	0.00	0.750	1.83	0.22	35.20	0.67	11.22	0.93	7.62	1.07	5.82
30	−5.71	0.666	1.71	0.04	36.48	0.65	10.27	0.91	7.14	1.04	5.57
30	−11.31	0.587	1.60	0.25	37.75	0.63	9.39	0.88	6.69	1.00	5.34
30	−14.03	0.549	1.54	0.45	38.52	0.61	8.98	0.86	6.49	0.98	5.25
35	0.00	0.671	1.84	0.43	35.33	0.71	9.64	0.99	6.20	1.13	4.48
35	−5.71	0.580	1.72	0.37	36.71	0.69	8.60	0.97	5.65	1.11	4.17
35	−11.31	0.496	1.58	0.28	38.76	0.67	7.69	0.94	5.21	1.07	3.98
35	−14.03	0.456	1.52	0.57	39.78	0.65	7.26	0.92	5.01	1.05	3.89
40	0.00	0.587	1.84	0.23	36.01	0.75	8.03	1.05	4.79	1.21	3.17
40	−5.71	0.492	1.70	0.11	37.78	0.74	6.97	1.04	4.24	1.19	2.88
40	−11.31	0.406	1.57	0.62	39.81	0.72	6.04	1.01	3.80	1.15	2.68
40	−14.03	0.366	1.50	1.11	41.21	0.70	5.64	0.99	3.64	1.13	2.64

B. Physical Dimensions for a Technoeconomic Wall Section

ϕ Degree	α Degree	Top Width, mm	Bottom Width, mm	Horizontal Angle θ Degree	Depth of Shear Key D_{sk}, mm $\mu = 0.50$	$\mu = 0.70$	$\mu = 0.80$	Volume of Wall, m³
25	0.00	200	1185	90.00	600	500	400	1.039
25	−5.71	200	1025	84.29	600	500	400	0.919
25	−11.31	200	865	78.69	700	600	500	0.799
25	−14.03	200	790	75.97	600	600	500	0.743
30	0.00	200	1130	90.00	400	400	300	0.998
30	−5.71	200	955	84.29	400	300	300	0.866
30	−11.31	200	790	78.69	500	400	400	0.743
30	−14.03	200	710	75.97	500	400	400	0.683
35	0.00	200	1070	90.00	300	200	200	0.953
35	−5.71	200	885	84.29	300	200	200	0.814
35	−11.31	200	705	78.69	300	200	200	0.679
35	−14.03	200	620	75.97	400	300	300	0.615

(Continued)

TABLE 5.11 (Continued)

H = 1.5 m, $\gamma_{masonry}$ = 22.56 kN/m³, γ_{sat} = 15.70 kN/m³ and $\iota = \phi$

ϕ Degree	α Degree	Top Width, mm	Bottom Width, mm	Horizontal Angle θ Degree	Depth of Shear Key D_{sk}, mm			Volume of Wall, m³
					$\mu = 0.50$	$\mu = 0.70$	$\mu = 0.80$	
40	0.00	200	995	90.00	200	200	100	0.896
40	−5.71	200	800	84.29	200	200	100	0.750
40	−11.31	200	615	78.69	200	100	100	0.611
40	−14.03	200	525	75.97	300	200	200	0.544

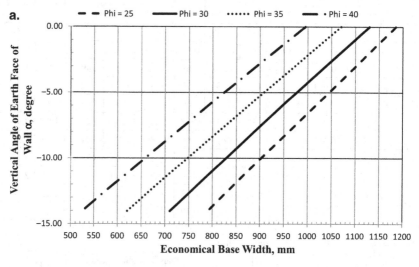

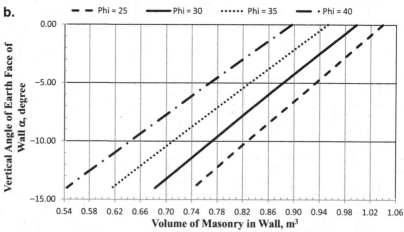

FIGURE 5.12 Design Charts for a Masonry Wall H = 1.5 m, $\gamma_{masonry}$ = 22.56 kN/m³, γ_{sat} = 15.70 kN/m³ and $\iota = \phi$: (a) Economical Base Width for a Wall, (b) Volume of a Masonry in Wall

TABLE 5.12

$H = 1.5$ m, $\gamma_{masonry} = 22.56$ kN/m^3, $\gamma_{sat} = 17.27$ kN/m^3 and $\iota = \phi$

A. Results of Stability Analysis

ϕ Degree	α Degree	K_{ah}	F_{OVT}	Pressure at Heel, kN/m^2	Toe, kN/m^2	$\mu = 0.50$ F_{SLD}	Net Force, kN	$\mu = 0.70$ F_{SLD}	Net Force, kN	$\mu = 0.80$ F_{SLD}	Net Force, kN
25	0.00	0.821	1.82	0.17	34.70	0.61	14.41	0.85	10.51	0.97	8.56
25	−5.71	0.746	1.72	0.06	35.64	0.59	13.44	0.82	9.97	0.94	8.24
25	−11.31	0.675	1.62	0.34	36.41	0.57	12.49	0.80	9.44	0.91	7.91
25	−14.03	0.641	1.57	0.10	37.34	0.56	12.08	0.78	9.23	0.89	7.81
30	0.00	0.750	1.82	0.03	35.11	0.63	12.84	0.89	9.09	1.01	7.22
30	−5.71	0.666	1.73	0.35	35.75	0.62	11.64	0.87	8.34	0.99	6.70
30	−11.31	0.587	1.61	0.46	36.98	0.60	10.58	0.84	7.73	0.96	6.31
30	−14.03	0.549	1.55	0.30	38.04	0.59	10.10	0.83	7.48	0.94	6.17
35	0.00	0.671	1.83	0.13	35.35	0.67	11.08	0.94	7.51	1.07	5.73
35	−5.71	0.580	1.72	0.22	36.45	0.66	9.80	0.92	6.72	1.06	5.18
35	−11.31	0.496	1.60	0.44	37.95	0.64	8.67	0.90	6.06	1.03	4.76
35	−14.03	0.456	1.53	0.33	39.28	0.63	8.17	0.88	5.80	1.01	4.62
40	0.00	0.587	1.83	0.14	35.76	0.72	9.25	1.00	5.88	1.14	4.19
40	−5.71	0.492	1.71	0.32	37.07	0.71	7.94	0.99	5.08	1.14	3.65
40	−11.31	0.406	1.57	0.12	39.65	0.69	6.85	0.97	4.51	1.11	3.34
40	−14.03	0.366	1.50	0.75	40.66	0.68	6.34	0.95	4.24	1.09	3.19

B. Physical Dimensions for a Technoeconomic Wall Section

ϕ Degree	α Degree	Top Width, mm	Bottom Width, mm	Horizontal Angle θ Degree	Depth of Shear Key D_{sk}, mm $\mu = 0.50$	$\mu = 0.70$	$\mu = 0.80$	Volume of Wall, m^3
25	0.00	200	1245	90.00	600	500	500	1.084
25	−5.71	200	1080	84.29	600	500	400	0.960
25	−11.31	200	925	78.69	700	600	500	0.844
25	−14.03	200	845	75.97	600	600	500	0.784
30	0.00	200	1185	90.00	500	400	300	1.039
30	−5.71	200	1015	84.29	400	400	300	0.911
30	−11.31	200	845	78.69	400	300	300	0.784
30	−14.03	200	760	75.97	500	400	400	0.720
35	0.00	200	1120	90.00	300	300	200	0.990
35	−5.71	200	935	84.29	300	200	200	0.851
35	−11.31	200	755	78.69	300	200	200	0.716
35	−14.03	200	665	75.97	400	300	300	0.649

(Continued)

TABLE 5.12 (Continued)

H = 1.5 m, $\gamma_{masonry}$ = 22.56 kN/m³, γ_{sat} = 17.27 kN/m³ and $\iota = \phi$

ϕ Degree	α Degree	Top Width, mm	Bottom Width, mm	Horizontal Angle θ Degree	Depth of Shear Key D_{sk}, mm			Volume of Wall, m³
					$\mu = 0.50$	$\mu = 0.70$	$\mu = 0.80$	
40	0.00	200	1045	90.00	200	200	100	0.934
40	−5.71	200	850	84.29	200	200	100	0.788
40	−11.31	200	655	78.69	200	200	100	0.641
40	−14.03	200	565	75.97	300	200	200	0.574

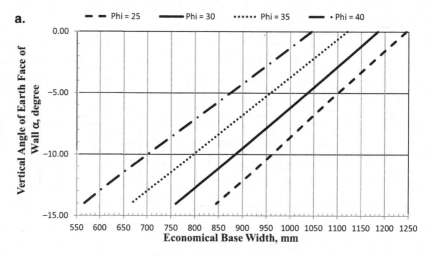

a.

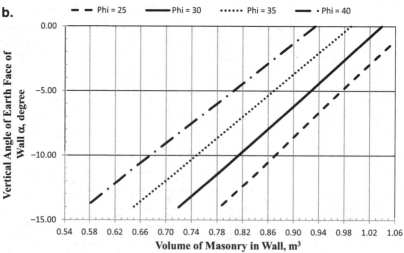

b.

FIGURE 5.13 Design Charts for a Masonry Wall H = 1.5 m, $\gamma_{masonry}$ = 22.56 kN/m³, γ_{sat} = 17.27 kN/m³ and $\iota = \phi$: (a) Economical Base Width for a Wall, (b) Volume of a Masonry in Wall

5.4 CHARTS FOR THE DESIGN OF A 2.0-M-HIGH MASONRY BREAST WALL

The charts/tables developed for a 2.0-m-high stone masonry breast wall for the previously discussed six combinations are presented in Tables 5.13 to 5.18 and Figures 5.14 to 5.19.

TABLE 5.13
$H = 2.0$ m, $\gamma_{masonry} = 22.56$ kN/m³, $\gamma_{sat} = 15.70$ kN/m³ and $\iota = 0°$

A. Results of Stability Analysis

ϕ Degree	α Degree	K_{ah}	F_{OVT}	Pressure at Heel, kN/m²	Toe, kN/m²	$\mu = 0.50$ F_{SLD}	Net Force, kN	$\mu = 0.70$ F_{SLD}	Net Force, kN	$\mu = 0.80$ F_{SLD}	Net Force, kN
25	0.00	0.346	1.84	0.35	47.66	0.93	6.87	1.30	2.43	1.48	0.20
25	−5.71	0.316	1.68	0.40	49.52	0.85	7.20	1.19	3.41	1.36	1.51
25	−11.31	0.286	1.51	0.16	52.65	0.77	7.44	1.08	4.29	1.24	2.71
25	−14.03	0.271	1.51	5.56	48.60	0.75	7.30	1.05	4.35	1.21	2.88
30	0.00	0.279	1.85	0.36	48.64	1.02	4.86	1.42	0.78	1.63	0.00
30	−5.71	0.250	1.67	0.66	50.86	0.93	5.25	1.30	1.85	1.48	0.14
30	−11.31	0.221	1.50	1.88	53.63	0.83	5.54	1.17	2.77	1.33	1.38
30	−14.03	0.207	1.50	10.10	47.37	0.81	5.42	1.14	2.86	1.30	1.58
35	0.00	0.226	1.85	0.19	49.92	1.11	3.32	1.55	0.00	1.77	0.00
35	−5.71	0.198	1.65	0.71	52.74	1.00	3.79	1.40	0.74	1.60	0.00
35	−11.31	0.170	1.51	6.80	51.85	0.90	4.05	1.26	1.60	1.45	0.37
35	−14.03	0.156	1.51	18.45	43.35	0.87	4.02	1.22	1.79	1.39	0.68
40	0.00	0.184	1.86	0.16	51.11	1.20	2.17	1.68	0.00	1.92	0.00
40	−5.71	0.156	1.62	0.23	55.57	1.07	2.75	1.50	0.02	1.71	0.00
40	−11.31	0.130	1.51	12.23	50.76	0.96	3.02	1.35	0.87	1.54	0.00
40	−14.03	0.117	1.50	30.24	38.11	0.91	3.07	1.28	1.16	1.46	0.21

B. Physical Dimensions for a Technoeconomic Wall Section

ϕ Degree	α Degree	Top Width, mm	Bottom Width, mm	Horizontal Angle θ Degree	Depth of Shear Key D_{sk}, mm $\mu = 0.50$	$\mu = 0.70$	$\mu = 0.80$	Volume of Wall, m³
25	0.00	200	1030	90.00	400	200	0	1.230
25	−5.71	200	845	84.29	500	400	300	1.045
25	−11.31	200	665	78.69	500	400	300	0.865
25	−14.03	200	605	75.97	500	400	400	0.805
30	0.00	200	925	90.00	300	100	0	1.125
30	−5.71	200	735	84.29	400	300	100	0.935

(Continued)

TABLE 5.13 (Continued)

H = 2.0 m, $\gamma_{masonry}$ = 22.56 kN/m³, γ_{sat} = 15.70 kN/m³ and ι = 0°

ϕ Degree	α Degree	Top Width, mm	Bottom Width, mm	Horizontal Angle θ Degree	Depth of Shear Key D_{sk}, mm			Volume of Wall, m³
					$\mu = 0.50$	$\mu = 0.70$	$\mu = 0.80$	
30	−11.31	200	555	78.69	400	300	200	0.755
30	−14.03	200	495	75.97	400	300	200	0.695
35	0.00	200	830	90.00	200	0	0	1.030
35	−5.71	200	635	84.29	300	200	0	0.835
35	−11.31	200	465	78.69	300	200	100	0.665
35	−14.03	200	400	75.97	300	200	100	0.600
40	0.00	200	750	90.00	100	0	0	0.950
40	−5.71	200	545	84.29	200	100	0	0.745
40	−11.31	200	380	78.69	200	100	0	0.580
40	−14.03	200	310	75.97	200	200	100	0.510

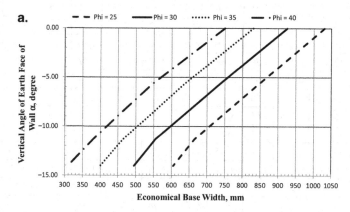

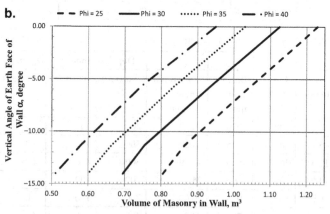

FIGURE 5.14 Design Charts for a Masonry Wall H = 2.0 m, $\gamma_{masonry}$ = 22.56 kN/m³, γ_{sat} = 15.70 kN/m³ and ι = 0°: (a) Economical Base Width for a Wall, (b) Volume of a Masonry in Wall

TABLE 5.14

$H = 2.0$ m, $\gamma_{masonry} = 22.56$ kN/m³, $\gamma_{sat} = 17.27$ kN/m³ and $\iota = 0°$

A. Results of Stability Analysis

ϕ Degree	α Degree	K_{ah}	F_{OVT}	Pressure at Heel, kN/m²	Pressure at Toe, kN/m²	$\mu = 0.50$ F_{SLD}	$\mu = 0.50$ Net Force, kN	$\mu = 0.70$ F_{SLD}	$\mu = 0.70$ Net Force, kN	$\mu = 0.80$ F_{SLD}	$\mu = 0.80$ Net Force, kN
25	0.00	0.346	1.83	0.17	47.42	0.88	8.06	1.24	3.43	1.41	1.12
25	−5.71	0.316	1.68	0.31	49.01	0.82	8.25	1.15	4.28	1.31	2.29
25	−11.31	0.286	1.53	0.39	51.48	0.75	8.35	1.05	5.01	1.20	3.35
25	−14.03	0.271	1.50	3.68	49.55	0.73	8.22	1.02	5.13	1.16	3.59
30	0.00	0.279	1.84	0.23	48.31	0.97	5.77	1.36	1.54	1.55	0.00
30	−5.71	0.250	1.66	0.00	50.88	0.89	6.08	1.25	2.53	1.42	0.76
30	−11.31	0.221	1.50	1.19	53.21	0.81	6.22	1.13	3.31	1.29	1.86
30	−14.03	0.207	1.50	7.64	48.63	0.79	6.08	1.10	3.40	1.26	2.06
35	0.00	0.226	1.84	0.07	49.54	1.06	4.03	1.48	0.15	1.70	0.00
35	−5.71	0.198	1.66	0.75	51.86	0.97	4.37	1.36	1.18	1.55	0.00
35	−11.31	0.170	1.50	4.39	53.08	0.87	4.58	1.22	2.03	1.40	0.75
35	−14.03	0.156	1.50	14.76	45.47	0.85	4.49	1.19	2.16	1.36	0.99
40	0.00	0.184	1.87	0.54	50.12	1.16	2.68	1.62	0.00	1.85	0.00
40	−5.71	0.156	1.63	0.28	54.52	1.04	3.18	1.45	0.32	1.66	0.00
40	−11.31	0.130	1.51	10.26	50.99	0.94	3.36	1.32	1.11	1.50	0.00
40	−14.03	0.117	1.51	26.06	39.73	0.90	3.35	1.26	1.34	1.44	0.33

B. Physical Dimensions for a Technoeconomic Wall Section

ϕ Degree	α Degree	Top Width, mm	Bottom Width, mm	Horizontal Angle θ Degree	Depth of Shear Key D_{sk}, mm $\mu = 0.50$	$\mu = 0.70$	$\mu = 0.80$	Volume of Wall, m³
25	0.00	200	1080	90.00	400	300	100	1.280
25	−5.71	200	895	84.29	500	400	300	1.095
25	−11.31	200	715	78.69	500	400	400	0.915
25	−14.03	200	645	75.97	500	400	400	0.845
30	0.00	200	970	90.00	300	100	0	1.170
30	−5.71	200	775	84.29	400	300	200	0.975
30	−11.31	200	595	78.69	400	300	200	0.795
30	−14.03	200	530	75.97	400	300	300	0.730
35	0.00	200	870	90.00	200	0	0	1.070
35	−5.71	200	675	84.29	300	200	0	0.875
35	−11.31	200	495	78.69	300	200	100	0.695
35	−14.03	200	430	75.97	300	200	200	0.630

(Continued)

TABLE 5.14 (Continued)

$H = 2.0$ m, $\gamma_{masonry} = 22.56$ kN/m³, $\gamma_{sat} = 17.27$ kN/m³ and $\iota = 0°$

					Depth of Shear Key D_{sk}, mm			
ϕ Degree	α Degree	Top Width, mm	Bottom Width, mm	Horizontal Angle θ Degree	$\mu = 0.50$	$\mu = 0.70$	$\mu = 0.80$	Volume of Wall, m³
40	0.00	200	790	90.00	100	0	0	0.990
40	−5.71	200	580	84.29	200	100	0	0.780
40	−11.31	200	410	78.69	200	200	0	0.610
40	−14.03	200	340	75.97	200	200	100	0.540

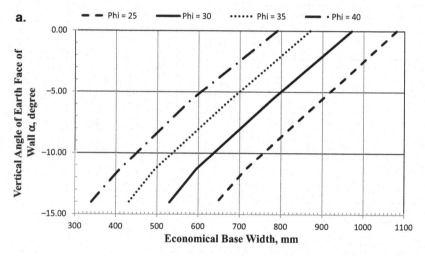

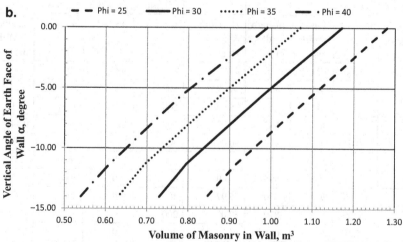

FIGURE 5.15 Design Charts for a Masonry Wall $H = 2.0$ m, $\gamma_{masonry} = 22.56$ kN/m³, $\gamma_{sat} = 17.27$ kN/m³ and $\iota = 0°$: (a) Economical Base Width for a Wall, (b) Volume of a Masonry in Wall

TABLE 5.15

$H = 2.0$ m, $\gamma_{masonry} = 22.56$ kN/m^3, $\gamma_{sat} = 15.70$ kN/m^3 and $\iota = \frac{1}{2}\phi$

A. Results of Stability Analysis

ϕ Degree	α Degree	K_{ah}	F_{OVT}	Pressure at Heel, kN/m^2	Toe, kN/m^2	$\mu = 0.50$ F_{SLD}	Net Force, kN	$\mu = 0.70$ F_{SLD}	Net Force, kN	$\mu = 0.80$ F_{SLD}	Net Force, kN
25	0.00	0.424	1.82	0.14	47.02	0.84	9.37	1.18	4.53	1.35	2.12
25	−5.71	0.385	1.69	0.36	48.38	0.79	9.32	1.11	5.15	1.27	3.07
25	−11.31	0.346	1.54	0.57	50.47	0.73	9.21	1.02	5.70	1.17	3.95
25	−14.03	0.326	1.50	2.35	50.06	0.71	9.06	0.99	5.83	1.13	4.22
30	0.00	0.348	1.83	0.01	47.99	0.92	6.99	1.29	2.54	1.47	0.32
30	−5.71	0.309	1.66	0.00	50.11	0.86	7.01	1.20	3.27	1.37	1.40
30	−11.31	0.270	1.51	0.56	52.78	0.79	6.97	1.10	3.90	1.26	2.37
30	−14.03	0.251	1.50	5.92	49.15	0.77	6.75	1.07	3.93	1.23	2.52
35	0.00	0.284	1.84	0.08	48.87	1.01	5.03	1.41	0.93	1.61	0.00
35	−5.71	0.246	1.66	0.50	51.19	0.93	5.12	1.31	1.75	1.49	0.07
35	−11.31	0.208	1.51	3.21	52.90	0.85	5.14	1.19	2.44	1.36	1.09
35	−14.03	0.190	1.50	11.97	46.68	0.83	4.96	1.16	2.51	1.33	1.29
40	0.00	0.232	1.85	0.12	49.87	1.09	3.49	1.53	0.00	1.75	0.00
40	−5.71	0.194	1.64	0.33	53.35	1.01	3.71	1.41	0.70	1.61	0.00
40	−11.31	0.158	1.51	7.51	52.24	0.92	3.76	1.28	1.40	1.47	0.22
40	−14.03	0.142	1.52	22.35	41.29	0.89	3.64	1.24	1.52	1.42	0.46

B. Physical Dimensions for a Technoeconomic Wall Section

ϕ Degree	α Degree	Top Width, mm	Bottom Width, mm	Horizontal Angle θ Degree	Depth of Shear Key D_{sk}, mm $\mu = 0.50$	$\mu = 0.70$	$\mu = 0.80$	Volume of Wall, m^3
25	0.00	200	1140	90.00	500	300	200	1.340
25	−5.71	200	950	84.29	600	500	400	1.150
25	−11.31	200	765	78.69	600	500	400	0.965
25	−14.03	200	685	75.97	600	500	400	0.885
30	0.00	200	1030	90.00	300	200	0	1.230
30	−5.71	200	830	84.29	300	200	100	1.030
30	−11.31	200	640	78.69	400	300	300	0.840
30	−14.03	200	570	75.97	400	300	300	0.770
35	0.00	200	930	90.00	200	100	0	1.130
35	−5.71	200	725	84.29	200	100	0	0.925
35	−11.31	200	535	78.69	300	200	200	0.735
35	−14.03	200	465	75.97	300	200	200	0.665

(Continued)

TABLE 5.15 (Continued)

$H = 2.0$ m, $\gamma_{masonry} = 22.56$ kN/m³, $\gamma_{sat} = 15.70$ kN/m³ and $\iota = \frac{1}{2}\phi$

					Depth of Shear Key D_{sk}, mm			
ϕ Degree	α Degree	Top Width, mm	Bottom Width, mm	Horizontal Angle θ Degree	$\mu = 0.50$	$\mu = 0.70$	$\mu = 0.80$	Volume of Wall, m³
40	0.00	200	840	90.00	100	0	0	1.040
40	−5.71	200	625	84.29	100	0	0	0.825
40	−11.31	200	440	78.69	200	200	100	0.640
40	−14.03	200	370	75.97	200	200	100	0.570

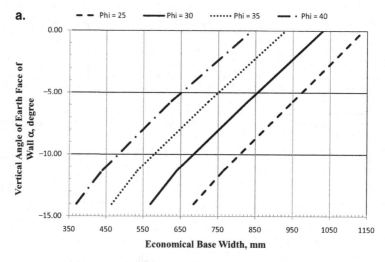

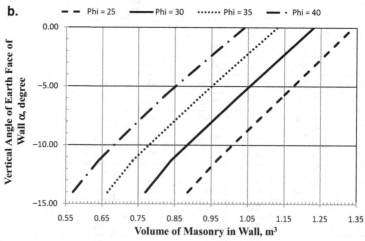

FIGURE 5.16 Design Charts for a Masonry Wall $H = 2.0$ m, $\gamma_{masonry} = 22.56$ kN/m³, $\gamma_{sat} = 15.70$ kN/m³ and $\iota = \frac{1}{2}\phi$: (a) Economical Base Width for a Wall, (b) Volume of a Masonry in Wall

TABLE 5.16

$H = 2.0$ m, $\gamma_{masonry} = 22.56$ kN/m³, $\gamma_{sat} = 17.27$ kN/m³ and $\iota = \frac{1}{2}\phi$

A. Results of Stability Analysis

ϕ Degree	α Degree	K_{ah}	F_{OVT}	Pressure at		$\mu = 0.50$		$\mu = 0.70$		$\mu = 0.80$	
				Heel, kN/m²	Toe, kN/m²	F_{SLD}	Net Force, kN	F_{SLD}	Net Force, kN	F_{SLD}	Net Force, kN
25	0.00	0.424	1.83	0.28	46.48	0.81	10.83	1.13	5.79	1.29	3.26
25	−5.71	0.385	1.69	0.24	47.98	0.76	10.64	1.06	6.28	1.21	4.10
25	−11.31	0.346	1.55	0.04	50.27	0.70	10.40	0.99	6.71	1.13	4.87
25	−14.03	0.326	1.51	1.81	49.71	0.68	10.16	0.96	6.75	1.09	5.05
30	0.00	0.348	1.83	0.24	47.32	0.88	8.14	1.23	3.50	1.41	1.18
30	−5.71	0.309	1.67	0.01	49.49	0.82	8.02	1.15	4.11	1.32	2.15
30	−11.31	0.270	1.52	0.14	52.27	0.76	7.84	1.06	4.62	1.22	3.00
30	−14.03	0.251	1.51	5.23	48.69	0.75	7.54	1.04	4.55	1.19	3.06
35	0.00	0.284	1.84	0.37	48.08	0.96	5.92	1.35	1.65	1.54	0.00
35	−5.71	0.246	1.67	0.56	50.40	0.90	5.88	1.26	2.35	1.44	0.59
35	−11.31	0.208	1.50	1.60	53.47	0.82	5.81	1.15	2.99	1.32	1.58
35	−14.03	0.190	1.50	9.60	47.69	0.81	5.55	1.13	2.97	1.29	1.69
40	0.00	0.232	1.85	0.44	49.00	1.05	4.18	1.47	0.25	1.68	0.00
40	−5.71	0.194	1.65	0.47	52.34	0.97	4.27	1.36	1.11	1.56	0.00
40	−11.31	0.158	1.50	5.25	53.19	0.89	4.24	1.24	1.77	1.42	0.54
40	−14.03	0.142	1.51	18.53	43.27	0.87	4.04	1.22	1.81	1.39	0.70

B. Physical Dimensions for a Technoeconomic Wall Section

ϕ Degree	α Degree	Top Width, mm	Bottom Width, mm	Horizontal Angle θ Degree	Depth of Shear Key D_{sk}, mm			Volume of Wall, m³
					$\mu = 0.50$	$\mu = 0.70$	$\mu = 0.80$	
25	0.00	200	1200	90.00	500	400	300	1.400
25	−5.71	200	1005	84.29	500	400	300	1.205
25	−11.31	200	815	78.69	600	500	400	1.015
25	−14.03	200	735	75.97	600	500	400	0.935
30	0.00	200	1085	90.00	400	200	100	1.285
30	−5.71	200	880	84.29	400	200	200	1.080
30	−11.31	200	685	78.69	400	400	300	0.885
30	−14.03	200	615	75.97	400	400	300	0.815
35	0.00	200	980	90.00	200	100	0	1.180
35	−5.71	200	770	84.29	200	100	0	0.970
35	−11.31	200	570	78.69	300	200	200	0.770
35	−14.03	200	500	75.97	300	200	200	0.700

(Continued)

TABLE 5.16 (Continued)

H = 2.0 m, $\gamma_{masonry}$ = 22.56 kN/m³, γ_{sat} = 17.27 kN/m³ and ι = ½ϕ

ϕ Degree	α Degree	Top Width, mm	Bottom Width, mm	Horizontal Angle θ Degree	Depth of Shear Key D_{sk}, mm			Volume of Wall, m³
					$\mu = 0.50$	$\mu = 0.70$	$\mu = 0.80$	
40	0.00	200	885	90.00	100	0	0	1.085
40	−5.71	200	665	84.29	100	100	0	0.865
40	−11.31	200	470	78.69	200	200	100	0.670
40	−14.03	200	400	75.97	200	200	100	0.600

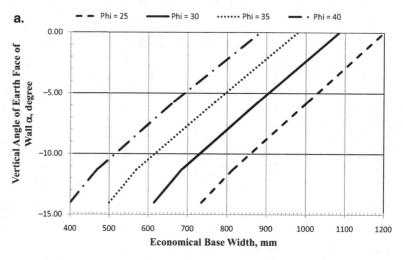

a.

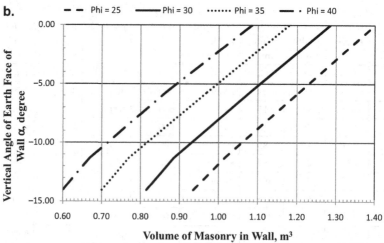

b.

FIGURE 5.17 Design Charts for a Masonry Wall H = 2.0 m, $\gamma_{masonry}$ = 22.56 kN/m³, γ_{sat} = 17.27 kN/m³ and ι = ½ϕ: (a) Economical Base Width for a Wall, (b) Volume of a Masonry in Wall

TABLE 5.17

H = 2.0 m, $\gamma_{masonry}$ = 22.56 kN/m³, γ_{sat} = 15.70 kN/m³ and $\iota = \phi$

A. Results of Stability Analysis

				Pressure at		$\mu = 0.50$		$\mu = 0.70$		$\mu = 0.80$	
ϕ Degree	α Degree	K_{ah}	F_{OVT}	Heel, kN/ m²	Toe, kN/ m²	F_{SLD}	Net Force, kN	F_{SLD}	Net Force, kN	F_{SLD}	Net Force, kN
25	0.00	0.821	1.80	0.05	44.83	0.62	22.96	0.87	16.51	0.99	13.28
25	−5.71	0.746	1.71	0.25	45.46	0.60	21.45	0.84	15.72	0.96	12.85
25	−11.31	0.675	1.61	0.40	46.39	0.58	20.05	0.81	15.02	0.92	12.51
25	−14.03	0.641	1.56	0.20	47.29	0.56	19.41	0.79	14.74	0.90	12.40
30	0.00	0.750	1.81	0.23	44.91	0.65	20.41	0.91	14.20	1.04	11.10
30	−5.71	0.666	1.70	0.13	46.03	0.63	18.66	0.88	13.24	1.01	10.54
30	−11.31	0.587	1.60	0.23	47.29	0.61	17.04	0.85	12.38	0.97	10.06
30	−14.03	0.549	1.54	0.20	48.20	0.60	16.30	0.83	12.01	0.95	9.87
35	0.00	0.671	1.81	0.24	45.24	0.68	17.65	0.96	11.74	1.09	8.79
35	−5.71	0.580	1.70	0.09	46.64	0.67	15.74	0.94	10.68	1.07	8.15
35	−11.31	0.496	1.58	0.07	48.42	0.65	14.03	0.91	9.78	1.03	7.66
35	−14.03	0.456	1.51	0.24	49.43	0.63	13.25	0.89	9.39	1.01	7.45
40	0.00	0.587	1.82	0.33	45.59	0.73	14.75	1.02	9.18	1.17	6.39
40	−5.71	0.492	1.69	0.06	47.42	0.72	12.81	1.00	8.13	1.15	5.80
40	−11.31	0.406	1.56	0.46	49.34	0.69	11.10	0.97	7.27	1.11	5.36
40	−14.03	0.366	1.51	1.78	49.58	0.68	10.30	0.96	6.86	1.09	5.14

B. Physical Dimensions for a Technoeconomic Wall Section

					Depth of Shear Key D_{sk}, mm			
ϕ Degree	α Degree	Top Width, mm	Bottom Width, mm	Horizontal Angle θ Degree	$\mu = 0.50$	$\mu = 0.70$	$\mu = 0.80$	Volume of Wall, m³
25	0.00	200	1600	90.00	800	700	600	1.800
25	−5.71	200	1395	84.29	800	700	600	1.595
25	−11.31	200	1195	78.69	800	600	600	1.395
25	−14.03	200	1095	75.97	700	600	600	1.295
30	0.00	200	1530	90.00	600	500	400	1.730
30	−5.71	200	1305	84.29	600	500	400	1.505
30	−11.31	200	1090	78.69	600	500	400	1.290
30	−14.03	200	985	75.97	600	500	400	1.185
35	0.00	200	1445	90.00	400	300	200	1.645
35	−5.71	200	1205	84.29	400	300	300	1.405
35	−11.31	200	975	78.69	400	300	300	1.175
35	−14.03	200	865	75.97	400	300	300	1.065

(Continued)

TABLE 5.17 (Continued)

$H = 2.0$ m, $\gamma_{masonry} = 22.56$ kN/m³, $\gamma_{sat} = 15.70$ kN/m³ and $\iota = \phi$

ϕ Degree	α Degree	Top Width, mm	Bottom Width, mm	Horizontal Angle θ Degree	Depth of Shear Key D_{sk}, mm $\mu = 0.50$	$\mu = 0.70$	$\mu = 0.80$	Volume of Wall, m³
40	0.00	200	1350	90.00	200	200	100	1.550
40	−5.71	200	1095	84.29	300	200	200	1.295
40	−11.31	200	855	78.69	300	200	200	1.055
40	−14.03	200	745	75.97	300	200	200	0.945

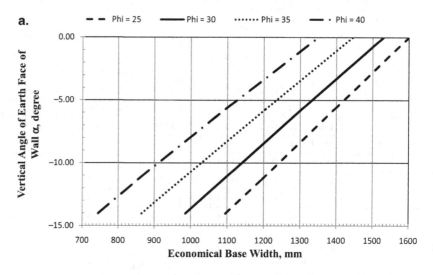

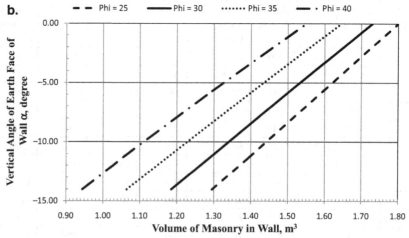

FIGURE 5.18 Design Charts for a Masonry Wall H = 2.0 m, $\gamma_{masonry}$ = 22.56 kN/m³, γ_{sat} = 15.70 kN/ m³ and $\iota = \phi$: (a) Economical Base Width for a Wall, (b) Volume of a Masonry in Wall

TABLE 5.18

$H = 2.0 \text{ m}, \gamma_{masonry} = 22.56 \text{ kN/m}^3, \gamma_{sat} = 17.27 \text{ kN/m}^3 \text{ and } \iota = \phi$

A. Results of Stability Analysis

ϕ Degree	α Degree	K_{ah}	F_{OVT}	Heel, kN/m²	Toe, kN/m²	F_{SLD}	Net Force, kN	F_{SLD}	Net Force, kN	F_{SLD}	Net Force, kN
				Pressure at		**$\mu = 0.50$**		**$\mu = 0.70$**		**$\mu = 0.80$**	
25	0.00	0.821	1.81	0.19	44.41	0.59	26.08	0.83	19.32	0.94	15.94
25	−5.71	0.746	1.72	0.31	45.05	0.57	24.25	0.80	18.24	0.92	15.23
25	−11.31	0.675	1.62	0.35	45.99	0.55	22.56	0.78	17.27	0.89	14.62
25	−14.03	0.641	1.57	0.33	46.63	0.54	21.76	0.76	16.82	0.87	14.35
30	0.00	0.750	1.80	0.04	44.82	0.62	23.28	0.86	16.80	0.98	13.57
30	−5.71	0.666	1.71	0.17	45.61	0.60	21.13	0.84	15.45	0.96	12.61
30	−11.31	0.587	1.60	0.21	46.81	0.58	19.18	0.82	14.28	0.94	11.83
30	−14.03	0.549	1.55	0.41	47.39	0.57	18.26	0.80	13.73	0.92	11.46
35	0.00	0.671	1.81	0.02	45.18	0.65	20.19	0.91	14.03	1.04	10.95
35	−5.71	0.580	1.70	0.13	46.19	0.64	17.85	0.90	12.54	1.02	9.89
35	−11.31	0.496	1.59	0.13	47.78	0.62	15.79	0.87	11.31	1.00	9.07
35	−14.03	0.456	1.52	0.01	48.99	0.61	14.87	0.85	10.79	0.98	8.75
40	0.00	0.587	1.81	0.08	45.53	0.69	16.94	0.97	11.13	1.11	8.23
40	−5.71	0.492	1.70	0.17	46.85	0.69	14.55	0.96	9.65	1.10	7.19
40	−11.31	0.406	1.56	0.05	49.10	0.67	12.53	0.94	8.50	1.07	6.49
40	−14.03	0.366	1.50	0.13	50.54	0.65	11.63	0.92	8.03	1.05	6.23

B. Physical Dimensions for a Technoeconomic Wall Section

ϕ Degree	α Degree	Top Width, mm	Bottom Width, mm	Horizontal Angle θ Degree	$\mu = 0.50$	$\mu = 0.70$	$\mu = 0.80$	Volume of Wall, m³
					Depth of Shear Key D_{sk}, mm			
25	0.00	200	1685	90.00	800	700	600	1.885
25	−5.71	200	1475	84.29	800	700	600	1.675
25	−11.31	200	1270	78.69	800	700	600	1.470
25	−14.03	200	1170	75.97	800	700	600	1.370
30	0.00	200	1605	90.00	600	500	500	1.805
30	−5.71	200	1380	84.29	600	500	500	1.580
30	−11.31	200	1160	78.69	600	500	400	1.360
30	−14.03	200	1055	75.97	600	500	400	1.255
35	0.00	200	1515	90.00	400	300	200	1.715
35	−5.71	200	1275	84.29	400	400	300	1.475
35	−11.31	200	1040	78.69	400	300	300	1.240
35	−14.03	200	925	75.97	400	300	300	1.125

(Continued)

TABLE 5.18 (Continued)

$H = 2.0$ m, $\gamma_{masonry} = 22.56$ kN/m³, $\gamma_{sat} = 17.27$ kN/m³ and $\iota = \phi$

					Depth of Shear Key D_{sk}, mm			
ϕ Degree	α Degree	Top Width, mm	Bottom Width, mm	Horizontal Angle θ Degree	$\mu = 0.50$	$\mu = 0.70$	$\mu = 0.80$	Volume of Wall, m³
40	0.00	200	1415	90.00	200	200	100	1.615
40	−5.71	200	1160	84.29	300	200	200	1.360
40	−11.31	200	910	78.69	300	200	200	1.110
40	−14.03	200	790	75.97	300	200	200	0.990

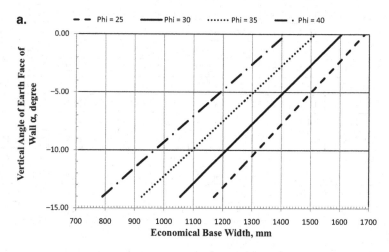

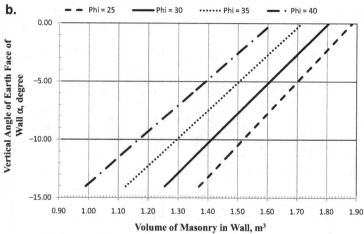

FIGURE 5.19 Design Charts for a Masonry Wall $H = 2.0$ m, $\gamma_{masonry} = 22.56$ kN/m³, $\gamma_{sat} = 17.27$ kN/m³ and $\iota = \phi$: (a) Economical Base Width for a Wall, (b) Volume of a Masonry in Wall

5.5 CHARTS FOR THE DESIGN OF A 2.5-M-HIGH MASONRY BREAST WALL

The charts/tables developed for a 2.5-m-high stone masonry breast wall for the previously discussed six combinations are presented in the Table 5.19 to Table 5.24 and Figures 5.20 to 5.25.

TABLE 5.19

$H = 2.5$ m, $\gamma_{masonry} = 22.56$ kN/m³, $\gamma_{sat} = 15.70$ kN/m³ and $\iota = 0°$

A. *Results of Stability Analysis*

				Pressure at		$\mu = 0.50$		$\mu = 0.70$		$\mu = 0.80$	
ϕ Degree	α Degree	K_{ah}	F_{OVT}	Heel, kN/ m²	Toe, kN/ m²	F_{SLD}	Net Force, kN	F_{SLD}	Net Force, kN	F_{SLD}	Net Force, kN
25	0.00	0.346	1.81	0.05	57.68	0.90	11.23	1.26	4.48	1.44	1.11
25	−5.71	0.316	1.67	0.45	59.04	0.83	11.62	1.16	5.85	1.33	2.96
25	−11.31	0.286	1.52	0.51	61.58	0.76	11.89	1.06	7.06	1.21	4.65
25	−14.03	0.271	1.50	5.02	58.40	0.73	11.72	1.03	7.25	1.17	5.01
30	0.00	0.279	1.82	0.24	58.45	0.99	8.07	1.38	1.89	1.58	0.00
30	−5.71	0.250	1.65	0.50	60.55	0.90	8.60	1.26	3.44	1.44	0.86
30	−11.31	0.221	1.50	1.96	62.65	0.81	8.94	1.14	4.73	1.30	2.62
30	−14.03	0.207	1.50	10.01	56.39	0.79	8.76	1.10	4.88	1.26	2.94
35	0.00	0.226	1.83	0.02	59.78	1.07	5.68	1.50	0.04	1.71	0.00
35	−5.71	0.198	1.64	0.60	62.30	0.97	6.33	1.36	1.72	1.55	0.00
35	−11.31	0.170	1.51	6.70	60.84	0.87	6.65	1.22	2.94	1.40	1.09
35	−14.03	0.156	1.50	17.84	52.49	0.84	6.58	1.18	3.23	1.34	1.55
40	0.00	0.184	1.84	0.12	60.81	1.16	3.88	1.62	0.00	1.85	0.00
40	−5.71	0.156	1.62	0.58	64.47	1.03	4.68	1.44	0.55	1.65	0.00
40	−11.31	0.130	1.51	12.48	58.91	0.93	5.02	1.30	1.78	1.48	0.16
40	−14.03	0.117	1.51	30.82	44.77	0.89	5.01	1.24	2.12	1.42	0.68

B. *Physical Dimensions for a Technoeconomic Wall Section*

					Depth of Shear Key D_{sk}, mm			
ϕ Degree	α Degree	Top Width, mm	Bottom Width, mm	Horizontal Angle θ Degree	$\mu = 0.50$	$\mu = 0.70$	$\mu = 0.80$	Volume of Wall, m³
25	0.00	200	1300	90.00	600	300	100	1.875
25	−5.71	200	1080	84.29	600	400	300	1.600
25	−11.31	200	865	78.69	700	500	400	1.331
25	−14.03	200	785	75.97	700	500	500	1.231
30	0.00	200	1170	90.00	400	200	0	1.713
30	−5.71	200	940	84.29	400	200	100	1.425
30	−11.31	200	725	78.69	500	400	300	1.156
30	−14.03	200	650	75.97	500	400	300	1.063
35	0.00	200	1050	90.00	200	0	0	1.563
35	−5.71	200	815	84.29	300	100	0	1.269

(Continued)

TABLE 5.19 (Continued)

$H = 2.5$ m, $\gamma_{masonry} = 22.56$ kN/m³, $\gamma_{sat} = 15.70$ kN/m³ and $\iota = 0°$

ϕ Degree	α Degree	Top Width, mm	Bottom Width, mm	Horizontal Angle θ Degree	Depth of Shear Key D_{sk}, mm			Volume of Wall, m³
					$\mu = 0.50$	$\mu = 0.70$	$\mu = 0.80$	
35	−11.31	200	610	78.69	400	300	200	1.013
35	−14.03	200	530	75.97	400	300	200	0.913
40	0.00	200	950	90.00	100	0	0	1.438
40	−5.71	200	705	84.29	200	0	0	1.131
40	−11.31	200	505	78.69	300	200	100	0.881
40	−14.03	200	425	75.97	300	200	100	0.781

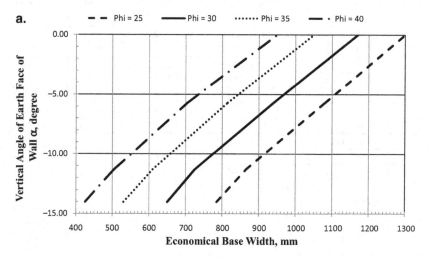

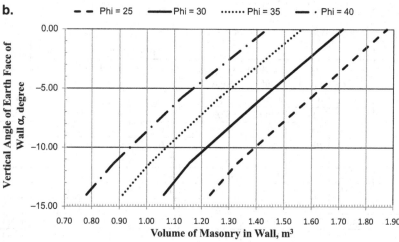

FIGURE 5.20 Design Charts for a Masonry Wall $H = 2.5$ m, $\gamma_{masonry} = 22.56$ kN/m³, $\gamma_{sat} = 15.70$ kN/m³ and $\iota = 0°$: (a) Economical Base Width for a Wall, (b) Volume of a Masonry in Wall

TABLE 5.20

$H = 2.5$ m, $\gamma_{masonry} = 22.56$ kN/m³, $\gamma_{sat} = 17.27$ kN/m³ and $\iota = 0°$

A. *Results of Stability Analysis*

ϕ Degree	α Degree	K_{ah}	F_{OVT}	Pressure at Heel, kN/m²	Toe, kN/m²	$\mu = 0.50$ F_{SLD}	Net Force, kN	$\mu = 0.70$ F_{SLD}	Net Force, kN	$\mu = 0.80$ F_{SLD}	Net Force, kN
25	0.00	0.346	1.81	0.01	57.30	0.86	13.06	1.21	6.02	1.38	2.50
25	−5.71	0.316	1.67	0.06	58.88	0.80	13.29	1.12	7.24	1.28	4.22
25	−11.31	0.286	1.53	0.39	60.85	0.73	13.33	1.03	8.24	1.17	5.69
25	−14.03	0.271	1.50	3.91	58.57	0.71	13.11	0.99	8.39	1.14	6.03
30	0.00	0.279	1.82	0.37	57.85	0.94	9.46	1.32	3.02	1.51	0.00
30	−5.71	0.250	1.66	0.33	60.06	0.87	9.84	1.21	4.44	1.39	1.73
30	−11.31	0.221	1.50	1.23	62.38	0.79	10.01	1.10	5.58	1.26	3.37
30	−14.03	0.207	1.50	7.74	57.54	0.77	9.78	1.07	5.70	1.23	3.66
35	0.00	0.226	1.83	0.32	58.94	1.03	6.73	1.44	0.85	1.65	0.00
35	−5.71	0.198	1.65	0.63	61.46	0.94	7.23	1.31	2.40	1.50	0.00
35	−11.31	0.170	1.50	4.69	61.71	0.85	7.45	1.19	3.57	1.36	1.63
35	−14.03	0.156	1.50	14.89	53.96	0.82	7.28	1.15	3.76	1.31	1.99
40	0.00	0.184	1.84	0.51	59.83	1.11	4.68	1.56	0.00	1.78	0.00
40	−5.71	0.156	1.63	0.78	63.31	1.00	5.33	1.41	1.01	1.61	0.00
40	−11.31	0.130	1.51	9.86	60.08	0.90	5.59	1.27	2.19	1.45	0.49
40	−14.03	0.117	1.50	24.73	49.11	0.87	5.54	1.21	2.52	1.38	1.01

B. *Physical Dimensions for a Technoeconomic Wall Section*

ϕ Degree	α Degree	Top Width, mm	Bottom Width, mm	Horizontal Angle θ Degree	Depth of Shear Key D_{sk}, mm $\mu = 0.50$	$\mu = 0.70$	$\mu = 0.80$	Volume of Wall, m³
25	0.00	200	1365	90.00	600	400	200	1.956
25	−5.71	200	1140	84.29	600	400	300	1.675
25	−11.31	200	925	78.69	700	500	500	1.406
25	−14.03	200	840	75.97	700	500	500	1.300
30	0.00	200	1230	90.00	400	200	0	1.788
30	−5.71	200	995	84.29	400	200	100	1.494
30	−11.31	200	775	78.69	500	400	300	1.219
30	−14.03	200	695	75.97	500	400	300	1.119
35	0.00	200	1105	90.00	200	100	0	1.631
35	−5.71	200	865	84.29	300	100	0	1.331
35	−11.31	200	650	78.69	400	300	200	1.063
35	−14.03	200	570	75.97	400	300	200	0.963

(Continued)

TABLE 5.20 (Continued)

$H = 2.5$ m, $\gamma_{masonry} = 22.56$ kN/m³, $\gamma_{sat} = 17.27$ kN/m³ and $\iota = 0°$

ϕ Degree	α Degree	Top Width, mm	Bottom Width, mm	Horizontal Angle θ Degree	Depth of Shear Key D_{sk}, mm			Volume of Wall, m³
					$\mu = 0.50$	$\mu = 0.70$	$\mu = 0.80$	
40	0.00	200	1000	90.00	200	0	0	1.500
40	−5.71	200	750	84.29	200	0	0	1.188
40	−11.31	200	540	78.69	300	200	100	0.925
40	−14.03	200	455	75.97	300	200	100	0.819

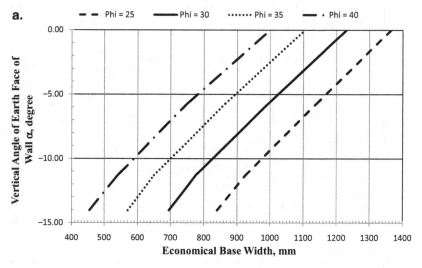

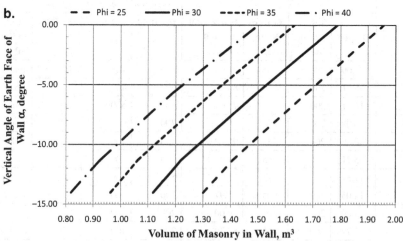

FIGURE 5.21 Design Charts for a Masonry Wall $H = 2.5$ m, $\gamma_{masonry} = 22.56$ kN/m³, $\gamma_{sat} = 17.27$ kN/m³ and $\iota = 0°$: (a) Economical Base Width for a Wall, (b) Volume of a Masonry in Wall

TABLE 5.21

$H = 2.5$ m, $\gamma_{masonry} = 22.56$ kN/m^3, $\gamma_{sat} = 15.70$ kN/m^3 and $\iota = \frac{1}{2}\phi$

A. Results of Stability Analysis

ϕ Degree	α Degree	K_{ah}	F_{OVT}	Pressure at		$\mu = 0.50$		$\mu = 0.70$		$\mu = 0.80$	
				Heel, kN/m^2	Toe, kN/m^2	F_{SLD}	Net Force, kN	F_{SLD}	Net Force, kN	F_{SLD}	Net Force, kN
25	0.00	0.424	1.81	0.31	56.55	0.83	15.05	1.16	7.66	1.32	3.97
25	−5.71	0.385	1.68	0.25	58.12	0.77	14.94	1.08	8.59	1.24	5.42
25	−11.31	0.346	1.54	0.28	60.22	0.71	14.71	1.00	9.35	1.14	6.68
25	−14.03	0.326	1.50	2.46	59.27	0.69	14.42	0.97	9.48	1.11	7.01
30	0.00	0.348	1.81	0.05	57.65	0.90	11.36	1.26	4.59	1.44	1.21
30	−5.71	0.309	1.67	0.42	59.23	0.84	11.29	1.17	5.58	1.34	2.72
30	−11.31	0.270	1.51	0.32	62.32	0.77	11.19	1.07	6.52	1.23	4.18
30	−14.03	0.251	1.50	5.99	58.20	0.75	10.82	1.05	6.52	1.20	4.37
35	0.00	0.284	1.82	0.24	58.37	0.98	8.29	1.37	2.07	1.57	0.00
35	−5.71	0.246	1.66	0.64	60.54	0.91	8.38	1.27	3.26	1.45	0.70
35	−11.31	0.208	1.51	3.45	61.72	0.83	8.32	1.16	4.22	1.32	2.17
35	−14.03	0.190	1.50	12.46	54.94	0.81	8.02	1.13	4.29	1.29	2.43
40	0.00	0.232	1.83	0.17	59.47	1.06	5.92	1.48	0.21	1.70	0.00
40	−5.71	0.194	1.64	0.55	62.51	0.97	6.17	1.36	1.61	1.56	0.00
40	−11.31	0.158	1.50	7.68	60.82	0.89	6.18	1.24	2.61	1.42	0.82
40	−14.03	0.142	1.50	20.74	51.34	0.85	6.01	1.19	2.84	1.37	1.25

B. Physical Dimensions for a Technoeconomic Wall Section

ϕ Degree	α Degree	Top Width, mm	Bottom Width, mm	Horizontal Angle θ Degree	Depth of Shear Key D_{sk}, mm			Volume of Wall, m^3
					$\mu = 0.50$	$\mu = 0.70$	$\mu = 0.80$	
25	0.00	200	1445	90.00	700	400	300	2.056
25	−5.71	200	1210	84.29	600	500	400	1.763
25	−11.31	200	985	78.69	700	600	500	1.481
25	−14.03	200	890	75.97	700	600	500	1.363
30	0.00	200	1305	90.00	500	300	100	1.881
30	−5.71	200	1065	84.29	500	300	200	1.581
30	−11.31	200	830	78.69	400	300	300	1.288
30	−14.03	200	745	75.97	500	400	400	1.181
35	0.00	200	1180	90.00	300	100	0	1.725
35	−5.71	200	930	84.29	300	200	100	1.413
35	−11.31	200	700	78.69	400	300	200	1.125
35	−14.03	200	615	75.97	400	300	200	1.019

(Continued)

TABLE 5.21 (Continued)

$H = 2.5$ m, $\gamma_{masonry} = 22.56$ kN/m³, $\gamma_{sat} = 15.70$ kN/m³ and $\iota = \frac{1}{2}\phi$

					Depth of Shear Key D_{sk}, mm			
ϕ Degree	α Degree	Top Width, mm	Bottom Width, mm	Horizontal Angle θ Degree	$\mu = 0.50$	$\mu = 0.70$	$\mu = 0.80$	Volume of Wall, m³
40	0.00	200	1065	90.00	200	0	0	1.581
40	−5.71	200	805	84.29	200	100	0	1.256
40	−11.31	200	580	78.69	300	200	100	0.975
40	−14.03	200	490	75.97	300	200	200	0.863

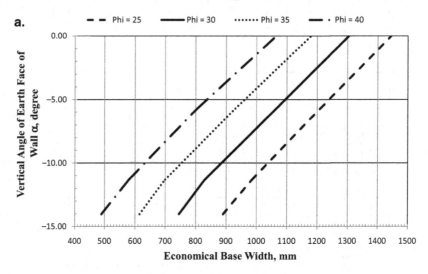

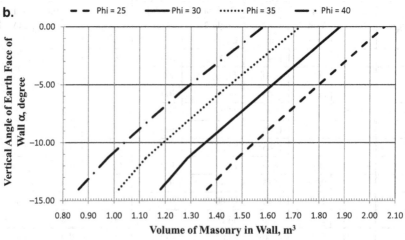

FIGURE 5.22 Design Charts for a Masonry Wall $H = 2.5$ m, $\gamma_{masonry} = 22.56$ kN/m³, $\gamma_{sat} = 15.70$ kN/m³ and $\iota = \frac{1}{2}\phi$: (a) Economical Base Width for a Wall, (b) Volume of Masonry in Wall

TABLE 5.22

$H = 2.5$ m, $\gamma_{masonry} = 22.56$ kN/m^3, $\gamma_{sat} = 17.27$ kN/m^3 and $\iota = \frac{1}{2}\phi$

A. Results of Stability Analysis

ϕ Degree	α Degree	K_{ah}	F_{OVT}	Pressure at Heel, kN/m^2	Toe, kN/m^2	$\mu = 0.50$ F_{SLD}	Net Force, kN	$\mu = 0.70$ F_{SLD}	Net Force, kN	$\mu = 0.80$ F_{SLD}	Net Force, kN
25	0.00	0.424	1.81	0.09	56.40	0.79	17.39	1.10	9.70	1.26	5.85
25	−5.71	0.385	1.68	0.22	57.65	0.74	17.00	1.04	10.34	1.19	7.01
25	−11.31	0.346	1.55	0.03	59.76	0.69	16.54	0.97	10.89	1.11	8.07
25	−14.03	0.326	1.50	1.53	59.39	0.67	16.16	0.94	10.95	1.07	8.35
30	0.00	0.348	1.82	0.38	56.88	0.86	13.15	1.20	6.07	1.38	2.53
30	−5.71	0.309	1.67	0.11	58.97	0.81	12.90	1.13	6.92	1.29	3.93
30	−11.31	0.270	1.52	0.49	61.24	0.74	12.52	1.04	7.58	1.19	5.11
30	−14.03	0.251	1.50	5.15	58.00	0.73	12.06	1.02	7.52	1.16	5.25
35	0.00	0.284	1.82	0.32	57.83	0.94	9.72	1.31	3.24	1.50	0.00
35	−5.71	0.246	1.66	0.53	59.97	0.87	9.57	1.22	4.22	1.40	1.54
35	−11.31	0.208	1.50	1.91	62.28	0.80	9.36	1.12	5.06	1.28	2.91
35	−14.03	0.190	1.51	10.34	55.80	0.79	8.92	1.10	5.00	1.26	3.03
40	0.00	0.232	1.83	0.38	58.74	1.02	7.02	1.43	1.07	1.63	0.00
40	−5.71	0.194	1.65	0.66	61.58	0.94	7.05	1.32	2.26	1.51	0.00
40	−11.31	0.158	1.50	5.89	61.36	0.86	6.90	1.21	3.15	1.38	1.28
40	−14.03	0.142	1.50	17.93	52.40	0.84	6.61	1.17	3.26	1.34	1.58

B. Physical Dimensions for a Technoeconomic Wall Section

ϕ Degree	α Degree	Top Width, mm	Bottom Width, mm	Horizontal Angle θ Degree	Depth of Shear Key D_{sk}, mm $\mu = 0.50$	$\mu = 0.70$	$\mu = 0.80$	Volume of Wall, m^3
25	0.00	200	1515	90.00	700	500	400	2.144
25	−5.71	200	1280	84.29	700	500	400	1.850
25	−11.31	200	1050	78.69	700	500	400	1.563
25	−14.03	200	950	75.97	700	600	500	1.438
30	0.00	200	1375	90.00	500	300	200	1.969
30	−5.71	200	1125	84.29	500	300	200	1.656
30	−11.31	200	890	78.69	500	300	300	1.363
30	−14.03	200	800	75.97	500	400	400	1.250
35	0.00	200	1240	90.00	300	200	0	1.800
35	−5.71	200	985	84.29	300	200	100	1.481
35	−11.31	200	745	78.69	300	200	100	1.181
35	−14.03	200	660	75.97	400	300	200	1.075

(Continued)

TABLE 5.22 (Continued)

H = 2.5 m, $\gamma_{masonry}$ = 22.56 kN/m³, γ_{sat} = 17.27 kN/m³ and ι = ½ϕ

ϕ Degree	α Degree	Top Width, mm	Bottom Width, mm	Horizontal Angle θ Degree	Depth of Shear Key D_{sk}, mm			Volume of Wall, m³
					μ = 0.50	μ = 0.70	μ = 0.80	
40	0.00	200	1120	90.00	200	0	0	1.650
40	−5.71	200	855	84.29	200	100	0	1.319
40	−11.31	200	620	78.69	200	100	100	1.025
40	−14.03	200	530	75.97	300	200	200	0.913

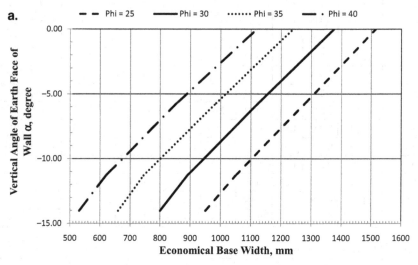

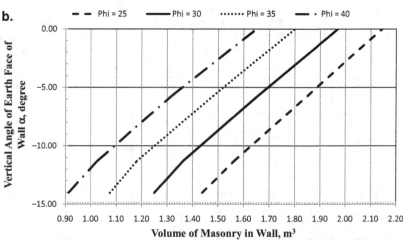

FIGURE 5.23 Design Charts for a Masonry Wall H = 2.5 m, $\gamma_{masonry}$ = 22.56 kN/m³, γ_{sat} = 17.27 kN/m³ and ι = ½ϕ: (a) Economical Base Width for a Wall, (b) Volume of a Masonry in Wall

TABLE 5.23

H = 2.5 m, $\gamma_{masonry}$ = 22.56 kN/m^3, γ_{sat} = 15.70 kN/m^3 and ι = ϕ

A. Results of Stability Analysis

				Pressure at		$\mu = 0.50$		$\mu = 0.70$		$\mu = 0.80$	
ϕ Degree	α Degree	K_{ah}	F_{OVT}	Heel, kN/m^2	Toe, kN/m^2	F_{SLD}	Net Force, kN	F_{SLD}	Net Force, kN	F_{SLD}	Net Force, kN
25	0.00	0.821	1.79	0.01	54.63	0.61	36.30	0.85	26.37	0.97	21.41
25	−5.71	0.746	1.70	0.03	55.41	0.59	33.91	0.83	25.12	0.94	20.72
25	−11.31	0.675	1.61	0.12	56.35	0.57	31.67	0.80	23.95	0.91	20.09
25	−14.03	0.641	1.55	0.02	57.09	0.55	30.62	0.78	23.43	0.89	19.84
30	0.00	0.750	1.80	0.11	54.78	0.64	32.33	0.89	22.81	1.02	18.04
30	−5.71	0.666	1.70	0.05	55.82	0.62	29.53	0.87	21.21	0.99	17.06
30	−11.31	0.587	1.60	0.42	56.72	0.60	26.90	0.84	19.74	0.96	16.15
30	−14.03	0.549	1.54	0.22	57.74	0.59	25.73	0.82	19.13	0.94	15.84
35	0.00	0.671	1.80	0.22	55.01	0.67	28.01	0.94	18.94	1.07	14.41
35	−5.71	0.580	1.70	0.37	56.03	0.66	24.91	0.92	17.13	1.05	13.24
35	−11.31	0.496	1.58	0.11	57.97	0.64	22.22	0.89	15.69	1.02	12.43
35	−14.03	0.456	1.52	0.20	58.96	0.62	20.96	0.87	15.03	0.99	12.07
40	0.00	0.587	1.81	0.27	55.39	0.71	23.49	1.00	14.95	1.14	10.69
40	−5.71	0.492	1.69	0.19	56.95	0.70	20.36	0.98	13.19	1.12	9.61
40	−11.31	0.406	1.56	0.53	58.77	0.68	17.64	0.95	11.77	1.09	8.84
40	−14.03	0.366	1.50	1.14	59.66	0.67	16.41	0.93	11.17	1.07	8.54

B. Physical Dimensions for a Technoeconomic Wall Section

					Depth of Shear Key D_{sk}, mm			
ϕ Degree	α Degree	Top Width, mm	Bottom Width, mm	Horizontal Angle θ Degree	$\mu = 0.50$	$\mu = 0.70$	$\mu = 0.80$	Volume of Wall, m^3
25	0.00	200	2020	90.00	900	800	700	2.775
25	−5.71	200	1765	84.29	1000	900	800	2.456
25	−11.31	200	1520	78.69	1000	800	800	2.150
25	−14.03	200	1400	75.97	900	800	800	2.000
30	0.00	200	1930	90.00	700	600	500	2.663
30	−5.71	200	1655	84.29	800	600	600	2.319
30	−11.31	200	1395	78.69	700	600	500	1.994
30	−14.03	200	1265	75.97	700	600	500	1.831
35	0.00	200	1825	90.00	500	400	300	2.531
35	−5.71	200	1535	84.29	500	300	300	2.169
35	−11.31	200	1250	78.69	500	400	400	1.813
35	−14.03	200	1115	75.97	500	400	400	1.644

(Continued)

TABLE 5.23 (Continued)

$H = 2.5$ m, $\gamma_{masonry} = 22.56$ kN/m³, $\gamma_{sat} = 15.70$ kN/m³ and $\iota = \phi$

ϕ Degree	α Degree	Top Width, mm	Bottom Width, mm	Horizontal Angle θ Degree	Depth of Shear Key D_{sk}, mm			Volume of Wall, m³
					$\mu = 0.50$	$\mu = 0.70$	$\mu = 0.80$	
40	0.00	200	1705	90.00	300	200	200	2.381
40	−5.71	200	1395	84.29	300	200	200	1.994
40	−11.31	200	1100	78.69	400	300	200	1.625
40	−14.03	200	960	75.97	400	300	200	1.450

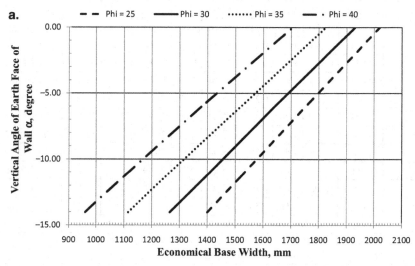

a.

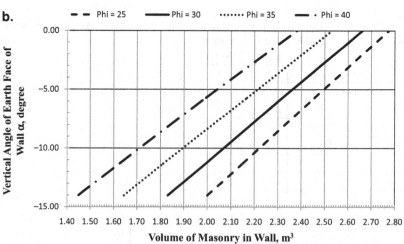

b.

FIGURE 5.24 Design Charts for a Masonry Wall $H = 2.5$ m, $\gamma_{masonry} = 22.56$ kN/m³, $\gamma_{sat} = 15.70$ kN/m³ and $\iota = \phi$: (a) Economical Base Width for a Wall and (b) Volume of a Masonry in Wall

TABLE 5.24

$H = 2.5$ m, $\gamma_{masonry} = 22.56$ kN/m³, $\gamma_{sat} = 17.27$ kN/m³ and $\iota = \phi$

A. Results of Stability Analysis

				Pressure at		$\mu = 0.50$		$\mu = 0.70$		$\mu = 0.80$	
ϕ Degree	α Degree	K_{ah}	F_{OVT}	Heel, kN/m²	Toe, kN/m²	F_{SLD}	Net Force, kN	F_{SLD}	Net Force, kN	F_{SLD}	Net Force, kN
25	0.00	0.821	1.80	0.11	54.25	0.58	41.17	0.81	30.79	0.93	25.60
25	−5.71	0.746	1.71	0.10	54.99	0.56	38.30	0.79	29.06	0.90	24.44
25	−11.31	0.675	1.61	0.19	55.85	0.55	35.58	0.76	27.45	0.87	23.38
25	−14.03	0.641	1.57	0.32	56.27	0.54	34.28	0.75	26.67	0.86	22.87
30	0.00	0.750	1.80	0.20	54.41	0.61	36.74	0.85	26.77	0.97	21.79
30	−5.71	0.666	1.70	0.17	55.33	0.59	33.37	0.83	24.64	0.95	20.27
30	−11.31	0.587	1.60	0.18	56.49	0.58	30.28	0.81	22.74	0.92	18.96
30	−14.03	0.549	1.55	0.25	57.16	0.57	28.82	0.79	21.86	0.90	18.37
35	0.00	0.671	1.80	0.06	54.88	0.64	31.94	0.89	22.49	1.02	17.76
35	−5.71	0.580	1.70	0.22	55.79	0.63	28.24	0.88	20.08	1.01	16.00
35	−11.31	0.496	1.58	0.07	57.46	0.61	24.98	0.86	18.10	0.98	14.66
35	−14.03	0.456	1.53	0.54	57.95	0.60	23.44	0.84	17.15	0.96	14.01
40	0.00	0.587	1.80	0.17	55.18	0.68	26.86	0.95	17.95	1.09	13.50
40	−5.71	0.492	1.69	0.20	56.50	0.67	23.09	0.94	15.57	1.08	11.81
40	−11.31	0.406	1.57	0.21	58.48	0.66	19.85	0.92	13.67	1.05	10.59
40	−14.03	0.366	1.51	0.53	59.52	0.64	18.39	0.90	12.85	1.03	10.09

B. Physical Dimensions for a Technoeconomic Wall Section

					Depth of Shear Key D_{sk}, mm			
ϕ Degree	α Degree	Top Width, mm	Bottom Width, mm	Horizontal Angle θ Degree	$\mu = 0.50$	$\mu = 0.70$	$\mu = 0.80$	Volume of Wall m³
25	0.00	200	2125	90.00	1000	800	700	2.906
25	−5.71	200	1865	84.29	1000	900	800	2.581
25	−11.31	200	1615	78.69	1000	900	800	2.269
25	−14.03	200	1495	75.97	1000	800	800	2.119
30	0.00	200	2030	90.00	700	600	500	2.788
30	−5.71	200	1750	84.29	700	600	500	2.438
30	−11.31	200	1480	78.69	700	600	600	2.100
30	−14.03	200	1350	75.97	700	600	600	1.938
35	0.00	200	1915	90.00	500	400	300	2.644
35	−5.71	200	1620	84.29	500	400	300	2.275
35	−11.31	200	1330	78.69	500	400	400	1.913
35	−14.03	200	1195	75.97	500	400	400	1.744

(Continued)

TABLE 5.24 (Continued)

$H = 2.5$ m, $\gamma_{masonry} = 22.56$ kN/m³, $\gamma_{sat} = 17.27$ kN/m³ and $\iota = \phi$

φ Degree	α Degree	Top Width, mm	Bottom Width, mm	Horizontal Angle θ Degree	Depth of Shear Key D_{sk}, mm			Volume of Wall m³
					$\mu = 0.50$	$\mu = 0.70$	$\mu = 0.80$	
40	0.00	200	1790	90.00	300	300	200	2.488
40	−5.71	200	1475	84.29	300	200	200	2.094
40	−11.31	200	1170	78.69	400	300	300	1.713
40	−14.03	200	1025	75.97	400	300	300	1.531

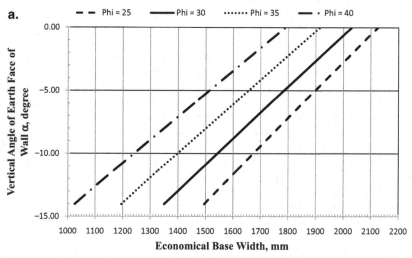

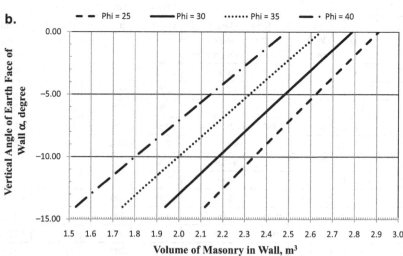

FIGURE 5.25 Design Charts for a Masonry Wall $H = 2.5$ m, $\gamma_{masonry} = 22.56$ kN/m³, $\gamma_{sat} = 17.27$ kN/m³ and $\iota = \phi$ (a) Economical Base Width for a Wall and (b) Volume of a Masonry in Wall

5.6 CHARTS FOR THE DESIGN OF A 3.0-M-HIGH MASONRY BREAST WALL

The charts/tables developed for a 3.0-m-high stone masonry breast wall for the previously discussed six combinations are presented in the Tables 5.25 to 5.30 and Figures 5.26 to 5.31.

TABLE 5.25

$H = 3.0$ m, $\gamma_{masonry} = 22.56$ kN/m³, $\gamma_{sat} = 15.70$ kN/m³ and $\iota = 0°$

A. Results of Stability Analysis

				Pressure at		$\mu = 0.50$		$\mu = 0.70$		$\mu = 0.80$	
ϕ Degree	α Degree	K_{ah}	F_{OVT}	Heel, kN/m²	Toe, kN/m²	F_{SLD}	Net Force, kN	F_{SLD}	Net Force, kN	F_{SLD}	Net Force, kN
25	0.00	0.346	1.83	0.10	71.95	0.92	15.53	1.29	5.55	1.48	0.56
25	−5.71	0.316	1.67	0.30	74.61	0.85	16.23	1.19	7.71	1.36	3.45
25	−11.31	0.286	1.52	0.69	78.48	0.77	16.72	1.08	9.60	1.24	6.04
25	−14.03	0.271	1.50	7.84	73.46	0.75	16.46	1.05	9.84	1.20	6.53
30	0.00	0.279	1.84	0.30	73.22	1.01	10.96	1.42	1.80	1.62	0.00
30	−5.71	0.250	1.66	0.64	76.68	0.93	11.85	1.30	4.20	1.48	0.38
30	−11.31	0.221	1.50	3.37	79.81	0.84	12.42	1.17	6.18	1.34	3.06
30	−14.03	0.207	1.50	14.53	71.77	0.81	12.23	1.13	6.49	1.30	3.62
35	0.00	0.226	1.85	0.29	74.88	1.11	7.46	1.55	0.00	1.77	0.00
35	−5.71	0.198	1.65	0.65	79.59	1.00	8.57	1.40	1.72	1.60	0.00
35	−11.31	0.170	1.51	9.53	78.55	0.90	9.15	1.26	3.64	1.44	0.89
35	−14.03	0.156	1.50	26.15	66.84	0.87	9.10	1.21	4.13	1.39	1.64
40	0.00	0.184	1.86	0.25	76.66	1.20	4.88	1.68	0.00	1.92	0.00
40	−5.71	0.156	1.63	0.86	82.77	1.07	6.16	1.50	0.00	1.71	0.00
40	−11.31	0.130	1.50	16.67	78.13	0.96	6.86	1.34	2.04	1.53	0.00
40	−14.03	0.117	1.50	45.37	57.16	0.91	6.90	1.28	2.61	1.46	0.47

B. Physical Dimensions for a Technoeconomic Wall Section

					Depth of Shear Key D_{sk}, mm			
ϕ Degree	α Degree	Top Width, mm	Bottom Width, mm	Horizontal Angle θ Degree	$\mu = 0.50$	$\mu = 0.70$	$\mu = 0.80$	Volume of Wall, m³
25	0.00	300	1540	90.00	700	400	100	2.760
25	−5.71	300	1265	84.29	700	500	300	2.348
25	−11.31	300	1000	78.69	800	600	500	1.950
25	−14.03	300	905	75.97	800	600	500	1.808
30	0.00	300	1385	90.00	400	200	0	2.528
30	−5.71	300	1100	84.29	500	300	0	2.100
30	−11.31	300	835	78.69	500	300	200	1.703
30	−14.03	300	740	75.97	600	400	300	1.560

(Continued)

TABLE 5.25 (Continued)

H = 3.0 m, $\gamma_{masonry}$ = 22.56 kN/m³, γ_{sat} = 15.70 kN/m³ and $\iota = 0°$

| φ Degree | α Degree | Top Width, mm | Bottom Width, mm | Horizontal Angle θ Degree | Depth of Shear Key D_{sk}, mm | | | Volume of Wall, m³ |
					μ = 0.50	μ = 0.70	μ = 0.80	
35	0.00	300	1245	90.00	300	0	0	2.318
35	−5.71	300	950	84.29	300	100	0	1.875
35	−11.31	300	695	78.69	400	300	200	1.493
35	−14.03	300	595	75.97	400	300	200	1.343
40	0.00	300	1125	90.00	200	0	0	2.138
40	−5.71	300	820	84.29	200	0	0	1.680
40	−11.31	300	565	78.69	300	200	0	1.298
40	−14.03	300	465	75.97	300	200	100	1.148

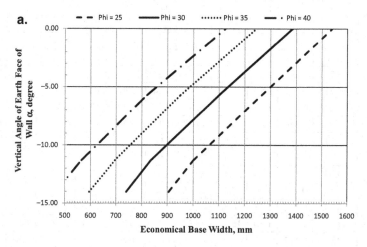

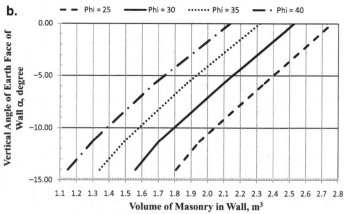

FIGURE 5.26 Design Charts for a Masonry Wall H = 3.0 m, $\gamma_{masonry}$ = 22.56 kN/m³, γ_{sat} = 15.70 kN/m³ and $\iota = 0°$: (a) Economical Base Width for a Wall, (b) Volume of a Masonry in Wall

TABLE 5.26

$H = 3.0$ m, $\gamma_{masonry} = 22.56$ kN/m³, $\gamma_{sat} = 17.27$ kN/m³ and $\iota = 0°$

A. Results of Stability Analysis

ϕ Degree	α Degree	K_{ah}	F_{OVT}	Pressure at Heel, kN/m²	Toe, kN/m²	$\mu = 0.50$ F_{SLD}	Net Force, kN	$\mu = 0.70$ F_{SLD}	Net Force, kN	$\mu = 0.80$ F_{SLD}	Net Force, kN
25	0.00	0.346	1.83	0.26	71.14	0.88	18.12	1.24	7.73	1.41	2.53
25	−5.71	0.316	1.67	0.19	73.82	0.82	18.59	1.15	9.67	1.31	5.21
25	−11.31	0.286	1.52	0.18	77.66	0.75	18.82	1.05	11.33	1.20	7.58
25	−14.03	0.271	1.50	5.06	74.84	0.72	18.53	1.01	11.60	1.16	8.13
30	0.00	0.279	1.84	0.35	72.47	0.97	12.99	1.36	3.46	1.55	0.00
30	−5.71	0.250	1.66	0.33	75.96	0.89	13.64	1.25	5.65	1.43	1.65
30	−11.31	0.221	1.50	1.28	80.40	0.81	14.04	1.13	7.50	1.29	4.23
30	−14.03	0.207	1.50	11.45	72.95	0.79	13.69	1.10	7.65	1.26	4.64
35	0.00	0.226	1.84	0.11	74.31	1.06	9.06	1.48	0.33	1.70	0.00
35	−5.71	0.198	1.65	0.74	78.22	0.97	9.87	1.35	2.70	1.55	0.00
35	−11.31	0.170	1.50	5.95	80.35	0.87	10.35	1.22	4.61	1.39	1.74
35	−14.03	0.156	1.50	22.14	68.20	0.85	10.09	1.19	4.85	1.36	2.23
40	0.00	0.184	1.86	0.25	75.82	1.15	6.10	1.61	0.00	1.84	0.00
40	−5.71	0.156	1.63	0.42	81.78	1.04	7.16	1.45	0.73	1.66	0.00
40	−11.31	0.130	1.50	13.85	78.30	0.94	7.63	1.31	2.58	1.50	0.05
40	−14.03	0.117	1.50	37.30	61.77	0.89	7.60	1.25	3.11	1.43	0.86

B. Physical Dimensions for a Technoeconomic Wall Section

ϕ Degree	α Degree	Top Width, mm	Bottom Width, mm	Horizontal Angle θ Degree	Depth of Shear Key D_{sk}, mm $\mu = 0.50$	$\mu = 0.70$	$\mu = 0.80$	Volume of Wall, m³
25	0.00	300	1620	90.00	700	400	200	2.880
25	−5.71	300	1340	84.29	700	500	300	2.460
25	−11.31	300	1070	78.69	700	500	400	2.055
25	−14.03	300	965	75.97	800	600	500	1.898
30	0.00	300	1455	90.00	500	200	0	2.633
30	−5.71	300	1165	84.29	500	300	100	2.198
30	−11.31	300	890	78.69	500	300	200	1.785
30	−14.03	300	795	75.97	600	400	400	1.643
35	0.00	300	1305	90.00	300	0	0	2.408
35	−5.71	300	1010	84.29	300	100	0	1.965
35	−11.31	300	740	78.69	300	200	100	1.560
35	−14.03	300	645	75.97	400	300	200	1.418

(Continued)

TABLE 5.26 (Continued)

$H = 3.0$ m, $\gamma_{masonry} = 22.56$ kN/m^3, $\gamma_{sat} = 17.27$ kN/m^3 and $\iota = 0°$

ϕ Degree	α Degree	Top Width, mm	Bottom Width, mm	Horizontal Angle θ Degree	Depth of Shear Key D_{sk}, mm			Volume of Wall, m^3
					$\mu = 0.50$	$\mu = 0.70$	$\mu = 0.80$	
40	0.00	300	1180	90.00	200	0	0	2.220
40	−5.71	300	870	84.29	200	0	0	1.755
40	−11.31	300	610	78.69	300	200	100	1.365
40	−14.03	300	505	75.97	300	200	100	1.208

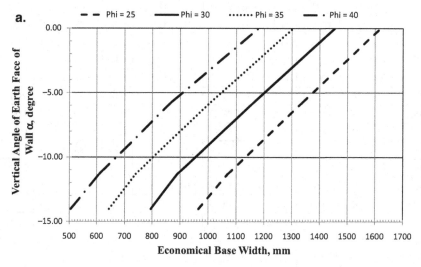

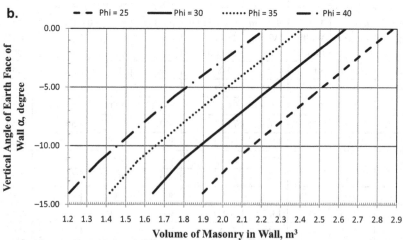

FIGURE 5.27 Design Charts for a Masonry Wall $H = 3.0$ m, $\gamma_{masonry} = 22.56$ kN/m^3, $\gamma_{sat} = 17.27$ kN/m^3 and $\iota = 0°$: (a) Economical Base Width for a Wall, (b) Volume of a Masonry in Wall

TABLE 5.27
$H = 3.0$ m, $\gamma_{masonry} = 22.56$ kN/m^3, $\gamma_{sat} = 15.70$ kN/m^3 and $\iota = \frac{1}{2}\phi$

A. *Results of Stability Analysis*

ϕ Degree	α Degree	K_{ah}	F_{OVT}	Pressure at Heel, kN/m^2	Toe, kN/m^2	$\mu = 0.50$ F_{SLD}	Net Force, kN	$\mu = 0.70$ F_{SLD}	Net Force, kN	$\mu = 0.80$ F_{SLD}	Net Force, kN
25	0.00	0.424	1.82	0.21	70.52	0.84	21.08	1.18	10.20	1.35	4.77
25	−5.71	0.385	1.68	0.03	73.13	0.79	21.03	1.10	11.69	1.26	7.02
25	−11.31	0.346	1.54	0.48	76.12	0.73	20.76	1.02	12.88	1.17	8.94
25	−14.03	0.326	1.50	3.97	74.61	0.71	20.36	0.99	13.08	1.13	9.44
30	0.00	0.348	1.83	0.01	71.99	0.92	15.72	1.29	5.72	1.47	0.72
30	−5.71	0.309	1.66	0.01	75.17	0.86	15.78	1.20	7.36	1.37	3.16
30	−11.31	0.270	1.51	0.84	79.18	0.79	15.68	1.10	8.78	1.26	5.32
30	−14.03	0.251	1.50	8.88	73.73	0.77	15.19	1.07	8.84	1.23	5.66
35	0.00	0.284	1.84	0.12	73.30	1.01	11.31	1.41	2.10	1.61	0.00
35	−5.71	0.246	1.66	0.39	77.18	0.93	11.56	1.30	3.99	1.49	0.21
35	−11.31	0.208	1.50	4.24	80.00	0.85	11.60	1.19	5.54	1.36	2.51
35	−14.03	0.190	1.50	17.29	70.78	0.83	11.20	1.16	5.70	1.32	2.95
40	0.00	0.232	1.85	0.17	74.80	1.09	7.85	1.53	0.00	1.75	0.00
40	−5.71	0.194	1.64	0.07	80.51	1.00	8.39	1.40	1.61	1.61	0.00
40	−11.31	0.158	1.51	11.27	78.36	0.92	8.46	1.28	3.14	1.47	0.49
40	−14.03	0.142	1.50	31.88	63.90	0.88	8.25	1.24	3.51	1.41	1.14

B. *Physical Dimensions for a Technoeconomic Wall Section*

ϕ Degree	α Degree	Top Width, mm	Bottom Width, mm	Horizontal Angle θ Degree	Depth of Shear Key D_{sk}, mm $\mu = 0.50$	$\mu = 0.70$	$\mu = 0.80$	Volume of Wall, m^3
25	0.00	300	1710	90.00	800	500	300	3.015
25	−5.71	300	1420	84.29	800	600	400	2.580
25	−11.31	300	1145	78.69	800	600	500	2.168
25	−14.03	300	1030	75.97	800	600	500	1.995
30	0.00	300	1545	90.00	500	300	100	2.768
30	−5.71	300	1245	84.29	500	400	200	2.318
30	−11.31	300	960	78.69	500	400	300	1.890
30	−14.03	300	855	75.97	500	400	300	1.733
35	0.00	300	1395	90.00	400	100	0	2.543
35	−5.71	300	1085	84.29	400	200	0	2.078
35	−11.31	300	800	78.69	400	200	100	1.650
35	−14.03	300	695	75.97	500	300	300	1.493

(Continued)

TABLE 5.27 (Continued)

H = 3.0 m, $\gamma_{masonry}$ = 22.56 kN/m³, γ_{sat} = 15.70 kN/m³ and ι = ½ϕ

| ϕ Degree | α Degree | Top Width, mm | Bottom Width, mm | Horizontal Angle θ Degree | Depth of Shear Key D_{sk}, mm | | | Volume of Wall, m³ |
					μ = 0.50	μ = 0.70	μ = 0.80	
40	0.00	300	1260	90.00	100	0	0	2.340
40	−5.71	300	935	84.29	200	100	0	1.853
40	−11.31	300	660	78.69	200	100	0	1.440
40	−14.03	300	550	75.97	300	200	200	1.275

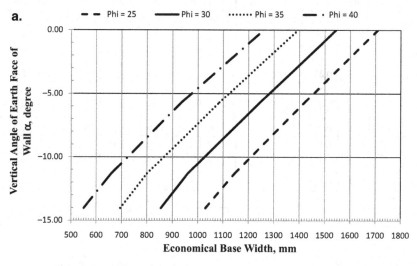

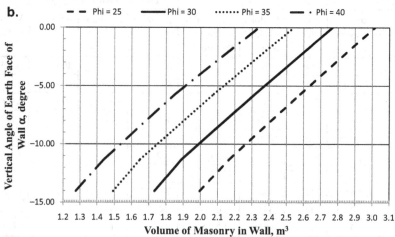

FIGURE 5.28 Design Charts for a Masonry Wall H = 3.0 m, $\gamma_{masonry}$ = 22.56 kN/m³, γ_{sat} = 15.70 kN/m³ and ι = ½ϕ: (a) Economical Base Width for a Wall, (b) Volume of a Masonry in Wall

TABLE 5.28

H = 3.0 m, $\gamma_{masonry}$ = 22.56 kN/m³, γ_{sat} = 17.27 kN/m³ and ι = ½ϕ

A. Results of Stability Analysis

ϕ Degree	α Degree	K_{ah}	F_{OVT}	Pressure at Heel, kN/ m²	Toe, kN/ m²	μ = 0.50 F_{SLD}	Net Force, kN	μ = 0.70 F_{SLD}	Net Force, kN	μ = 0.80 F_{SLD}	Net Force, kN
25	0.00	0.424	1.83	0.42	69.72	0.81	24.37	1.13	13.02	1.29	7.34
25	−5.71	0.385	1.68	0.13	72.23	0.76	23.97	1.06	14.18	1.21	9.29
25	−11.31	0.346	1.55	0.40	75.03	0.71	23.36	0.99	15.05	1.13	10.90
25	−14.03	0.326	1.50	2.31	75.01	0.68	22.88	0.96	15.24	1.09	11.41
30	0.00	0.348	1.83	0.15	71.20	0.88	18.35	1.23	7.93	1.41	2.71
30	−5.71	0.309	1.67	0.01	74.23	0.82	18.06	1.15	9.24	1.32	4.84
30	−11.31	0.270	1.52	0.64	77.93	0.76	17.62	1.07	10.34	1.22	6.70
30	−14.03	0.251	1.50	7.35	73.58	0.74	16.99	1.04	10.29	1.19	6.95
35	0.00	0.284	1.83	0.11	72.61	0.96	13.39	1.35	3.81	1.54	0.00
35	−5.71	0.246	1.66	0.19	76.33	0.90	13.30	1.26	5.39	1.44	1.43
35	−11.31	0.208	1.50	2.40	80.21	0.82	13.08	1.15	6.73	1.32	3.55
35	−14.03	0.190	1.50	14.40	71.53	0.81	12.49	1.13	6.69	1.29	3.80
40	0.00	0.232	1.85	0.41	73.77	1.05	9.45	1.47	0.61	1.68	0.00
40	−5.71	0.194	1.65	0.31	78.96	0.97	9.64	1.36	2.55	1.55	0.00
40	−11.31	0.158	1.50	7.87	79.79	0.89	9.54	1.24	3.99	1.42	1.21
40	−14.03	0.142	1.50	26.27	66.72	0.86	9.15	1.21	4.18	1.38	1.69

B. Physical Dimensions for a Technoeconomic Wall Section

ϕ Degree	α Degree	Top Width, mm	Bottom Width, mm	Horizontal Angle θ Degree	Depth of Shear Key D_{sk}, mm μ = 0.50	μ = 0.70	μ = 0.80	Volume of Wall, m³
25	0.00	300	1800	90.00	800	600	400	3.150
25	−5.71	300	1505	84.29	800	600	500	2.708
25	−11.31	300	1225	78.69	800	600	500	2.288
25	−14.03	300	1100	75.97	800	600	500	2.100
30	0.00	300	1625	90.00	600	400	200	2.888
30	−5.71	300	1320	84.29	600	400	300	2.430
30	−11.31	300	1030	78.69	500	400	300	1.995
30	−14.03	300	920	75.97	500	400	300	1.830
35	0.00	300	1465	90.00	300	100	0	2.648
35	−5.71	300	1150	84.29	400	200	100	2.175
35	−11.31	300	855	78.69	400	200	200	1.733
35	−14.03	300	750	75.97	400	200	200	1.575

(Continued)

TABLE 5.28 (Continued)

H = 3.0 m, $\gamma_{masonry}$ = 22.56 kN/m³, γ_{sat} = 17.27 kN/m³ and $\iota = \frac{1}{2}\phi$

					Depth of Shear Key D_{sk}, mm			
ϕ Degree	α Degree	Top Width, mm	Bottom Width, mm	Horizontal Angle θ Degree	$\mu = 0.50$	$\mu = 0.70$	$\mu = 0.80$	Volume of Wall, m³
40	0.00	300	1325	90.00	100	0	0	2.438
40	−5.71	300	995	84.29	200	100	0	1.943
40	−11.31	300	705	78.69	200	100	100	1.508
40	−14.03	300	595	75.97	300	200	200	1.343

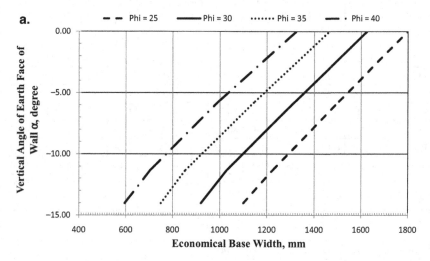

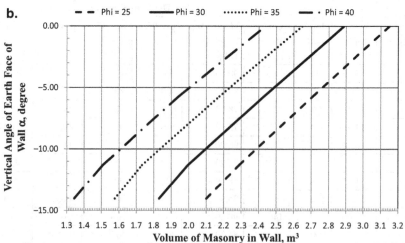

FIGURE 5.29 Design Charts for a Masonry Wall H = 3.0 m, $\gamma_{masonry}$ = 22.56 kN/m³, γ_{sat} = 17.27 kN/m³ and $\iota = \frac{1}{2}\phi$: (a) Economical Base Width for a Wall, (b) Volume of a Masonry in Wall

TABLE 5.29

$H = 3.0$ m, $\gamma_{masonry} = 22.56$ kN/m³, $\gamma_{sat} = 15.70$ kN/m³ and $\iota = \phi$

A. Results of Stability Analysis

ϕ Degree	α Degree	K_{ah}	F_{OVT}	Pressure at Heel, kN/m²	Toe, kN/m²	$\mu = 0.50$ F_{SLD}	Net Force, kN	$\mu = 0.70$ F_{SLD}	Net Force, kN	$\mu = 0.80$ F_{SLD}	Net Force, kN
25	0.00	0.821	1.80	0.07	67.25	0.62	51.67	0.87	37.14	0.99	29.88
25	−5.71	0.746	1.71	0.21	68.36	0.60	48.29	0.84	35.41	0.96	28.96
25	−11.31	0.675	1.61	0.39	69.81	0.58	45.14	0.81	33.84	0.92	28.19
25	−14.03	0.641	1.55	0.05	71.19	0.56	43.71	0.79	33.20	0.90	27.95
30	0.00	0.750	1.81	0.07	67.66	0.65	45.99	0.91	32.04	1.03	25.07
30	−5.71	0.666	1.70	0.02	69.23	0.63	42.02	0.88	29.84	1.01	23.76
30	−11.31	0.587	1.60	0.35	70.93	0.61	38.34	0.85	27.86	0.97	22.63
30	−14.03	0.549	1.53	0.02	72.60	0.59	36.70	0.83	27.07	0.95	22.26
35	0.00	0.671	1.81	0.22	68.02	0.68	39.75	0.96	26.47	1.09	19.82
35	−5.71	0.580	1.70	0.32	69.76	0.67	35.38	0.94	23.98	1.07	18.28
35	−11.31	0.496	1.58	0.38	72.34	0.65	31.54	0.91	21.96	1.04	17.17
35	−14.03	0.456	1.51	0.03	74.51	0.63	29.84	0.88	21.17	1.01	16.83
40	0.00	0.587	1.81	0.18	68.73	0.73	33.25	1.02	20.74	1.16	14.48
40	−5.71	0.492	1.69	0.31	70.90	0.72	28.79	1.00	18.26	1.15	12.99
40	−11.31	0.406	1.56	0.37	74.36	0.69	25.01	0.97	16.41	1.11	12.11
40	−14.03	0.366	1.50	2.27	74.80	0.68	23.21	0.95	15.49	1.09	11.62

B. Physical Dimensions for a Technoeconomic Wall Section

ϕ Degree	α Degree	Top Width, mm	Bottom Width, mm	Horizontal Angle θ Degree	Depth of Shear Key D_{sk}, mm $\mu = 0.50$	$\mu = 0.70$	$\mu = 0.80$	Volume of Wall, m³
25	0.00	300	2400	90.00	1100	1000	800	4.050
25	−5.71	300	2090	84.29	1100	900	800	3.585
25	−11.31	300	1790	78.69	1200	1000	900	3.135
25	−14.03	300	1640	75.97	1100	1000	900	2.910
30	0.00	300	2290	90.00	900	700	600	3.885
30	−5.71	300	1955	84.29	800	700	600	3.383
30	−11.31	300	1635	78.69	900	700	700	2.903
30	−14.03	300	1475	75.97	900	700	700	2.663
35	0.00	300	2165	90.00	500	400	300	3.698
35	−5.71	300	1810	84.29	600	400	400	3.165
35	−11.31	300	1465	78.69	500	400	300	2.648
35	−14.03	300	1295	75.97	600	500	400	2.393

(Continued)

TABLE 5.29 (Continued)

H = 3.0 m, $\gamma_{masonry}$ = 22.56 kN/m³, γ_{sat} = 15.70 kN/m³ and $\iota = \phi$

ϕ Degree	α Degree	Top Width, mm	Bottom Width, mm	Horizontal Angle θ Degree	Depth of Shear Key D_{sk}, mm			Volume of Wall, m³
					$\mu = 0.50$	$\mu = 0.70$	$\mu = 0.80$	
40	0.00	300	2020	90.00	300	200	100	3.480
40	−5.71	300	1645	84.29	400	300	200	2.918
40	−11.31	300	1280	78.69	300	300	200	2.370
40	−14.03	300	1115	75.97	400	300	300	2.123

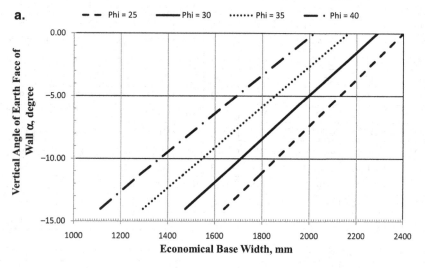

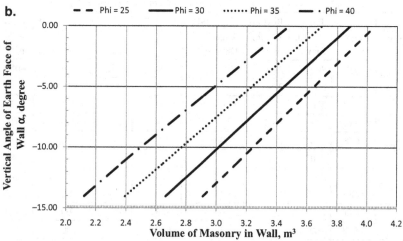

FIGURE 5.30 Design Charts for a Masonry Wall H = 3.0 m, $\gamma_{masonry}$ = 22.56 kN/m³, γ_{sat} = 15.70 kN/m³ and $\iota = \phi$: (a) Economical Base Width for a Wall and (b) Volume of a Masonry in Wall

TABLE 5.30

$H = 3.0$ m, $\gamma_{masonry} = 22.56$ kN/m³, $\gamma_{sat} = 17.27$ kN/m³ and $\iota = \phi$

A. Results of Stability Analysis

ϕ Degree	α Degree	K_{ah}	F_{OVT}	Pressure at Heel, kN/m²	Pressure at Toe, kN/m²	$\mu = 0.50$ F_{SLD}	$\mu = 0.50$ Net Force, kN	$\mu = 0.70$ F_{SLD}	$\mu = 0.70$ Net Force, kN	$\mu = 0.80$ F_{SLD}	$\mu = 0.80$ Net Force, kN
25	0.00	0.821	1.80	0.16	66.74	0.59	58.71	0.82	43.52	0.94	35.93
25	−5.71	0.746	1.71	0.31	67.74	0.57	54.61	0.80	41.08	0.92	34.32
25	−11.31	0.675	1.61	0.14	69.41	0.55	50.83	0.77	38.95	0.89	33.01
25	−14.03	0.641	1.56	0.04	70.42	0.54	49.03	0.76	37.95	0.87	32.40
30	0.00	0.750	1.81	0.20	67.09	0.62	52.34	0.86	37.76	0.99	30.47
30	−5.71	0.666	1.71	0.25	68.42	0.60	47.54	0.84	34.76	0.96	28.37
30	−11.31	0.587	1.60	0.31	70.22	0.58	43.16	0.82	32.13	0.94	26.61
30	−14.03	0.549	1.55	0.36	71.35	0.57	41.12	0.80	30.94	0.92	25.84
35	0.00	0.671	1.81	0.17	67.62	0.65	45.39	0.91	31.53	1.04	24.59
35	−5.71	0.580	1.70	0.02	69.47	0.64	40.20	0.89	28.27	1.02	22.30
35	−11.31	0.496	1.59	0.20	71.68	0.62	35.53	0.87	25.45	1.00	20.41
35	−14.03	0.456	1.53	0.32	73.15	0.61	33.42	0.86	24.24	0.98	19.65
40	0.00	0.587	1.81	0.28	68.14	0.69	38.07	0.97	25.00	1.11	18.46
40	−5.71	0.492	1.70	0.25	70.28	0.69	32.74	0.96	21.70	1.10	16.19
40	−11.31	0.406	1.56	0.08	73.66	0.67	28.19	0.94	19.14	1.07	14.61
40	−14.03	0.366	1.50	0.93	74.99	0.66	26.10	0.92	17.97	1.05	13.91

B. Physical Dimensions for a Technoeconomic Wall Section

ϕ Degree	α Degree	Top Width, mm	Bottom Width, mm	Horizontal Angle θ Degree	Depth of Shear Key D_{sk}, mm $\mu = 0.50$	Depth of Shear Key D_{sk}, mm $\mu = 0.70$	Depth of Shear Key D_{sk}, mm $\mu = 0.80$	Volume of Wall, m³
25	0.00	300	2525	90.00	1200	1000	900	4.238
25	−5.71	300	2210	84.29	1100	1000	900	3.765
25	−11.31	300	1900	78.69	1200	1000	900	3.300
25	−14.03	300	1750	75.97	1200	1000	900	3.075
30	0.00	300	2410	90.00	900	700	600	4.065
30	−5.71	300	2070	84.29	800	700	600	3.555
30	−11.31	300	1740	78.69	800	700	600	3.060
30	−14.03	300	1580	75.97	900	700	700	2.820
35	0.00	300	2275	90.00	500	400	300	3.863
35	−5.71	300	1910	84.29	600	500	400	3.315
35	−11.31	300	1560	78.69	500	400	400	2.790
35	−14.03	300	1390	75.97	600	500	500	2.535

(Continued)

TABLE 5.30 (Continued)

H = 3.0 m, $\gamma_{masonry}$ = 22.56 kN/m³, γ_{sat} = 17.27 kN/m³ and $\iota = \phi$

| ϕ Degree | α Degree | Top Width, mm | Bottom Width, mm | Horizontal Angle θ Degree | Depth of Shear Key D_{sk}, mm | | | Volume of Wall, m³ |
					$\mu = 0.50$	$\mu = 0.70$	$\mu = 0.80$	
40	0.00	300	2125	90.00	300	200	200	3.638
40	−5.71	300	1740	84.29	400	300	200	3.060
40	−11.31	300	1365	78.69	400	300	200	2.498
40	−14.03	300	1190	75.97	400	400	300	2.235

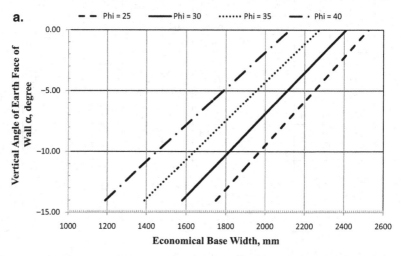

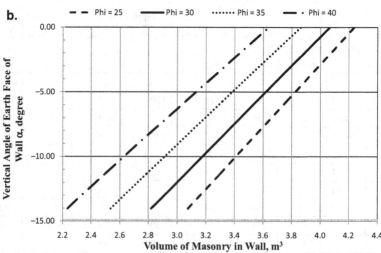

FIGURE 5.31 Design Charts for Masonry Wall H = 3.0 m, $\gamma_{masonry}$ = 22.56 kN/m³, γ_{sat} = 17.27 kN/m³ and $\iota = \phi$: (a) Economical Base Width for a Wall, (b) Volume of a Masonry in Wall

5.7 CHARTS FOR THE DESIGN OF A 3.5-M-HIGH MASONRY BREAST WALL

The charts/tables developed for a 3.5-m-high stone masonry breast wall for the previously discussed six combinations are presented in Tables 5.31 to 5.36 and Figures 5.32 to 5.37.

TABLE 5.31
$H = 3.5$ m, $\gamma_{masonry} = 22.56$ kN/m^3, $\gamma_{sat} = 15.70$ kN/m^3 and $\iota = 0°$

A. Results of Stability Analysis

				Pressure at		$\mu = 0.50$		$\mu = 0.70$		$\mu = 0.80$	
ϕ Degree	α Degree	K_{ah}	F_{OVT}	Heel, kN/m^2	Toe, kN/m^2	F_{SLD}	Net Force, kN	F_{SLD}	Net Force, kN	F_{SLD}	Net Force, kN
25	0.00	0.346	1.82	0.25	81.48	0.91	21.74	1.27	8.40	1.45	1.73
25	−5.71	0.316	1.67	0.41	84.05	0.84	22.61	1.17	11.22	1.34	5.52
25	−11.31	0.286	1.51	0.25	88.25	0.76	23.20	1.06	13.70	1.21	8.94
25	−14.03	0.271	1.50	7.37	83.13	0.74	22.84	1.03	14.01	1.18	9.59
30	0.00	0.279	1.83	0.23	82.98	0.99	15.58	1.39	3.39	1.59	0.00
30	−5.71	0.250	1.65	0.55	86.28	0.91	16.68	1.27	6.49	1.45	1.40
30	−11.31	0.221	1.51	4.54	87.59	0.82	17.25	1.15	8.88	1.32	4.70
30	−14.03	0.207	1.51	15.71	79.29	0.80	16.97	1.12	9.28	1.27	5.44
35	0.00	0.226	1.83	0.16	84.68	1.08	10.85	1.51	0.00	1.73	0.00
35	−5.71	0.198	1.64	0.62	89.01	0.98	12.22	1.37	3.11	1.56	0.00
35	−11.31	0.170	1.51	9.57	87.30	0.88	12.89	1.23	5.57	1.41	1.91
35	−14.03	0.156	1.51	27.11	74.02	0.85	12.73	1.19	6.09	1.36	2.77
40	0.00	0.184	1.84	0.25	86.30	1.17	7.33	1.63	0.00	1.87	0.00
40	−5.71	0.156	1.62	0.33	92.64	1.04	8.99	1.46	0.84	1.66	0.00
40	−11.31	0.130	1.50	17.17	85.84	0.93	9.75	1.30	3.36	1.49	0.16
40	−14.03	0.117	1.50	44.16	65.71	0.89	9.77	1.24	4.09	1.42	1.25

B. Physical Dimensions for a Technoeconomic Wall Section

					Depth of Shear Key D_{sk}, mm			
ϕ Degree	α Degree	Top Width, mm	Bottom Width, mm	Horizontal Angle θ Degree	$\mu = 0.50$	$\mu = 0.70$	$\mu = 0.80$	Volume of Wall, m^3
25	0.00	300	1815	90.00	800	500	200	3.701
25	−5.71	300	1500	84.29	800	600	400	3.150
25	−11.31	300	1195	78.69	800	600	500	2.616
25	−14.03	300	1085	75.97	800	600	500	2.424
30	0.00	300	1630	90.00	500	200	0	3.378
30	−5.71	300	1305	84.29	600	300	100	2.809
30	−11.31	300	1010	78.69	600	400	300	2.293
30	−14.03	300	900	75.97	600	400	300	2.100

(Continued)

TABLE 5.31 (Continued)

$H = 3.5$ m, $\gamma_{masonry} = 22.56$ kN/m³, $\gamma_{sat} = 15.70$ kN/m³ and $\iota = 0°$

ϕ Degree	α Degree	Top Width, mm	Bottom Width, mm	Horizontal Angle θ Degree	Depth of Shear Key D_{sk}, mm			Volume of Wall, m³
					$\mu = 0.50$	$\mu = 0.70$	$\mu = 0.80$	
35	0.00	300	1465	90.00	200	0	0	3.089
35	−5.71	300	1130	84.29	400	200	0	2.503
35	−11.31	300	840	78.69	400	200	100	1.995
35	−14.03	300	730	75.97	400	200	200	1.803
40	0.00	300	1325	90.00	100	0	0	2.844
40	−5.71	300	975	84.29	200	0	0	2.231
40	−11.31	300	690	78.69	300	100	0	1.733
40	−14.03	300	575	75.97	400	300	200	1.531

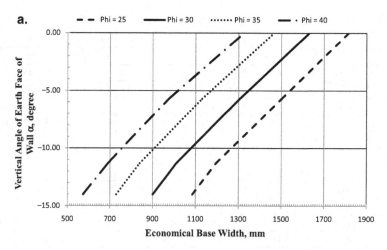

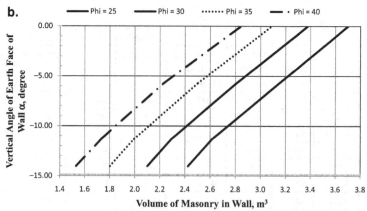

FIGURE 5.32 Design Charts for a Masonry Wall $H = 3.5$ m, $\gamma_{masonry} = 22.56$ kN/m³, $\gamma_{sat} = 15.70$ kN/m³ and $\iota = 0°$: (a) Economical Base Width for a Wall, (b) Volume of a Masonry in Wall

TABLE 5.32

H = 3.5 m, $\gamma_{masonry}$ = 22.56 kN/m³, γ_{sat} = 17.27 kN/m³ and ι = 0°

A. Results of Stability Analysis

ϕ Degree	α Degree	K_{ah}	F_{OVT}	Pressure at Heel, kN/m²	Pressure at Toe, kN/m²	$\mu = 0.50$ F_{SLD}	$\mu = 0.50$ Net Force, kN	$\mu = 0.70$ F_{SLD}	$\mu = 0.70$ Net Force, kN	$\mu = 0.80$ F_{SLD}	$\mu = 0.80$ Net Force, kN
25	0.00	0.346	1.82	0.51	80.56	0.87	25.25	1.22	11.33	1.39	4.36
25	−5.71	0.316	1.67	0.52	83.05	0.81	25.78	1.13	13.83	1.29	7.86
25	−11.31	0.286	1.52	0.26	86.92	0.74	26.01	1.03	15.98	1.18	10.96
25	−14.03	0.271	1.50	5.37	83.72	0.71	25.58	1.00	16.29	1.14	11.64
30	0.00	0.279	1.82	0.08	82.44	0.95	18.37	1.33	5.68	1.52	0.00
30	−5.71	0.250	1.65	0.07	85.76	0.87	19.13	1.22	8.48	1.40	3.15
30	−11.31	0.221	1.50	1.42	89.39	0.79	19.50	1.11	10.77	1.27	6.40
30	−14.03	0.207	1.50	11.63	81.72	0.77	19.02	1.08	10.96	1.23	6.93
35	0.00	0.226	1.83	0.40	83.66	1.04	12.95	1.45	1.31	1.66	0.00
35	−5.71	0.198	1.65	0.70	87.71	0.95	13.99	1.32	4.45	1.51	0.00
35	−11.31	0.170	1.50	6.40	88.74	0.85	14.48	1.20	6.83	1.37	3.00
35	−14.03	0.156	1.50	22.33	76.52	0.83	14.12	1.16	7.15	1.33	3.66
40	0.00	0.184	1.84	0.27	85.44	1.12	8.98	1.57	0.00	1.80	0.00
40	−5.71	0.156	1.62	0.13	91.43	1.01	10.32	1.41	1.80	1.62	0.00
40	−11.31	0.130	1.50	13.65	87.04	0.91	10.84	1.27	4.14	1.46	0.79
40	−14.03	0.117	1.50	37.76	68.82	0.88	10.68	1.23	4.70	1.40	1.70

B. Physical Dimensions for a Technoeconomic Wall Section

ϕ Degree	α Degree	Top Width, mm	Bottom Width, mm	Horizontal Angle θ Degree	Depth of Shear Key D_{sk}, mm $\mu = 0.50$	$\mu = 0.70$	$\mu = 0.80$	Volume of Wall, m³
25	0.00	300	1910	90.00	800	500	300	3.868
25	−5.71	300	1590	84.29	800	600	400	3.308
25	−11.31	300	1280	78.69	800	600	500	2.765
25	−14.03	300	1160	75.97	800	600	500	2.555
30	0.00	300	1710	90.00	500	200	0	3.518
30	−5.71	300	1380	84.29	600	400	200	2.940
30	−11.31	300	1070	78.69	600	400	300	2.398
30	−14.03	300	960	75.97	600	400	300	2.205
35	0.00	300	1540	90.00	300	0	0	3.220
35	−5.71	300	1200	84.29	400	200	0	2.625
35	−11.31	300	895	78.69	400	300	100	2.091
35	−14.03	300	785	75.97	400	300	200	1.899

(Continued)

TABLE 5.32 (Continued)

H = 3.5 m, $\gamma_{masonry}$ = 22.56 kN/m³, γ_{sat} = 17.27 kN/m³ and ι = 0°

ϕ Degree	α Degree	Top Width, mm	Bottom Width, mm	Horizontal Angle θ Degree	Depth of Shear Key D_{sk}, mm			Volume of Wall, m³
					$\mu = 0.50$	$\mu = 0.70$	$\mu = 0.80$	
40	0.00	300	1390	90.00	100	0	0	2.958
40	−5.71	300	1035	84.29	300	100	0	2.336
40	−11.31	300	740	78.69	300	100	0	1.820
40	−14.03	300	625	75.97	300	200	100	1.619

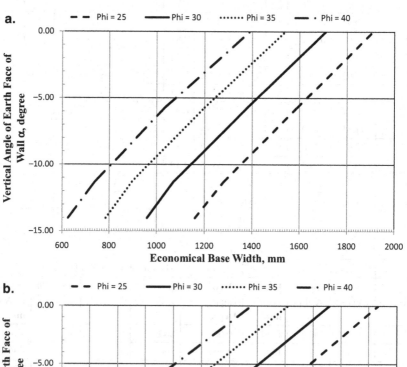

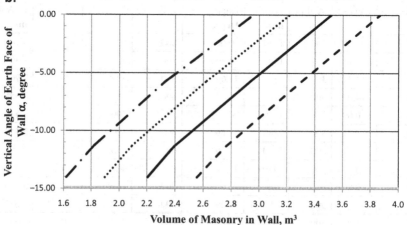

FIGURE 5.33 Design Charts for Masonry Wall H = 3.5 m, $\gamma_{masonry}$ = 22.56 kN/m³, γ_{sat} = 17.27 kN/m³ and ι = 0°: (a) Economical Base Width for a Wall, (b) Volume of a Masonry in Wall

TABLE 5.33

$H = 3.5$ m, $\gamma_{masonry} = 22.56$ kN/m^3, $\gamma_{sat} = 15.70$ kN/m^3 and $\iota = \frac{1}{2}\phi$

A. *Results of Stability Analysis*

ϕ Degree	α Degree	K_{ah}	F_{OVT}	Pressure at Heel, kN/m^2	Toe, kN/m^2	$\mu = 0.50$ F_{SLD}	Net Force, kN	$\mu = 0.70$ F_{SLD}	Net Force, kN	$\mu = 0.80$ F_{SLD}	Net Force, kN
25	0.00	0.424	1.81	0.04	80.42	0.83	29.35	1.16	14.81	1.33	7.54
25	−5.71	0.385	1.68	0.47	82.26	0.78	29.07	1.09	16.54	1.24	10.27
25	−11.31	0.346	1.54	0.25	85.77	0.72	28.70	1.01	18.14	1.15	12.86
25	−14.03	0.326	1.50	3.27	84.66	0.69	28.16	0.97	18.43	1.11	13.57
30	0.00	0.348	1.82	0.08	81.61	0.90	22.02	1.27	8.65	1.45	1.97
30	−5.71	0.309	1.67	0.47	84.21	0.84	21.94	1.18	10.67	1.35	5.04
30	−11.31	0.270	1.51	0.67	88.60	0.77	21.77	1.08	12.54	1.23	7.92
30	−14.03	0.251	1.50	9.02	82.65	0.75	21.06	1.05	12.57	1.20	8.32
35	0.00	0.284	1.83	0.31	82.76	0.99	16.01	1.38	3.72	1.58	0.00
35	−5.71	0.246	1.65	0.61	86.42	0.91	16.25	1.28	6.15	1.46	1.11
35	−11.31	0.208	1.50	4.61	88.61	0.83	16.19	1.17	8.10	1.33	4.06
35	−14.03	0.190	1.50	17.89	78.81	0.81	15.61	1.14	8.26	1.30	4.59
40	0.00	0.232	1.83	0.27	84.35	1.07	11.35	1.50	0.05	1.71	0.00
40	−5.71	0.194	1.63	0.39	89.53	0.98	11.93	1.37	2.91	1.57	0.00
40	−11.31	0.158	1.51	11.55	86.73	0.90	11.94	1.25	4.87	1.43	1.34
40	−14.03	0.142	1.50	31.92	71.84	0.86	11.60	1.21	5.30	1.38	2.15

B. *Physical Dimensions for a Technoeconomic Wall Section*

ϕ Degree	α Degree	Top Width, mm	Bottom Width, mm	Horizontal Angle θ Degree	Depth of Shear Key D_{sk}, mm $\mu = 0.50$	$\mu = 0.70$	$\mu = 0.80$	Volume of Wall, m^3
25	0.00	300	2010	90.00	800	500	300	4.043
25	−5.71	300	1685	84.29	900	700	500	3.474
25	−11.31	300	1365	78.69	900	700	600	2.914
25	−14.03	300	1230	75.97	900	700	600	2.678
30	0.00	300	1820	90.00	500	300	100	3.710
30	−5.71	300	1480	84.29	600	400	300	3.115
30	−11.31	300	1150	78.69	600	500	400	2.538
30	−14.03	300	1030	75.97	600	500	400	2.328
35	0.00	300	1645	90.00	300	100	0	3.404
35	−5.71	300	1290	84.29	400	200	100	2.783
35	−11.31	300	965	78.69	400	300	200	2.214
35	−14.03	300	845	75.97	400	300	200	2.004

(Continued)

TABLE 5.33 (Continued)

H = 3.5 m, $\gamma_{masonry}$ = 22.56 kN/m³, γ_{sat} = 15.70 kN/m³ and ι = ½φ

ϕ Degree	α Degree	Top Width, mm	Bottom Width, mm	Horizontal Angle θ Degree	Depth of Shear Key D_{sk}, mm			Volume of Wall, m³
					$\mu = 0.50$	$\mu = 0.70$	$\mu = 0.80$	
40	0.00	300	1485	90.00	200	0	0	3.124
40	−5.71	300	1115	84.29	300	100	0	2.476
40	−11.31	300	800	78.69	300	200	100	1.925
40	−14.03	300	675	75.97	300	200	100	1.706

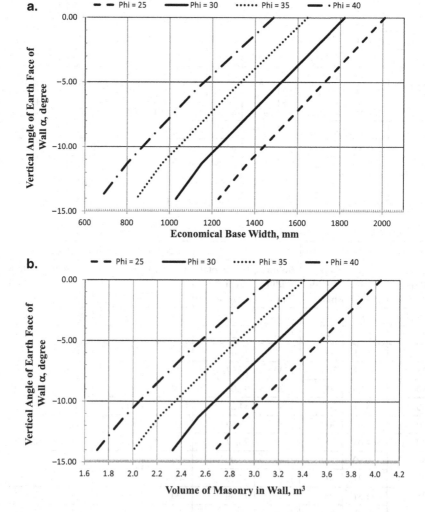

FIGURE 5.34 Design Charts for a Masonry Wall H = 3.5 m, $\gamma_{masonry}$ = 22.56 kN/m³, γ_{sat} = 15.70 kN/m³ and ι = ½φ: (a) Economical Base Width for a Wall, (b) Volume of a Masonry in Wall

TABLE 5.34

$H = 3.5$ m, $\gamma_{masonry} = 22.56$ kN/m³, $\gamma_{sat} = 17.27$ kN/m³ and $\iota = \frac{1}{2}\phi$

A. Results of Stability Analysis

				Pressure at		$\mu = 0.50$		$\mu = 0.70$		$\mu = 0.80$	
ϕ Degree	α Degree	K_{ah}	F_{OVT}	Heel, kN/m²	Toe, kN/m²	F_{SLD}	Net Force, kN	F_{SLD}	Net Force, kN	F_{SLD}	Net Force, kN
25	0.00	0.424	1.81	0.25	79.62	0.79	33.83	1.11	18.64	1.27	11.05
25	−5.71	0.385	1.68	0.13	81.84	0.75	33.15	1.04	20.03	1.19	13.47
25	−11.31	0.346	1.55	0.43	84.47	0.70	32.21	0.97	21.07	1.11	15.50
25	−14.03	0.326	1.50	2.10	84.59	0.67	31.55	0.94	21.30	1.08	16.17
30	0.00	0.348	1.82	0.33	80.71	0.87	25.57	1.21	11.62	1.38	4.64
30	−5.71	0.309	1.67	0.15	83.65	0.81	25.08	1.13	13.28	1.30	7.39
30	−11.31	0.270	1.52	0.23	87.70	0.75	24.43	1.05	14.70	1.20	9.84
30	−14.03	0.251	1.50	7.35	82.76	0.73	23.52	1.02	14.57	1.17	10.09
35	0.00	0.284	1.82	0.10	82.30	0.94	18.86	1.32	6.08	1.51	0.00
35	−5.71	0.246	1.65	0.23	85.79	0.88	18.63	1.23	8.07	1.41	2.79
35	−11.31	0.208	1.50	2.80	88.87	0.81	18.18	1.13	9.69	1.29	5.45
35	−14.03	0.190	1.51	15.22	79.49	0.79	17.34	1.11	9.60	1.27	5.72
40	0.00	0.232	1.83	0.40	83.46	1.03	13.53	1.44	1.76	1.64	0.00
40	−5.71	0.194	1.64	0.59	88.07	0.95	13.64	1.33	4.19	1.52	0.00
40	−11.31	0.158	1.50	8.64	87.74	0.87	13.37	1.22	5.97	1.39	2.26
40	−14.03	0.142	1.50	25.78	75.56	0.84	12.87	1.18	6.26	1.35	2.96

B. Physical Dimensions for a Technoeconomic Wall Section

					Depth of Shear Key D_{sk}, mm			
ϕ Degree	α Degree	Top Width, mm	Bottom Width, mm	Horizontal Angle θ Degree	$\mu = 0.50$	$\mu = 0.70$	$\mu = 0.80$	Volume of Wall, m³
25	0.00	300	2115	90.00	900	600	400	4.226
25	−5.71	300	1780	84.29	900	700	600	3.640
25	−11.31	300	1460	78.69	900	700	600	3.080
25	−14.03	300	1315	75.97	900	700	600	2.826
30	0.00	300	1915	90.00	600	300	200	3.876
30	−5.71	300	1565	84.29	700	500	300	3.264
30	−11.31	300	1230	78.69	700	500	400	2.678
30	−14.03	300	1105	75.97	600	500	400	2.459
35	0.00	300	1725	90.00	300	100	0	3.544
35	−5.71	300	1365	84.29	400	300	100	2.914
35	−11.31	300	1030	78.69	400	300	200	2.328
35	−14.03	300	910	75.97	400	300	200	2.118

(Continued)

TABLE 5.34 (Continued)

$H = 3.5$ m, $\gamma_{masonry} = 22.56$ kN/m³, $\gamma_{sat} = 17.27$ kN/m³ and $\iota = \frac{1}{2}\phi$

ϕ Degree	α Degree	Top Width, mm	Bottom Width, mm	Horizontal Angle θ Degree	Depth of Shear Key D_{sk}, mm			Volume of Wall, m³
					$\mu = 0.50$	$\mu = 0.70$	$\mu = 0.80$	
40	0.00	300	1560	90.00	200	0	0	3.255
40	−5.71	300	1185	84.29	300	100	0	2.599
40	−11.31	300	855	78.69	300	200	100	2.021
40	−14.03	300	725	75.97	300	200	100	1.794

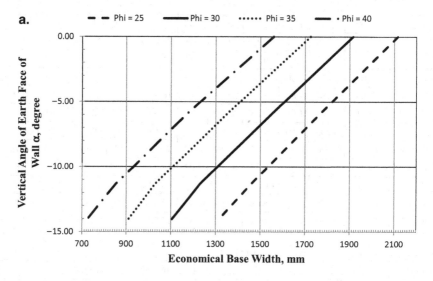

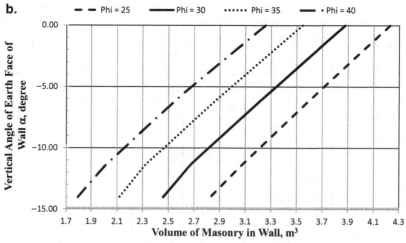

FIGURE 5.35 Design Charts for a Masonry Wall $H = 3.5$ m, $\gamma_{masonry} = 22.56$ kN/m³, $\gamma_{sat} = 17.27$ kN/m³ and $\iota = \frac{1}{2}\phi$: (a) Economical Base Width for a Wall, (b) Volume of Masonry in Wall

TABLE 5.35

$H = 3.5$ m, $\gamma_{masonry} = 22.56$ kN/m^3, $\gamma_{sat} = 15.70$ kN/m^3 and $\iota = \phi$

A. *Results of Stability Analysis*

ϕ Degree	α Degree	K_{ah}	F_{OVT}	Pressure at Heel, kN/m^2	Toe, kN/m^2	$\mu = 0.50$ F_{SLD}	Net Force, kN	$\mu = 0.70$ F_{SLD}	Net Force, kN	$\mu = 0.80$ F_{SLD}	Net Force, kN
25	0.00	0.821	1.80	0.04	77.03	0.61	70.91	0.86	51.37	0.98	41.60
25	−5.71	0.746	1.70	0.01	78.29	0.59	66.29	0.83	48.97	0.95	40.31
25	−11.31	0.675	1.60	0.13	79.74	0.57	61.92	0.80	46.73	0.91	39.14
25	−14.03	0.641	1.56	0.38	80.44	0.56	59.81	0.78	45.64	0.89	38.56
30	0.00	0.750	1.80	0.24	77.22	0.64	63.12	0.90	44.35	1.02	34.97
30	−5.71	0.666	1.70	0.30	78.63	0.62	57.63	0.87	41.23	1.00	33.04
30	−11.31	0.587	1.59	0.11	80.81	0.60	52.65	0.84	38.58	0.96	31.54
30	−14.03	0.549	1.54	0.09	82.08	0.59	50.32	0.82	37.35	0.94	30.87
35	0.00	0.671	1.80	0.21	77.77	0.67	54.69	0.94	36.85	1.08	27.93
35	−5.71	0.580	1.69	0.25	79.52	0.66	48.69	0.92	33.38	1.06	25.72
35	−11.31	0.496	1.58	0.44	81.84	0.64	43.34	0.89	30.47	1.02	24.03
35	−14.03	0.456	1.51	0.05	83.96	0.62	40.98	0.87	29.31	1.00	23.48
40	0.00	0.587	1.81	0.15	78.50	0.72	45.87	1.00	29.07	1.15	20.68
40	−5.71	0.492	1.68	0.04	80.84	0.71	39.75	0.99	25.64	1.13	18.59
40	−11.31	0.406	1.56	0.50	83.71	0.68	34.44	0.96	22.90	1.09	17.13
40	−14.03	0.366	1.50	1.69	84.80	0.67	32.03	0.94	21.69	1.07	16.52

B. *Physical Dimensions for a Technoeconomic Wall Section*

ϕ Degree	α Degree	Top Width, mm	Bottom Width, mm	Horizontal Angle θ Degree	Depth of Shear Key D_{sk}, mm $\mu = 0.50$	$\mu = 0.70$	$\mu = 0.80$	Volume of Wall, m^3
25	0.00	300	2820	90.00	1400	1100	1000	5.460
25	−5.71	300	2460	84.29	1300	1100	1000	4.830
25	−11.31	300	2115	78.69	1300	1100	1000	4.226
25	−14.03	300	1950	75.97	1300	1200	1100	3.938
30	0.00	300	2695	90.00	900	700	600	5.241
30	−5.71	300	2310	84.29	1000	800	700	4.568
30	−11.31	300	1935	78.69	900	800	700	3.911
30	−14.03	300	1755	75.97	900	800	700	3.596
35	0.00	300	2545	90.00	600	500	400	4.979
35	−5.71	300	2135	84.29	600	400	400	4.261
35	−11.31	300	1740	78.69	600	500	400	3.570
35	−14.03	300	1545	75.97	600	500	400	3.229

(Continued)

TABLE 5.35 (Continued)

H = 3.5 m, $\gamma_{masonry}$ = 22.56 kN/m³, γ_{sat} = 15.70 kN/m³ and ι = φ

ϕ Degree	α Degree	Top Width, mm	Bottom Width, mm	Horizontal Angle θ Degree	Depth of Shear Key D_{sk}, mm			Volume of Wall, m³
					$\mu = 0.50$	$\mu = 0.70$	$\mu = 0.80$	
40	0.00	300	2375	90.00	400	300	200	4.681
40	−5.71	300	1940	84.29	400	300	200	3.920
40	−11.31	300	1525	78.69	400	300	300	3.194
40	−14.03	300	1330	75.97	400	300	300	2.853

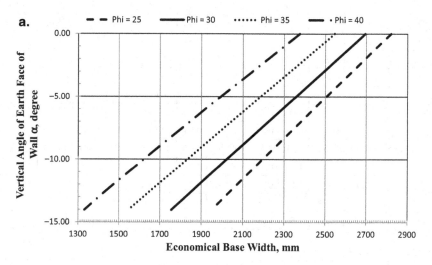

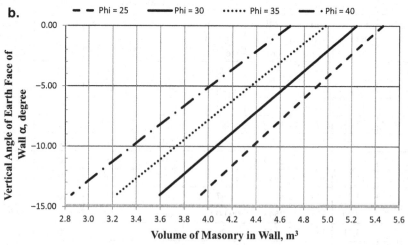

FIGURE 5.36 Design Charts for a Masonry Wall H = 3.5 m, $\gamma_{masonry}$ = 22.56 kN/m³, γ_{sat} = 15.70 kN/m³ and ι = φ: (a) Economical Base Width for a Wall, (b) Volume of a Masonry in Wall

TABLE 5.36

$H = 3.5$ m, $\gamma_{masonry} = 22.56$ kN/m³, $\gamma_{sat} = 17.27$ kN/m³ and $\iota = \phi$

A. Results of Stability Analysis

				Pressure at		$\mu = 0.50$		$\mu = 0.70$		$\mu = 0.80$	
ϕ Degree	α Degree	K_{ah}	F_{OVT}	Heel, kN/ m²	Toe, kN/ m²	F_{SLD}	Net Force, kN	F_{SLD}	Net Force, kN	F_{SLD}	Net Force, kN
25	0.00	0.821	1.80	0.09	76.57	0.58	80.50	0.82	60.06	0.93	49.85
25	−5.71	0.746	1.71	0.12	77.66	0.57	74.88	0.79	56.70	0.91	47.60
25	−11.31	0.675	1.62	0.38	78.82	0.55	69.56	0.77	53.54	0.88	45.53
25	−14.03	0.641	1.56	0.07	80.01	0.54	67.11	0.75	52.17	0.86	44.70
30	0.00	0.750	1.80	0.10	76.94	0.61	71.84	0.85	52.24	0.97	42.44
30	−5.71	0.666	1.70	0.26	78.11	0.60	65.21	0.83	48.01	0.95	39.42
30	−11.31	0.587	1.60	0.31	79.86	0.58	59.17	0.81	44.32	0.93	36.90
30	−14.03	0.549	1.55	0.23	81.07	0.57	56.37	0.79	42.67	0.91	35.81
35	0.00	0.671	1.80	0.22	77.31	0.64	62.35	0.90	43.70	1.03	34.38
35	−5.71	0.580	1.70	0.13	79.04	0.63	55.19	0.88	39.14	1.01	31.11
35	−11.31	0.496	1.59	0.16	81.31	0.61	48.79	0.86	35.24	0.98	28.46
35	−14.03	0.456	1.53	0.30	82.69	0.60	45.86	0.84	33.51	0.97	27.34
40	0.00	0.587	1.81	0.38	77.77	0.68	52.39	0.96	34.82	1.09	26.04
40	−5.71	0.492	1.69	0.30	79.89	0.68	45.06	0.95	30.24	1.08	22.84
40	−11.31	0.406	1.57	0.28	82.97	0.66	38.75	0.92	26.59	1.06	20.50
40	−14.03	0.366	1.50	0.65	84.70	0.65	35.92	0.91	25.02	1.04	19.57

B. Physical Dimensions for a Technoeconomic Wall Section

					Depth of Shear Key D_{sk}, mm			
ϕ Degree	α Degree	Top Width, mm	Bottom Width, mm	Horizontal Angle θ Degree	$\mu = 0.50$	$\mu = 0.70$	$\mu = 0.80$	Volume of Wall, m³
25	0.00	300	2965	90.00	1400	1200	1100	5.714
25	−5.71	300	2600	84.29	1300	1100	1000	5.075
25	−11.31	300	2250	78.69	1300	1100	1000	4.463
25	−14.03	300	2075	75.97	1300	1100	1000	4.156
30	0.00	300	2830	90.00	1000	800	700	5.478
30	−5.71	300	2440	84.29	1000	800	700	4.795
30	−11.31	300	2060	78.69	900	800	700	4.130
30	−14.03	300	1875	75.97	900	800	700	3.806
35	0.00	300	2675	90.00	700	500	400	5.206
35	−5.71	300	2255	84.29	600	500	400	4.471
35	−11.31	300	1850	78.69	700	500	500	3.763
35	−14.03	300	1655	75.97	600	500	500	3.421

(Continued)

TABLE 5.36 (Continued)

H = 3.5 m, $\gamma_{masonry}$ = 22.56 kN/m³, γ_{sat} = 17.27 kN/m³ and $\iota = \phi$

ϕ Degree	α Degree	Top Width, mm	Bottom Width, mm	Horizontal Angle θ Degree	Depth of Shear Key D_{sk}, mm			Volume of Wall, m³
					μ = 0.50	μ = 0.70	μ = 0.80	
40	0.00	300	2500	90.00	300	200	100	4.900
40	−5.71	300	2055	84.29	400	300	200	4.121
40	−11.31	300	1625	78.69	400	300	300	3.369
40	−14.03	300	1420	75.97	400	300	300	3.010

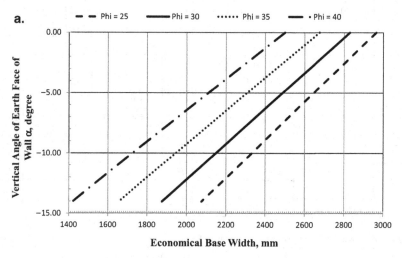

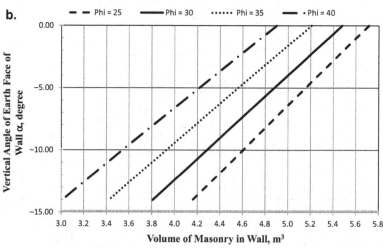

FIGURE 5.37 Design Charts for a Masonry Wall H = 3.5 m, $\gamma_{masonry}$ = 22.56 kN/m³, γ_{sat} = 17.27 kN/m³ and $\iota = \phi$: (a) Economical Base Width for a Wall, (b) Volume of a Masonry in Wall

5.8 CHARTS FOR THE DESIGN OF A 4.0-M-HIGH MASONRY BREAST WALL

The charts/tables developed for a 4.0-m-high stone masonry breast wall for the previously discussed six combinations are presented in the Tables 5.37 to 5.42 and Figures 5.38 to 5.43.

TABLE 5.37

$H = 4.0$ m, $\gamma_{masonry} = 22.56$ kN/m³, $\gamma_{sat} = 15.70$ kN/m³ and $\iota = 0°$

A. Results of Stability Analysis

ϕ Degree	α Degree	K_{ah}	F_{OVT}	Pressure at Heel, kN/m²	Pressure at Toe, kN/m²	$\mu = 0.50$ F_{SLD}	$\mu = 0.50$ Net Force, kN	$\mu = 0.70$ F_{SLD}	$\mu = 0.70$ Net Force kN	$\mu = 0.80$ F_{SLD}	$\mu = 0.80$ Net Force, kN
25	0.00	0.346	1.81	0.28	91.15	0.90	28.98	1.25	11.80	1.43	3.21
25	−5.71	0.316	1.66	0.35	93.73	0.82	30.04	1.15	15.36	1.32	8.02
25	−11.31	0.286	1.51	0.41	97.48	0.75	30.66	1.05	18.38	1.20	12.24
25	−14.03	0.271	1.50	7.65	92.15	0.73	30.17	1.02	18.77	1.17	13.07
30	0.00	0.279	1.81	0.02	92.90	0.98	21.01	1.37	5.34	1.56	0.00
30	−5.71	0.250	1.65	0.25	96.16	0.89	22.33	1.25	9.24	1.43	2.69
30	−11.31	0.221	1.50	3.32	98.20	0.81	23.03	1.13	12.30	1.29	6.94
30	−14.03	0.207	1.50	15.44	88.70	0.78	22.56	1.10	12.68	1.25	7.75
35	0.00	0.226	1.83	0.38	94.12	1.06	14.77	1.49	0.41	1.70	0.00
35	−5.71	0.198	1.63	0.32	98.83	0.96	16.50	1.34	4.82	1.53	0.00
35	−11.31	0.170	1.50	9.24	96.73	0.86	17.26	1.21	7.88	1.38	3.18
35	−14.03	0.156	1.50	26.40	83.57	0.83	17.04	1.17	8.53	1.33	4.28
40	0.00	0.184	1.83	0.06	96.18	1.14	10.26	1.60	0.00	1.83	0.00
40	−5.71	0.156	1.63	2.24	99.89	1.03	12.08	1.44	1.56	1.64	0.00
40	−11.31	0.130	1.50	17.16	94.54	0.91	13.13	1.28	4.95	1.46	0.86
40	−14.03	0.117	1.50	44.81	72.91	0.87	13.06	1.22	5.75	1.40	2.10

B. Physical Dimensions for a Technoeconomic Wall Section

ϕ Degree	α Degree	Top Width, mm	Bottom Width, mm	Horizontal Angle θ Degree	Depth of Shear Key D_{sk}, mm $\mu = 0.50$	Depth of Shear Key D_{sk}, mm $\mu = 0.70$	Depth of Shear Key D_{sk}, mm $\mu = 0.80$	Volume of Wall, m³
25	0.00	300	2090	90.00	800	500	200	4.780
25	−5.71	300	1735	84.29	900	700	500	4.070
25	−11.31	300	1395	78.69	900	700	600	3.390
25	−14.03	300	1270	75.97	900	700	600	3.140
30	0.00	300	1875	90.00	500	200	0	4.350
30	−5.71	300	1510	84.29	700	400	200	3.620
30	−11.31	300	1175	78.69	700	500	300	2.950
30	−14.03	300	1055	75.97	700	500	400	2.710

(Continued)

TABLE 5.37 (Continued)

H = 4.0 m, $\gamma_{masonry}$ = 22.56 kN/m³, γ_{sat} = 15.70 kN/m³ and ι = 0°

ϕ Degree	α Degree	Top Width, mm	Bottom Width, mm	Horizontal Angle θ Degree	Depth of Shear Key D_{sk}, mm			Volume of Wall, m³
					μ = 0.50	μ = 0.70	μ = 0.80	
35	0.00	300	1690	90.00	300	0	0	3.980
35	−5.71	300	1310	84.29	400	200	0	3.220
35	−11.31	300	985	78.69	500	300	200	2.570
35	−14.03	300	860	75.97	400	300	200	2.320
40	0.00	300	1525	90.00	200	0	0	3.650
40	−5.71	300	1145	84.29	300	100	0	2.890
40	−11.31	300	815	78.69	300	200	0	2.230
40	−14.03	300	690	75.97	300	200	100	1.980

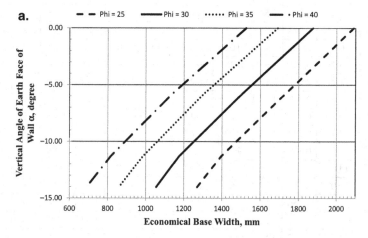

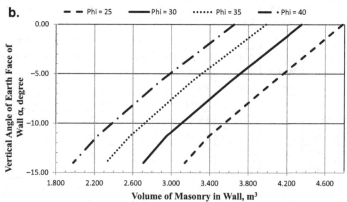

FIGURE 5.38 Design Charts for a Masonry Wall H = 4.0 m, $\gamma_{masonry}$ = 22.56 kN/m³, γ_{sat} = 15.70 kN/m³ and ι = 0°: (a) Economical Base Width for a Wall, (b) Volume of a Masonry in Wall

TABLE 5.38

H = 4.0 m, $\gamma_{masonry}$ = 22.56 kN/m³, γ_{sat} = 17.27 kN/m³ and ι = 0°

A. Results of Stability Analysis

				Pressure at		$\mu = 0.50$		$\mu = 0.70$		$\mu = 0.80$	
ϕ Degree	α Degree	K_{ah}	F_{OVT}	Heel, kN/m²	Toe, kN/m²	F_{SLD}	Net Force, kN	F_{SLD}	Net Force, kN	F_{SLD}	Net Force, kN
25	0.00	0.346	1.81	0.28	90.54	0.86	33.63	1.20	15.71	1.37	6.75
25	−5.71	0.316	1.67	0.17	93.06	0.79	34.22	1.11	18.84	1.27	11.15
25	−11.31	0.286	1.52	0.12	96.54	0.73	34.36	1.02	21.42	1.16	14.94
25	−14.03	0.271	1.50	5.45	93.01	0.71	33.76	0.99	21.77	1.13	15.77
30	0.00	0.279	1.81	0.14	92.08	0.94	24.59	1.31	8.25	1.50	0.09
30	−5.71	0.250	1.65	0.25	95.15	0.86	25.45	1.21	11.73	1.38	4.86
30	−11.31	0.221	1.50	1.26	98.88	0.78	25.86	1.09	14.61	1.25	8.98
30	−14.03	0.207	1.50	12.61	89.83	0.76	25.13	1.07	14.72	1.22	9.51
35	0.00	0.226	1.82	0.04	93.74	1.02	17.61	1.43	2.68	1.63	0.00
35	−5.71	0.198	1.64	0.41	97.55	0.93	18.81	1.30	6.57	1.49	0.44
35	−11.31	0.170	1.50	6.45	97.82	0.84	19.31	1.18	9.46	1.34	4.54
35	−14.03	0.156	1.50	22.29	85.46	0.82	18.83	1.14	9.86	1.30	5.38
40	0.00	0.184	1.83	0.12	95.28	1.10	12.39	1.54	0.00	1.76	0.00
40	−5.71	0.156	1.63	1.25	99.60	1.00	13.87	1.39	2.90	1.59	0.00
40	−11.31	0.130	1.50	14.42	95.00	0.90	14.51	1.25	5.90	1.43	1.60
40	−14.03	0.117	1.50	38.06	76.76	0.86	14.28	1.20	6.59	1.38	2.75

B. Physical Dimensions for a Technoeconomic Wall Section

					Depth of Shear Key D_{sk}, mm			
ϕ Degree	α Degree	Top Width, mm	Bottom Width, mm	Horizontal Angle θ Degree	$\mu = 0.50$	$\mu = 0.70$	$\mu = 0.80$	Volume of Wall, m³
25	0.00	300	2195	90.00	800	500	300	4.990
25	−5.71	300	1835	84.29	1000	700	500	4.270
25	−11.31	300	1490	78.69	1000	700	600	3.580
25	−14.03	300	1355	75.97	1000	800	600	3.310
30	0.00	300	1970	90.00	600	300	0	4.540
30	−5.71	300	1600	84.29	700	400	300	3.800
30	−11.31	300	1250	78.69	700	500	400	3.100
30	−14.03	300	1130	75.97	700	500	400	2.860
35	0.00	300	1770	90.00	300	0	0	4.140
35	−5.71	300	1390	84.29	400	200	0	3.380
35	−11.31	300	1050	78.69	500	300	200	2.700
35	−14.03	300	925	75.97	400	300	200	2.450

(Continued)

TABLE 5.32 (Continued)

H = 4.0 m, $\gamma_{masonry}$ = 22.56 kN/m³, γ_{sat} = 17.27 kN/m³ and ι = 0°

ϕ Degree	α Degree	Top Width, mm	Bottom Width, mm	Horizontal Angle θ Degree	Depth of Shear Key D_{sk}, mm			Volume of Wall, m³
					$\mu = 0.50$	$\mu = 0.70$	$\mu = 0.80$	
40	0.00	300	1600	90.00	200	0	0	3.800
40	−5.71	300	1210	84.29	300	100	0	3.020
40	−11.31	300	875	78.69	300	200	100	2.350
40	−14.03	300	745	75.97	300	200	100	2.090

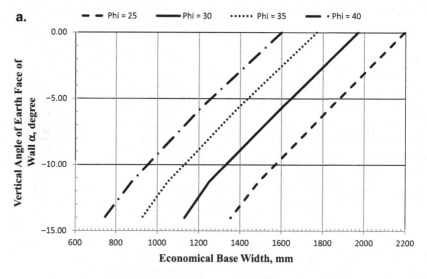

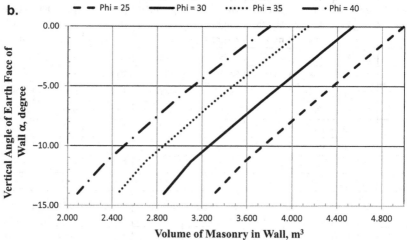

FIGURE 5.39 Design Charts for a Masonry Wall H = 4.0 m, $\gamma_{masonry}$ = 22.56 kN/m³, γ_{sat} = 17.27 kN/m³ and ι = 0°: (a) Economical Base Width for a Wall, (b) Volume of a Masonry in Wall

TABLE 5.39

$H = 4.0$ m, $\gamma_{masonry} = 22.56$ kN/m^3, $\gamma_{sat} = 15.70$ kN/m^3 and $\iota = \frac{1}{2}\phi$

A. Results of Stability Analysis

ϕ Degree	α Degree	K_{ah}	F_{OVT}	Pressure at Heel, kN/m^2	Toe, kN/m^2	$\mu = 0.50$ F_{SLD}	Net Force, kN	$\mu = 0.70$ F_{SLD}	Net Force, kN	$\mu = 0.80$ F_{SLD}	Net Force, kN
25	0.00	0.424	1.81	0.15	90.03	0.82	38.88	1.15	20.11	1.31	10.73
25	−5.71	0.385	1.67	0.26	92.14	0.77	38.50	1.08	22.34	1.23	14.26
25	−11.31	0.346	1.54	0.54	94.97	0.71	37.82	1.00	24.17	1.14	17.34
25	−14.03	0.326	1.50	3.24	94.11	0.69	37.13	0.96	24.57	1.10	18.29
30	0.00	0.348	1.81	0.04	91.37	0.89	29.37	1.25	12.15	1.43	3.54
30	−5.71	0.309	1.66	0.18	94.14	0.83	29.20	1.16	14.70	1.33	7.45
30	−11.31	0.270	1.51	0.25	98.44	0.76	28.85	1.07	16.96	1.22	11.01
30	−14.03	0.251	1.50	9.90	90.96	0.75	27.79	1.04	16.82	1.19	11.34
35	0.00	0.284	1.82	0.37	92.40	0.97	21.51	1.36	5.71	1.55	0.00
35	−5.71	0.246	1.65	0.60	95.99	0.90	21.72	1.26	8.74	1.44	2.25
35	−11.31	0.208	1.50	4.61	97.83	0.82	21.53	1.15	11.13	1.31	5.92
35	−14.03	0.190	1.50	18.21	87.48	0.80	20.74	1.12	11.28	1.28	6.55
40	0.00	0.232	1.82	0.21	94.11	1.05	15.46	1.47	0.96	1.68	0.00
40	−5.71	0.194	1.63	0.40	98.98	0.96	16.08	1.35	4.51	1.54	0.00
40	−11.31	0.158	1.50	11.42	95.86	0.88	16.02	1.23	6.95	1.41	2.42
40	−14.03	0.142	1.50	31.80	80.52	0.85	15.51	1.19	7.44	1.36	3.40

B. Physical Dimensions for a Technoeconomic Wall Section

ϕ Degree	α Degree	Top Width, mm	Bottom Width, mm	Horizontal Angle θ Degree	Depth of Shear Key D_{sk}, mm $\mu = 0.50$	$\mu = 0.70$	$\mu = 0.80$	Volume of Wall, m^3
25	0.00	300	2315	90.00	1000	700	400	5.230
25	−5.71	300	1945	84.29	1100	800	600	4.490
25	−11.31	300	1590	78.69	1100	800	700	3.780
25	−14.03	300	1435	75.97	1100	800	700	3.470
30	0.00	300	2095	90.00	700	400	100	4.790
30	−5.71	300	1710	84.29	700	400	300	4.020
30	−11.31	300	1340	78.69	800	600	400	3.280
30	−14.03	300	1210	75.97	700	600	500	3.020
35	0.00	300	1895	90.00	400	100	0	4.390
35	−5.71	300	1495	84.29	400	200	0	3.590
35	−11.31	300	1130	78.69	500	400	200	2.860
35	−14.03	300	995	75.97	500	400	300	2.590

(Continued)

TABLE 5.39 (Continued)

$H = 4.0$ m, $\gamma_{masonry} = 22.56$ kN/m³, $\gamma_{sat} = 15.70$ kN/m³ and $\iota = \frac{1}{2}\phi$

ϕ Degree	α Degree	Top Width, mm	Bottom Width, mm	Horizontal Angle θ Degree	Depth of Shear Key D_{sk}, mm			Volume of Wall, m³
					$\mu = 0.50$	$\mu = 0.70$	$\mu = 0.80$	
40	0.00	300	1710	90.00	200	0	0	4.020
40	−5.71	300	1295	84.29	300	100	0	3.190
40	−11.31	300	940	78.69	300	200	100	2.480
40	−14.03	300	800	75.97	300	200	100	2.200

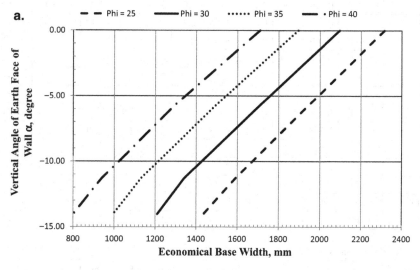

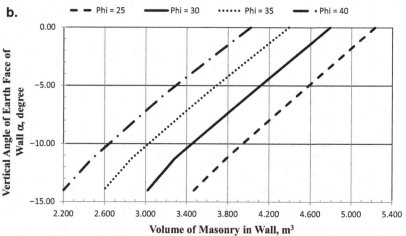

FIGURE 5.40 Design Charts for a Masonry Wall $H = 4.0$ m, $\gamma_{masonry} = 22.56$ kN/m³, $\gamma_{sat} = 15.70$ kN/m³ and $\iota = \frac{1}{2}\phi$: (a) Economical Base Width for a Wall, (b) Volume of a Masonry in Wall

TABLE 5.40

$H = 4.0$ m, $\gamma_{masonry} = 22.56$ kN/m^3, $\gamma_{sat} = 17.27$ kN/m^3 and $\iota = \frac{1}{2}\phi$

A. Results of Stability Analysis

				Pressure at		$\mu = 0.50$		$\mu = 0.70$		$\mu = 0.80$	
ϕ Degree	α Degree	K_{ah}	F_{OVT}	Heel, kN/m^2	Toe, kN/m^2	F_{SLD}	Net Force, kN	F_{SLD}	Net Force, kN	F_{SLD}	Net Force, kN
25	0.00	0.424	1.80	0.01	89.61	0.78	44.83	1.10	25.25	1.25	15.46
25	−5.71	0.385	1.67	0.03	91.63	0.74	43.80	1.03	26.86	1.18	18.39
25	−11.31	0.346	1.55	0.28	94.17	0.69	42.48	0.96	28.08	1.10	20.88
25	−14.03	0.326	1.50	1.71	94.47	0.66	41.60	0.93	28.37	1.06	21.75
30	0.00	0.348	1.81	0.01	90.78	0.85	34.06	1.19	16.11	1.36	7.13
30	−5.71	0.309	1.66	0.13	93.30	0.80	33.24	1.12	18.04	1.28	10.44
30	−11.31	0.270	1.52	0.39	96.96	0.74	32.26	1.04	19.70	1.18	13.42
30	−14.03	0.251	1.50	7.14	92.32	0.72	31.11	1.01	19.58	1.15	13.81
35	0.00	0.284	1.82	0.40	91.69	0.93	25.18	1.30	8.70	1.49	0.47
35	−5.71	0.246	1.65	0.08	95.54	0.87	24.84	1.21	11.26	1.39	4.47
35	−11.31	0.208	1.50	1.89	99.16	0.79	24.22	1.11	13.31	1.27	7.86
35	−14.03	0.190	1.50	14.64	89.26	0.78	23.08	1.09	13.13	1.24	8.16
40	0.00	0.232	1.82	0.24	93.32	1.01	18.32	1.41	3.22	1.62	0.00
40	−5.71	0.194	1.64	0.61	97.57	0.94	18.32	1.31	6.18	1.50	0.12
40	−11.31	0.158	1.50	7.76	97.80	0.85	17.93	1.20	8.44	1.37	3.70
40	−14.03	0.142	1.50	26.54	83.42	0.83	17.12	1.16	8.62	1.33	4.37

B. Physical Dimensions for a Technoeconomic Wall Section

					Depth of Shear Key D_{sk}, mm			
ϕ Degree	α Degree	Top Width, mm	Bottom Width, mm	Horizontal Angle θ Degree	$\mu = 0.50$	$\mu = 0.70$	$\mu = 0.80$	Volume of Wall, m^3
25	0.00	300	2430	90.00	1000	700	500	5.460
25	−5.71	300	2055	84.29	1000	700	600	4.710
25	−11.31	300	1695	78.69	1100	900	700	3.990
25	−14.03	300	1530	75.97	1100	900	800	3.660
30	0.00	300	2200	90.00	700	400	200	5.000
30	−5.71	300	1810	84.29	700	500	300	4.220
30	−11.31	300	1435	78.69	800	600	500	3.470
30	−14.03	300	1290	75.97	700	600	500	3.180
35	0.00	300	1990	90.00	400	200	0	4.580
35	−5.71	300	1580	84.29	400	200	100	3.760
35	−11.31	300	1200	78.69	500	400	300	3.000
35	−14.03	300	1065	75.97	500	400	300	2.730

(Continued)

TABLE 5.40 (Continued)

$H = 4.0$ m, $\gamma_{masonry} = 22.56$ kN/m³, $\gamma_{sat} = 17.27$ kN/m³ and $\iota = \frac{1}{2}\phi$

ϕ Degree	α Degree	Top Width, mm	Bottom Width, mm	Horizontal Angle θ Degree	Depth of Shear Key D_{sk}, mm			Volume of Wall, m³
					$\mu = 0.50$	$\mu = 0.70$	$\mu = 0.80$	
40	0.00	300	1795	90.00	300	0	0	4.190
40	−5.71	300	1375	84.29	300	100	0	3.350
40	−11.31	300	1000	78.69	400	200	100	2.600
40	−14.03	300	860	75.97	300	200	100	2.320

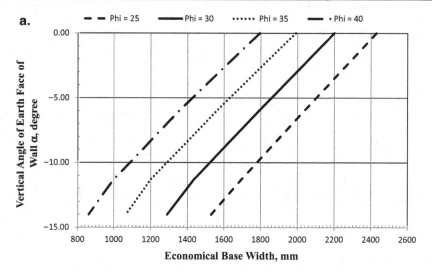

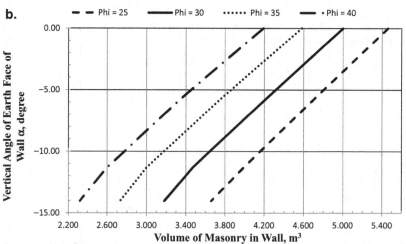

FIGURE 5.41 Design Charts for a Masonry Wall $H = 4.0$ m, $\gamma_{masonry} = 22.56$ kN/m³, $\gamma_{sat} = 17.27$ kN/m³ and $\iota = \frac{1}{2}\phi$: (a) Economical Base Width for a Wall, (b) Volume of a Masonry in Wall

TABLE 5.41

$H = 4.0$ m, $\gamma_{masonry} = 22.56$ kN/m^3, $\gamma_{sat} = 15.70$ kN/m^3 and $\iota = \phi$

A. *Results of Stability Analysis*

				Pressure at		$\mu = 0.50$		$\mu = 0.70$		$\mu = 0.80$	
ϕ Degree	α Degree	K_{ah}	F_{OVT}	Heel, kN/m^2	Toe, kN/m^2	F_{SLD}	Net Force, kN	F_{SLD}	Net Force, kN	F_{SLD}	Net Force, kN
25	0.00	0.821	1.80	0.22	86.60	0.61	93.09	0.85	67.76	0.97	55.10
25	−5.71	0.746	1.70	0.06	87.97	0.59	87.03	0.82	64.59	0.94	53.37
25	−11.31	0.675	1.61	0.21	89.35	0.57	81.25	0.79	61.56	0.91	51.72
25	−14.03	0.641	1.56	0.13	90.35	0.55	78.52	0.77	60.18	0.88	51.00
30	0.00	0.750	1.79	0.09	87.14	0.63	83.05	0.89	58.78	1.01	46.64
30	−5.71	0.666	1.70	0.16	88.50	0.62	75.78	0.86	54.58	0.99	43.98
30	−11.31	0.587	1.59	0.23	90.34	0.60	69.14	0.83	50.90	0.95	41.78
30	−14.03	0.549	1.54	0.03	91.76	0.58	66.07	0.82	49.28	0.93	40.88
35	0.00	0.671	1.80	0.15	87.59	0.67	72.02	0.93	48.94	1.07	37.41
35	−5.71	0.580	1.69	0.10	89.39	0.65	64.11	0.91	44.32	1.05	34.42
35	−11.31	0.496	1.58	0.38	91.54	0.63	57.01	0.89	40.36	1.01	32.04
35	−14.03	0.456	1.52	0.53	93.00	0.62	53.79	0.87	38.66	0.99	31.09
40	0.00	0.587	1.80	0.05	88.35	0.71	60.50	0.99	38.80	1.13	27.95
40	−5.71	0.492	1.68	0.09	90.47	0.70	52.38	0.98	34.14	1.12	25.02
40	−11.31	0.406	1.56	0.47	93.31	0.68	45.38	0.95	30.45	1.08	22.99
40	−14.03	0.366	1.50	1.70	94.25	0.66	42.18	0.93	28.81	1.06	22.12

B. *Physical Dimensions for a Technoeconomic Wall Section*

					Depth of Shear Key D_{sk}, mm			
ϕ Degree	α Degree	Top Width, mm	Bottom Width, mm	Horizontal Angle θ Degree	$\mu = 0.50$	$\mu = 0.70$	$\mu = 0.80$	Volume of Wall, m^3
25	0.00	300	3245	90.00	1500	1200	1100	7.090
25	−5.71	300	2835	84.29	1500	1300	1200	6.270
25	−11.31	300	2445	78.69	1500	1300	1100	5.490
25	−14.03	300	2255	75.97	1500	1300	1100	5.110
30	0.00	300	3095	90.00	1100	900	800	6.790
30	−5.71	300	2660	84.29	1000	900	700	5.920
30	−11.31	300	2240	78.69	1100	900	800	5.080
30	−14.03	300	2035	75.97	1100	900	800	4.670
35	0.00	300	2925	90.00	700	500	400	6.450
35	−5.71	300	2460	84.29	700	600	500	5.520
35	−11.31	300	2015	78.69	800	600	500	4.630
35	−14.03	300	1800	75.97	700	600	500	4.200

(Continued)

TABLE 5.41 (Continued)

$H = 4.0$ m, $\gamma_{masonry} = 22.56$ kN/m³, $\gamma_{sat} = 15.70$ kN/m³ and $\iota = \phi$

ϕ Degree	α Degree	Top Width, mm	Bottom Width, mm	Horizontal Angle θ Degree	Depth of Shear Key D_{sk}, mm			Volume of Wall, m³
					$\mu = 0.50$	$\mu = 0.70$	$\mu = 0.80$	
40	0.00	300	2730	90.00	400	300	200	6.060
40	−5.71	300	2240	84.29	500	300	200	5.080
40	−11.31	300	1770	78.69	500	400	300	4.140
40	−14.03	300	1550	75.97	500	400	300	3.700

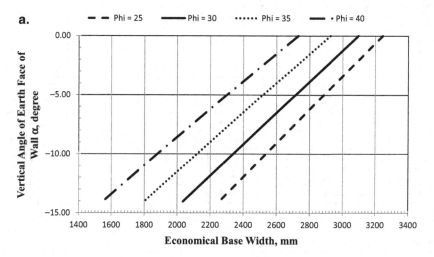

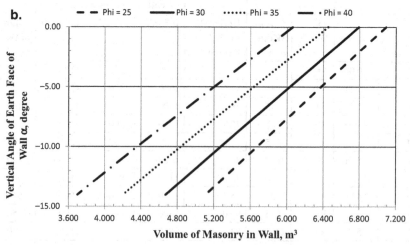

FIGURE 5.42 Design Charts for a Masonry Wall $H = 4.0$ m, $\gamma_{masonry} = 22.56$ kN/m³, $\gamma_{sat} = 15.70$ kN/m³ and $\iota = \phi$: (a) Economical Base Width for a Wall, (b) Volume of a Masonry in Wall

TABLE 5.42

H = 4.0 m, $\gamma_{masonry}$ = 22.56 kN/m³, γ_{sat} = 17.27 kN/m³ and $\iota = \phi$

A. Results of Stability Analysis

				Pressure at		$\mu = 0.50$		$\mu = 0.70$		$\mu = 0.80$	
ϕ Degree	α Degree	K_{ah}	F_{OVT}	Heel, kN/m²	Toe, kN/m²	F_{SLD}	Net Force, kN	F_{SLD}	Net Force, kN	F_{SLD}	Net Force, kN
25	0.00	0.821	1.79	0.22	86.20	0.58	105.63	0.81	79.14	0.92	65.89
25	−5.71	0.746	1.70	0.18	87.34	0.56	98.25	0.79	74.69	0.90	62.90
25	−11.31	0.675	1.61	0.17	88.75	0.54	91.31	0.76	70.56	0.87	60.19
25	−14.03	0.641	1.57	0.02	89.74	0.53	88.02	0.75	68.66	0.85	58.97
30	0.00	0.750	1.80	0.22	86.58	0.60	94.34	0.84	68.94	0.97	56.24
30	−5.71	0.666	1.70	0.22	87.90	0.59	85.66	0.83	63.40	0.95	52.27
30	−11.31	0.587	1.60	0.22	89.64	0.57	77.70	0.80	58.47	0.92	48.86
30	−14.03	0.549	1.55	0.01	90.94	0.56	74.02	0.79	56.27	0.90	47.40
35	0.00	0.671	1.80	0.22	87.07	0.64	81.99	0.89	57.86	1.02	45.79
35	−5.71	0.580	1.69	0.17	88.72	0.63	72.55	0.88	51.77	1.00	41.38
35	−11.31	0.496	1.58	0.02	91.11	0.61	64.14	0.85	46.61	0.97	37.84
35	−14.03	0.456	1.53	0.15	92.43	0.60	60.26	0.84	44.28	0.96	36.29
40	0.00	0.587	1.80	0.13	87.78	0.68	69.06	0.95	46.38	1.08	35.04
40	−5.71	0.492	1.69	0.27	89.63	0.67	59.35	0.94	40.19	1.07	30.61
40	−11.31	0.406	1.57	0.33	92.52	0.65	50.99	0.91	35.25	1.05	27.39
40	−14.03	0.366	1.50	0.22	94.67	0.64	47.31	0.90	33.23	1.02	26.19

B. Physical Dimensions for a Technoeconomic Wall Section

					Depth of Shear Key D_{sk}, mm			
ϕ Degree	α Degree	Top Width, mm	Bottom Width, mm	Horizontal Angle θ Degree	$\mu = 0.50$	$\mu = 0.70$	$\mu = 0.80$	Volume of Wall, m³
25	0.00	300	3410	90.00	1500	1300	1100	7.420
25	−5.71	300	2995	84.29	1600	1300	1200	6.590
25	−11.31	300	2595	78.69	1500	1300	1200	5.790
25	−14.03	300	2400	75.97	1500	1300	1200	5.400
30	0.00	300	3255	90.00	1100	900	800	7.110
30	−5.71	300	2810	84.29	1100	900	800	6.220
30	−11.31	300	2380	78.69	1100	900	800	5.360
30	−14.03	300	2170	75.97	1100	900	800	4.940
35	0.00	300	3075	90.00	700	500	400	6.750
35	−5.71	300	2600	84.29	700	600	500	5.800
35	−11.31	300	2140	78.69	700	500	500	4.880
35	−14.03	300	1920	75.97	700	600	500	4.440

(Continued)

TABLE 5.42 (Continued)

H = 4.0 m, $\gamma_{masonry}$ = 22.56 kN/m³, γ_{sat} = 17.27 kN/m³ and ι = ϕ

ϕ Degree	α Degree	Top Width, mm	Bottom Width, mm	Horizontal Angle θ Degree	Depth of Shear Key D_{sk}, mm			Volume of Wall, m³
					μ = 0.50	μ = 0.70	μ = 0.80	
40	0.00	300	2870	90.00	400	300	200	6.340
40	−5.71	300	2370	84.29	500	400	300	5.340
40	−11.31	300	1885	78.69	400	300	200	4.370
40	−14.03	300	1650	75.97	500	400	300	3.900

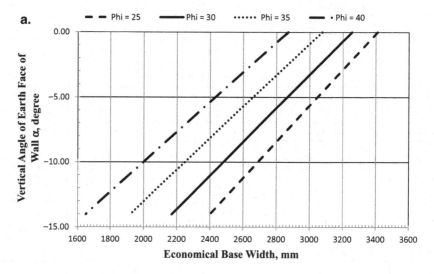

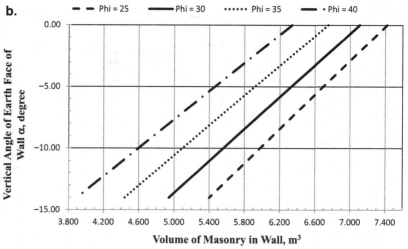

FIGURE 5.43 Design Charts for a Masonry Wall H = 4.0 m, $\gamma_{masonry}$ = 22.56 kN/m³, γ_{sat} = 17.27 kN/m³ and ι = ϕ: (a) Economical Base Width for a Wall, (b) Volume of a Masonry in Wall

5.9 CHARTS FOR THE DESIGN OF A 4.5-M-HIGH MASONRY BREAST WALL

The charts/tables developed for a 4.5-m-high stone masonry breast wall for the previously discussed six combinations are presented in the Tables 5.43 to 5.48 and Figures 5.44 to 5.49.

TABLE 5.43

$H = 4.5$ m, $\gamma_{masonry} = 22.56$ kN/m³, $\gamma_{sat} = 15.70$ kN/m³ and $\iota = 0°$

A. Results of Stability Analysis

				Pressure at		$\mu = 0.50$		$\mu = 0.70$		$\mu = 0.80$	
ϕ Degree	α Degree	K_{ah}	F_{OVT}	Heel, kN/m²	Toe, kN/m²	F_{SLD}	Net Force, kN	F_{SLD}	Net Force, kN	F_{SLD}	Net Force, kN
25	0.00	0.346	1.82	1.43	99.65	0.89	36.97	1.25	15.34	1.43	4.52
25	−5.71	0.316	1.71	4.41	99.03	0.83	37.72	1.16	19.02	1.33	9.68
25	−11.31	0.286	1.57	7.19	99.67	0.76	38.26	1.07	22.50	1.22	14.62
25	−14.03	0.271	1.51	9.55	99.53	0.73	38.32	1.02	23.95	1.16	16.77
30	0.00	0.279	1.85	2.81	99.61	0.98	26.54	1.37	6.70	1.56	0.00
30	−5.71	0.250	1.71	5.98	99.57	0.90	27.80	1.26	11.05	1.44	2.67
30	−11.31	0.221	1.56	10.35	99.94	0.82	28.75	1.14	15.02	1.31	8.15
30	−14.03	0.207	1.50	16.15	97.21	0.78	28.85	1.09	16.47	1.24	10.28
35	0.00	0.226	1.88	4.36	99.51	1.07	18.48	1.49	0.23	1.71	0.00
35	−5.71	0.198	1.71	8.46	99.48	0.98	20.22	1.37	5.18	1.56	0.00
35	−11.31	0.170	1.55	15.47	99.07	0.87	21.67	1.22	9.73	1.39	3.75
35	−14.03	0.156	1.50	26.99	91.91	0.82	21.87	1.15	11.23	1.32	5.92
40	0.00	0.184	1.90	5.65	99.75	1.16	12.56	1.62	0.00	1.85	0.00
40	−5.71	0.156	1.72	11.11	99.57	1.05	14.71	1.46	1.18	1.67	0.00
40	−11.31	0.130	1.52	21.11	99.07	0.91	16.72	1.27	6.40	1.46	1.24
40	−14.03	0.117	1.51	47.08	78.81	0.87	16.71	1.21	7.54	1.39	2.96

B. Physical Dimensions for a Technoeconomic Wall Section

					Depth of Shear Key D_{sk}, mm			
ϕ Degree	α Degree	Top Width, mm	Bottom Width, mm	Horizontal Angle θ Degree	$\mu = 0.50$	$\mu = 0.70$	$\mu = 0.80$	Volume of Wall, m³
25	0.00	300	2380	90.00	900	600	200	6.030
25	−5.71	300	2010	84.29	1000	600	400	5.198
25	−11.31	300	1640	78.69	1100	800	600	4.365
25	−14.03	300	1465	75.97	1100	800	700	3.971
30	0.00	300	2155	90.00	600	200	0	5.524
30	−5.71	300	1765	84.29	600	300	100	4.646
30	−11.31	300	1385	78.69	700	500	400	3.791
30	−14.03	300	1215	75.97	800	600	400	3.409

(Continued)

TABLE 5.43 (Continued)

H = 4.5 m, $\gamma_{masonry}$ = 22.56 kN/m³, γ_{sat} = 15.70 kN/m³ and ι = 0°

ϕ Degree	α Degree	Top Width, mm	Bottom Width, mm	Horizontal Angle θ Degree	Depth of Shear Key D_{sk}, mm			Volume of Wall, m³
					μ = 0.50	μ = 0.70	μ = 0.80	
35	0.00	300	1955	90.00	400	0	0	5.074
35	−5.71	300	1550	84.29	400	100	0	4.163
35	−11.31	300	1160	78.69	500	300	200	3.285
35	−14.03	300	995	75.97	500	400	200	2.914
40	0.00	300	1780	90.00	200	0	0	4.680
40	−5.71	300	1360	84.29	200	0	0	3.735
40	−11.31	300	955	78.69	400	200	100	2.824
40	−14.03	300	810	75.97	400	200	100	2.498

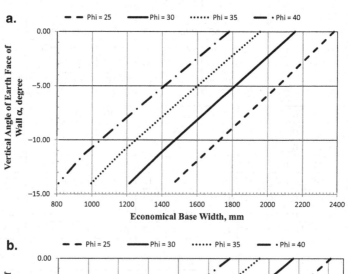

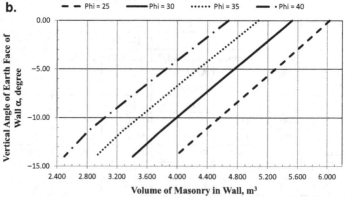

FIGURE 5.44 Design Charts for a Masonry Wall H = 4.5 m, $\gamma_{masonry}$ = 22.56 kN/m³, γ_{sat} = 15.70 kN/m³ and ι = 0°: (a) Economical Base Width for a Wall, (b) Volume of a Masonry in Wall

TABLE 5.44

H = 4.5 m, $\gamma_{masonry}$ = 22.56 kN/m³, γ_{sat} = 17.27 kN/m³ and ι = 0°

A. Results of Stability Analysis

ϕ Degree	α Degree	K_{ah}	F_{OVT}	Pressure at Heel, kN/m²	Toe, kN/m²	μ = 0.50 F_{SLD}	Net Force, kN	μ = 0.70 F_{SLD}	Net Force, kN	μ = 0.80 F_{SLD}	Net Force, kN
25	0.00	0.346	1.81	0.74	99.78	0.85	43.01	1.19	20.51	1.36	9.26
25	−5.71	0.316	1.69	2.75	99.97	0.79	43.26	1.11	23.77	1.27	14.03
25	−11.31	0.286	1.58	6.14	99.60	0.74	42.98	1.03	26.39	1.18	18.10
25	−14.03	0.271	1.52	7.87	99.89	0.70	42.78	0.99	27.62	1.13	20.04
30	0.00	0.279	1.84	2.22	99.58	0.93	31.21	1.31	10.57	1.50	0.25
30	−5.71	0.250	1.70	4.92	99.75	0.87	31.88	1.21	14.37	1.39	5.62
30	−11.31	0.221	1.57	9.49	99.38	0.79	32.13	1.11	17.64	1.27	10.40
30	−14.03	0.207	1.50	12.38	99.43	0.75	32.18	1.06	19.17	1.21	12.66
35	0.00	0.226	1.86	3.33	99.89	1.02	22.17	1.43	3.24	1.63	0.00
35	−5.71	0.198	1.71	7.37	99.55	0.94	23.25	1.32	7.53	1.51	0.00
35	−11.31	0.170	1.55	13.60	99.25	0.85	24.13	1.19	11.54	1.36	5.25
35	−14.03	0.156	1.50	23.39	93.37	0.81	24.10	1.13	12.87	1.29	7.25
40	0.00	0.184	1.89	4.98	99.69	1.11	15.35	1.55	0.00	1.78	0.00
40	−5.71	0.156	1.72	9.83	99.68	1.01	16.95	1.42	2.82	1.62	0.00
40	−11.31	0.130	1.53	18.76	99.25	0.89	18.42	1.25	7.54	1.43	2.11
40	−14.03	0.117	1.50	38.33	85.18	0.85	18.40	1.19	8.79	1.36	3.99

B. Physical Dimensions for a Technoeconomic Wall Section

ϕ Degree	α Degree	Top Width, mm	Bottom Width, mm	Horizontal Angle θ Degree	Depth of Shear Key D_{sk}, mm μ = 0.50	μ = 0.70	μ = 0.80	Volume of Wall, m³
25	0.00	300	2490	90.00	1000	600	400	6.278
25	−5.71	300	2110	84.29	1000	700	500	5.423
25	−11.31	300	1745	78.69	1100	800	700	4.601
25	−14.03	300	1565	75.97	1100	900	700	4.196
30	0.00	300	2255	90.00	600	300	0	5.749
30	−5.71	300	1860	84.29	700	400	200	4.860
30	−11.31	300	1480	78.69	800	500	400	4.005
30	−14.03	300	1295	75.97	800	600	500	3.589
35	0.00	300	2040	90.00	400	100	0	5.265
35	−5.71	300	1635	84.29	400	200	0	4.354
35	−11.31	300	1240	78.69	500	300	200	3.465
35	−14.03	300	1070	75.97	500	400	300	3.083

(Continued)

TABLE 5.44 (Continued)

H = 4.5 m, $\gamma_{masonry}$ = 22.56 kN/m³, γ_{sat} = 17.27 kN/m³ and ι = 0°

ϕ Degree	α Degree	Top Width, mm	Bottom Width, mm	Horizontal Angle θ Degree	Depth of Shear Key D_{sk}, mm			Volume of Wall, m³
					μ = 0.50	μ = 0.70	μ = 0.80	
40	0.00	300	1860	90.00	100	0	0	4.860
40	−5.71	300	1435	84.29	200	0	0	3.904
40	−11.31	300	1025	78.69	400	200	100	2.981
40	−14.03	300	865	75.97	400	200	100	2.621

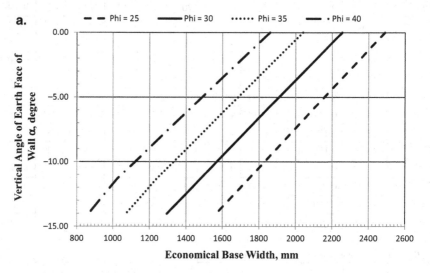

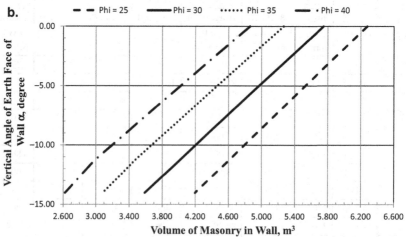

FIGURE 5.45 Design Charts for a Masonry Wall H = 4.5 m, $\gamma_{masonry}$ = 22.56 kN/m³, γ_{sat} = 17.27 kN/m³ and ι = 0°: (a) Economical Base Width for a Wall, (b) Volume of a Masonry in Wall

TABLE 5.45

$H = 4.5$ m, $\gamma_{masonry} = 22.56$ kN/m^3, $\gamma_{sat} = 15.70$ kN/m^3 and $\iota = \frac{1}{2}\phi$

A. *Results of Stability Analysis*

ϕ Degree	α Degree	K_{ah}	F_{OVT}	Pressure at Heel, kN/m^2	Toe, kN/m^2	$\mu = 0.50$ F_{SLD}	Net Force, kN	$\mu = 0.70$ F_{SLD}	Net Force, kN	$\mu = 0.80$ F_{SLD}	Net Force, kN
25	0.00	0.424	1.80	0.19	99.73	0.81	49.76	1.14	26.22	1.30	14.45
25	−5.71	0.385	1.69	2.33	99.61	0.77	48.75	1.07	28.31	1.23	18.09
25	−11.31	0.346	1.58	5.21	99.54	0.72	47.51	1.00	30.09	1.15	21.38
25	−14.03	0.326	1.53	6.93	99.66	0.69	46.85	0.97	30.89	1.10	22.91
30	0.00	0.348	1.82	1.50	99.53	0.89	37.36	1.24	15.65	1.42	4.80
30	−5.71	0.309	1.70	4.10	99.57	0.84	36.69	1.17	18.23	1.34	9.01
30	−11.31	0.270	1.57	7.64	99.93	0.77	35.93	1.08	20.65	1.24	13.01
30	−14.03	0.251	1.51	10.58	99.62	0.74	35.46	1.03	21.69	1.18	14.80
35	0.00	0.284	1.84	2.52	99.81	0.97	27.32	1.36	7.36	1.55	0.00
35	−5.71	0.246	1.71	6.64	99.05	0.91	27.00	1.27	10.37	1.46	2.05
35	−11.31	0.208	1.56	12.10	99.01	0.83	26.84	1.16	13.51	1.33	6.84
35	−14.03	0.190	1.50	18.38	96.51	0.79	26.59	1.11	14.77	1.26	8.85
40	0.00	0.232	1.87	3.89	99.82	1.06	19.41	1.48	0.99	1.69	0.00
40	−5.71	0.194	1.71	8.23	99.97	0.98	19.76	1.37	4.88	1.57	0.00
40	−11.31	0.158	1.54	17.01	98.85	0.88	20.18	1.23	8.67	1.41	2.92
40	−14.03	0.142	1.50	33.07	87.97	0.84	19.90	1.18	9.78	1.34	4.72

B. *Physical Dimensions for a Technoeconomic Wall Section*

ϕ Degree	α Degree	Top Width, mm	Bottom Width, mm	Horizontal Angle θ Degree	Depth of Shear Key D_{sk}, mm $\mu = 0.50$	$\mu = 0.70$	$\mu = 0.80$	Volume of Wall, m^3
25	0.00	300	2620	90.00	1100	800	500	6.570
25	−5.71	300	2230	84.29	1100	800	600	5.693
25	−11.31	300	1850	78.69	1200	900	800	4.838
25	−14.03	300	1665	75.97	1200	1000	800	4.421
30	0.00	300	2390	90.00	800	400	200	6.053
30	−5.71	300	1980	84.29	800	500	300	5.130
30	−11.31	300	1580	78.69	800	600	500	4.230
30	−14.03	300	1390	75.97	800	600	500	3.803
35	0.00	300	2170	90.00	400	100	0	5.558
35	−5.71	300	1750	84.29	500	200	0	4.613
35	−11.31	300	1335	78.69	600	400	300	3.679
35	−14.03	300	1145	75.97	600	400	300	3.251

(Continued)

TABLE 5.45 (Continued)

H = 4.5 m, $\gamma_{masonry}$ = 22.56 kN/m³, γ_{sat} = 15.70 kN/m³ and ι = ½ϕ

ϕ Degree	α Degree	Top Width, mm	Bottom Width, mm	Horizontal Angle θ Degree	Depth of Shear Key D_{sk}, mm			Volume of Wall, m³
					μ = 0.50	μ = 0.70	μ = 0.80	
40	0.00	300	1975	90.00	200	0	0	5.119
40	−5.71	300	1530	84.29	300	100	0	4.118
40	−11.31	300	1105	78.69	400	200	100	3.161
40	−14.03	300	930	75.97	400	300	200	2.768

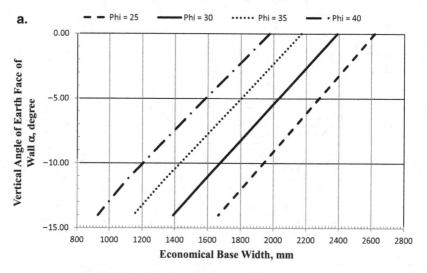

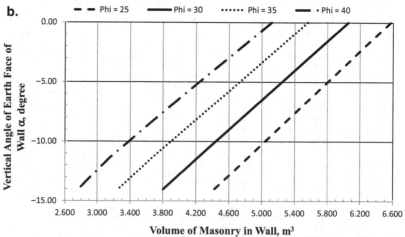

FIGURE 5.46 Design Charts for a Masonry Wall H = 4.5 m, $\gamma_{masonry}$ = 22.56 kN/m³, γ_{sat} = 15.70 kN/m³ and ι = ½ϕ: (a) Economical Base Width for a Wall, (b) Volume of a Masonry in Wall

TABLE 5.46

H = 4.5 m, $\gamma_{masonry}$ = 22.56 kN/m³, γ_{sat} = 17.27 kN/m³ and ι = ½ϕ

A. Results of Stability Analysis

ϕ Degree	α Degree	K_{ah}	F_{OVT}	Pressure at Heel, kN/ m²	Toe, kN/ m²	μ = 0.50 F_{SLD}	Net Force, kN	μ = 0.70 F_{SLD}	Net Force, kN	μ = 0.80 F_{SLD}	Net Force, kN
25	0.00	0.424	1.80	0.06	99.30	0.78	57.27	1.09	32.70	1.24	20.41
25	−5.71	0.385	1.69	1.62	99.63	0.74	55.53	1.03	34.14	1.18	23.44
25	−11.31	0.346	1.59	4.25	99.54	0.69	53.46	0.97	35.12	1.11	25.95
25	−14.03	0.326	1.53	5.51	99.94	0.67	52.43	0.93	35.61	1.07	27.19
30	0.00	0.348	1.81	0.77	99.70	0.85	43.47	1.19	20.89	1.36	9.60
30	−5.71	0.309	1.70	3.56	99.30	0.80	41.86	1.13	22.54	1.29	12.87
30	−11.31	0.270	1.58	6.96	99.41	0.75	40.28	1.05	24.17	1.20	16.11
30	−14.03	0.251	1.52	9.43	99.30	0.72	39.46	1.01	24.90	1.15	17.61
35	0.00	0.284	1.83	1.85	99.85	0.93	32.10	1.30	11.35	1.48	0.97
35	−5.71	0.246	1.70	5.09	99.75	0.87	31.06	1.22	13.71	1.40	5.04
35	−11.31	0.208	1.56	9.89	99.85	0.81	30.11	1.13	16.10	1.29	9.10
35	−14.03	0.190	1.50	15.05	98.11	0.77	29.54	1.08	17.07	1.23	10.84
40	0.00	0.232	1.86	3.66	99.35	1.01	23.04	1.42	3.87	1.62	0.00
40	−5.71	0.194	1.71	7.26	99.89	0.95	22.69	1.33	7.13	1.52	0.00
40	−11.31	0.158	1.54	14.53	99.57	0.86	22.46	1.21	10.36	1.38	4.30
40	−14.03	0.142	1.50	27.14	91.76	0.82	21.98	1.15	11.34	1.31	6.02

B. Physical Dimensions for a Technoeconomic Wall Section

ϕ Degree	α Degree	Top Width, mm	Bottom Width, mm	Horizontal Angle θ Degree	Depth of Shear Key D_{sk}, mm μ = 0.50	μ = 0.70	μ = 0.80	Volume of Wall, m³
25	0.00	300	2750	90.00	1200	800	600	6.863
25	−5.71	300	2350	84.29	1100	900	700	5.963
25	−11.31	300	1965	78.69	1200	1000	800	5.096
25	−14.03	300	1775	75.97	1200	1000	800	4.669
30	0.00	300	2500	90.00	700	400	200	6.300
30	−5.71	300	2090	84.29	800	500	400	5.378
30	−11.31	300	1685	78.69	800	500	400	4.466
30	−14.03	300	1490	75.97	800	700	500	4.028
35	0.00	300	2270	90.00	400	100	0	5.783
35	−5.71	300	1840	84.29	500	300	100	4.815
35	−11.31	300	1420	78.69	500	300	200	3.870
35	−14.03	300	1225	75.97	600	400	300	3.431

(Continued)

TABLE 5.46 (Continued)

H = 4.5 m, $\gamma_{masonry}$ = 22.56 kN/m³, γ_{sat} = 17.27 kN/m³ and ι = ½ϕ

ϕ Degree	α Degree	Top Width, mm	Bottom Width, mm	Horizontal Angle θ Degree	Depth of Shear Key D_{sk}, mm			Volume of Wall, m³
					μ = 0.50	μ = 0.70	μ = 0.80	
40	0.00	300	2070	90.00	200	0	0	5.333
40	−5.71	300	1615	84.29	300	100	0	4.309
40	−11.31	300	1180	78.69	400	300	100	3.330
40	−14.03	300	995	75.97	400	300	200	2.914

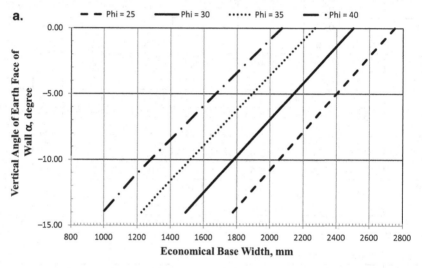

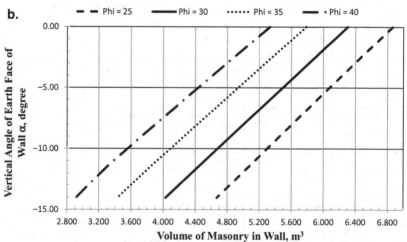

FIGURE 5.47 Design Charts for a Masonry Wall H = 4.5 m, $\gamma_{masonry}$ = 22.56 kN/m³, γ_{sat} = 17.27 kN/m³ and ι = ½ϕ: (a) Economical Base Width for a Wall, (b) Volume of a Masonry in Wall

TABLE 5.47

$H = 4.5$ m, $\gamma_{masonry} = 22.56$ kN/m³, $\gamma_{sat} = 15.70$ kN/m³ and $\iota = \phi$

A. Results of Stability Analysis

				Pressure at		$\mu = 0.50$		$\mu = 0.70$		$\mu = 0.80$	
ϕ Degree	α Degree	K_{ah}	F_{OVT}	Heel, kN/m²	Toe, kN/m²	F_{SLD}	Net Force, kN	F_{SLD}	Net Force, kN	F_{SLD}	Net Force, kN
25	0.00	0.821	1.79	0.12	96.48	0.60	118.39	0.84	86.56	0.96	70.64
25	−5.71	0.746	1.70	0.08	97.70	0.58	110.60	0.82	82.38	0.93	68.27
25	−11.31	0.675	1.61	0.22	99.04	0.56	103.20	0.79	78.44	0.90	66.05
25	−14.03	0.641	1.56	0.74	99.40	0.55	99.58	0.77	76.44	0.88	64.87
30	0.00	0.750	1.79	0.17	96.82	0.63	105.61	0.88	75.09	1.01	59.83
30	−5.71	0.666	1.70	0.32	98.07	0.61	96.32	0.86	69.65	0.98	56.31
30	−11.31	0.587	1.60	0.73	99.51	0.59	87.76	0.83	64.78	0.95	53.29
30	−14.03	0.549	1.55	1.43	99.94	0.58	83.67	0.82	62.44	0.93	51.82
35	0.00	0.671	1.80	0.32	97.17	0.66	91.62	0.93	62.60	1.06	48.10
35	−5.71	0.580	1.69	0.27	98.94	0.65	81.55	0.91	56.66	1.04	44.22
35	−11.31	0.496	1.59	1.72	99.78	0.63	72.26	0.88	51.22	1.01	40.71
35	−14.03	0.456	1.54	3.23	99.74	0.62	67.94	0.87	48.73	0.99	39.13
40	0.00	0.587	1.80	0.21	97.93	0.70	77.05	0.99	49.78	1.13	36.15
40	−5.71	0.492	1.69	0.89	99.35	0.69	66.56	0.97	43.57	1.11	32.08
40	−11.31	0.406	1.59	3.77	99.41	0.68	57.22	0.95	38.25	1.09	28.76
40	−14.03	0.366	1.53	5.79	99.42	0.67	53.12	0.93	36.10	1.07	27.58

B. Physical Dimensions for a Technoeconomic Wall Section

					Depth of Shear Key D_{sk}, mm			
ϕ Degree	α Degree	Top Width, mm	Bottom Width, mm	Horizontal Angle θ Degree	$\mu = 0.50$	$\mu = 0.70$	$\mu = 0.80$	Volume of Wall, m³
25	0.00	300	3665	90.00	1700	1400	1300	8.921
25	−5.71	300	3210	84.29	1600	1400	1200	7.898
25	−11.31	300	2775	78.69	1700	1400	1300	6.919
25	−14.03	300	2570	75.97	1700	1400	1300	6.458
30	0.00	300	3500	90.00	1200	900	800	8.550
30	−5.71	300	3015	84.29	1200	1000	900	7.459
30	−11.31	300	2550	78.69	1100	900	800	6.413
30	−14.03	300	2330	75.97	1200	1000	900	5.918
35	0.00	300	3310	90.00	800	600	500	8.123
35	−5.71	300	2790	84.29	800	700	600	6.953
35	−11.31	300	2305	78.69	800	600	500	5.861
35	−14.03	300	2075	75.97	800	700	600	5.344

(Continued)

TABLE 5.47 (Continued)

$H = 4.5$ m, $\gamma_{masonry} = 22.56$ kN/m³, $\gamma_{sat} = 15.70$ kN/m³ and $\iota = \phi$

ϕ Degree	α Degree	Top Width, mm	Bottom Width, mm	Horizontal Angle θ Degree	Depth of Shear Key D_{sk}, mm			Volume of Wall, m³
					$\mu = 0.50$	$\mu = 0.70$	$\mu = 0.80$	
40	0.00	300	3090	90.00	400	300	200	7.628
40	−5.71	300	2550	84.29	500	300	200	6.413
40	−11.31	300	2045	78.69	500	400	300	5.276
40	−14.03	300	1800	75.97	600	500	400	4.725

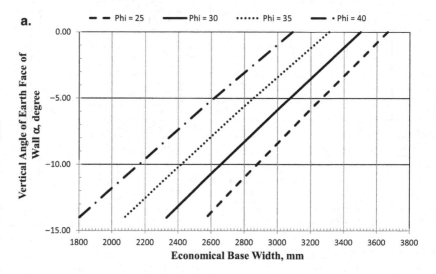

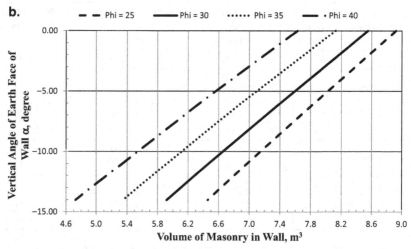

FIGURE 5.48 Design Charts for a Masonry Wall $H = 4.5$ m, $\gamma_{masonry} = 22.56$ kN/m³, $\gamma_{sat} = 15.70$ kN/m³ and $\iota = \phi$: (a) Economical Base Width for a Wall, (b) Volume of a Masonry in Wall

TABLE 5.48

$H = 4.5$ m, $\gamma_{masonry} = 22.56$ kN/m³, $\gamma_{sat} = 17.27$ kN/m³ and $\iota = \phi$

A. Results of Stability Analysis

				Pressure at		$\mu = 0.50$		$\mu = 0.70$		$\mu = 0.80$	
ϕ Degree	α Degree	K_{ah}	F_{OVT}	Heel, kN/ m²	Toe, kN/ m²	F_{SLD}	Net Force, kN	F_{SLD}	Net Force, kN	F_{SLD}	Net Force, kN
25	0.00	0.821	1.79	0.08	96.12	0.57	134.28	0.80	100.98	0.92	84.33
25	−5.71	0.746	1.70	0.21	97.07	0.56	124.80	0.78	95.15	0.89	80.33
25	−11.31	0.675	1.61	0.28	98.35	0.54	115.91	0.76	89.79	0.86	76.73
25	−14.03	0.641	1.57	0.33	99.12	0.53	111.67	0.74	87.26	0.85	75.05
30	0.00	0.750	1.79	0.30	96.27	0.60	119.89	0.84	87.94	0.96	71.97
30	−5.71	0.666	1.70	0.13	97.74	0.59	108.90	0.82	80.92	0.94	66.93
30	−11.31	0.587	1.60	0.08	99.49	0.57	98.75	0.80	74.58	0.91	62.49
30	−14.03	0.549	1.56	0.68	99.91	0.56	93.87	0.78	71.48	0.90	60.29
35	0.00	0.671	1.79	0.19	96.87	0.63	104.32	0.88	73.99	1.01	58.83
35	−5.71	0.580	1.69	0.16	98.47	0.62	92.28	0.87	66.17	0.99	53.11
35	−11.31	0.496	1.60	1.20	99.55	0.61	81.29	0.85	59.14	0.97	48.07
35	−14.03	0.456	1.55	2.65	99.42	0.60	76.13	0.84	55.85	0.96	45.71
40	0.00	0.587	1.79	0.14	97.53	0.67	87.94	0.94	59.45	1.07	45.20
40	−5.71	0.492	1.69	0.18	99.46	0.66	75.60	0.93	51.54	1.06	39.52
40	−11.31	0.406	1.59	2.42	99.94	0.65	64.51	0.92	44.58	1.05	34.62
40	−14.03	0.366	1.54	4.28	99.91	0.64	59.56	0.90	41.62	1.03	32.65

B. Physical Dimensions for a Technoeconomic Wall Section

					Depth of Shear Key D_{sk}, mm			
ϕ Degree	α Degree	Top Width, mm	Bottom Width, mm	Horizontal Angle θ Degree	$\mu = 0.50$	$\mu = 0.70$	$\mu = 0.80$	Volume of Wall, m³
25	0.00	300	3850	90.00	1700	1500	1300	9.338
25	−5.71	300	3390	84.29	1700	1400	1300	8.303
25	−11.31	300	2945	78.69	1700	1500	1400	7.301
25	−14.03	300	2730	75.97	1700	1500	1300	6.818
30	0.00	300	3680	90.00	1200	1000	900	8.955
30	−5.71	300	3180	84.29	1200	1000	900	7.830
30	−11.31	300	2700	78.69	1200	1000	900	6.750
30	−14.03	300	2475	75.97	1200	1100	1000	6.244
35	0.00	300	3475	90.00	800	600	500	8.494
35	−5.71	300	2945	84.29	800	600	500	7.301
35	−11.31	300	2445	78.69	800	600	500	6.176
35	−14.03	300	2210	75.97	800	600	500	5.648

(Continued)

TABLE 5.48 (Continued)

H = 4.5 m, $\gamma_{masonry}$ = 22.56 kN/m³, γ_{sat} = 17.27 kN/m³ and $\iota = \phi$

ϕ Degree	α Degree	Top Width, mm	Bottom Width, mm	Horizontal Angle θ Degree	Depth of Shear Key D_{sk}, mm			Volume of Wall, m³
					$\mu = 0.50$	$\mu = 0.70$	$\mu = 0.80$	
40	0.00	300	3245	90.00	400	300	200	7.976
40	−5.71	300	2685	84.29	500	300	200	6.716
40	−11.31	300	2165	78.69	500	400	300	5.546
40	−14.03	300	1915	75.97	500	400	300	4.984

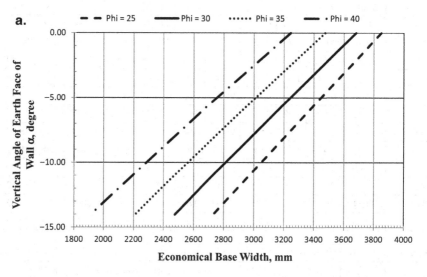

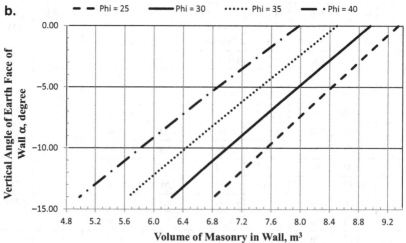

FIGURE 5.49 Design Charts for a Masonry Wall H = 4.5 m, $\gamma_{masonry}$ = 22.56 kN/m³, γ_{sat} = 17.27 kN/m³ and $\iota = \phi$: (a) Economical Base Width for a Wall, (b) Volume of a Masonry in Wall

5.10 CHARTS FOR THE DESIGN OF A 5.0-M-HIGH MASONRY BREAST WALL

The charts/tables developed for a 5.0-m-high stone masonry breast wall for the previously discussed six combinations are presented in Tables 5.49 to 5.54 and Figures 5.50 to 5.55.

TABLE 5.49
$H = 5.0$ m, $\gamma_{masonry} = 22.56$ kN/m³, $\gamma_{sat} = 15.70$ kN/m³ and $\iota = 0°$

A. Results of Stability Analysis

				Pressure at		$\mu = 0.50$		$\mu = 0.70$		$\mu = 0.80$	
ϕ Degree	α Degree	K_{ah}	F_{OVT}	Heel, kN/m²	Toe, kN/m²	F_{SLD}	Net Force, kN	F_{SLD}	Net Force, kN	F_{SLD}	Net Force, kN
25	0.00	0.346	1.94	10.39	99.88	0.92	43.50	1.29	15.93	1.47	2.15
25	−5.71	0.316	1.79	12.73	99.86	0.85	45.17	1.19	21.53	1.36	9.71
25	−11.31	0.286	1.65	16.33	99.49	0.78	46.14	1.09	26.25	1.24	16.31
25	−14.03	0.271	1.58	18.48	99.47	0.74	46.45	1.04	28.37	1.18	19.33
30	0.00	0.279	1.96	11.61	99.95	1.01	30.95	1.41	5.73	1.61	0.00
30	−5.71	0.250	1.80	14.75	99.82	0.92	33.09	1.29	11.93	1.48	1.34
30	−11.31	0.221	1.63	19.56	99.50	0.83	34.66	1.17	17.37	1.33	8.73
30	−14.03	0.207	1.55	23.82	98.29	0.79	35.14	1.10	19.66	1.26	11.93
35	0.00	0.226	1.99	13.04	99.90	1.10	21.29	1.54	0.00	1.75	0.00
35	−5.71	0.198	1.80	17.24	99.61	1.00	23.99	1.39	5.03	1.59	0.00
35	−11.31	0.170	1.60	23.79	99.38	0.88	26.25	1.23	11.30	1.41	3.83
35	−14.03	0.156	1.50	28.80	99.12	0.82	27.20	1.15	14.14	1.31	7.62
40	0.00	0.184	2.02	14.72	99.66	1.19	14.10	1.66	0.00	1.90	0.00
40	−5.71	0.156	1.80	20.00	99.45	1.07	17.38	1.49	0.36	1.70	0.00
40	−11.31	0.130	1.57	29.51	98.94	0.92	20.30	1.29	7.42	1.47	0.98
40	−14.03	0.117	1.50	46.12	88.76	0.86	21.04	1.20	9.88	1.37	4.30

B. Physical Dimensions for a Technoeconomic Wall Section

					Depth of Shear Key D_{sk}, mm			
ϕ Degree	α Degree	Top Width, mm	Bottom Width, mm	Horizontal Angle θ Degree	$\mu = 0.50$	$\mu = 0.70$	$\mu = 0.80$	Volume of Wall, m³
25	0.00	300	2780	90.00	1000	600	100	7.700
25	−5.71	300	2335	84.29	1100	700	400	6.588
25	−11.31	300	1910	78.69	1200	900	700	5.525
25	−14.03	300	1705	75.97	1200	900	700	5.013
30	0.00	300	2515	90.00	600	100	0	7.038
30	−5.71	300	2055	84.29	700	400	0	5.888
30	−11.31	300	1615	78.69	800	600	400	4.788
30	−14.03	300	1410	75.97	800	600	500	4.275

(Continued)

TABLE 5.49 (Continued)

$H = 5.0$ m, $\gamma_{masonry} = 22.56$ kN/m³, $\gamma_{sat} = 15.70$ kN/m³ and $\iota = 0°$

φ Degree	α Degree	Top Width, mm	Bottom Width, mm	Horizontal Angle θ Degree	Depth of Shear Key D_{sk}, mm			Volume of Wall, m³
					μ = 0.50	μ = 0.70	μ = 0.80	
35	0.00	300	2280	90.00	300	0	0	6.450
35	−5.71	300	1805	84.29	400	100	0	5.263
35	−11.31	300	1350	78.69	600	400	200	4.125
35	−14.03	300	1135	75.97	600	400	300	3.588
40	0.00	300	2080	90.00	100	0	0	5.950
40	−5.71	300	1585	84.29	300	0	0	4.713
40	−11.31	300	1115	78.69	400	200	0	3.538
40	−14.03	300	920	75.97	400	300	200	3.050

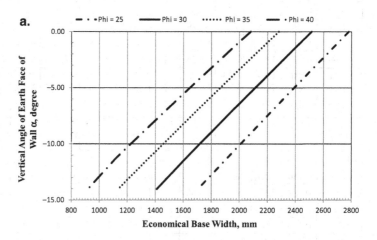

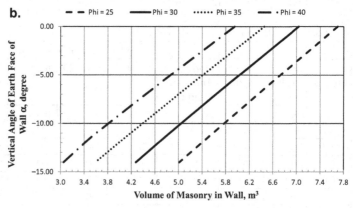

FIGURE 5.50 Design Charts for a Masonry Wall H = 5.0 m, $\gamma_{masonry}$ = 22.56 kN/m³, γ_{sat} = 15.70 kN/m³ and $\iota = 0°$: (a) Economical Base Width for a Wall, (b) Volume of a Masonry in Wall

TABLE 5.50
$H = 5.0$ m, $\gamma_{masonry} = 22.56$ kN/m^3, $\gamma_{sat} = 17.27$ kN/m^3 and $\iota = 0°$

A. Results of Stability Analysis

ϕ Degree	α Degree	K_{ah}	F_{OVT}	Pressure at Heel, kN/m²	Toe, kN/m²	$\mu = 0.50$ F_{SLD}	Net Force, kN	$\mu = 0.70$ F_{SLD}	Net Force, kN	$\mu = 0.80$ F_{SLD}	Net Force, kN
25	0.00	0.346	1.93	9.80	99.93	0.88	50.79	1.23	22.08	1.41	7.72
25	−5.71	0.316	1.80	12.24	99.59	0.82	51.60	1.15	26.81	1.31	14.42
25	−11.31	0.286	1.65	15.06	99.71	0.75	51.90	1.05	30.95	1.21	20.48
25	−14.03	0.271	1.58	16.72	99.98	0.72	51.87	1.01	32.78	1.15	23.23
30	0.00	0.279	1.96	11.24	99.70	0.96	36.52	1.35	10.23	1.54	0.00
30	−5.71	0.250	1.80	14.12	99.58	0.89	37.92	1.25	15.74	1.43	4.65
30	−11.31	0.221	1.63	17.99	99.79	0.81	38.84	1.13	20.63	1.30	11.52
30	−14.03	0.207	1.55	20.62	99.97	0.77	39.12	1.07	22.80	1.23	14.65
35	0.00	0.226	1.98	12.46	99.83	1.05	25.61	1.47	1.53	1.68	0.00
35	−5.71	0.198	1.81	16.82	99.00	0.97	27.49	1.35	7.60	1.55	0.00
35	−11.31	0.170	1.61	22.43	99.09	0.86	29.15	1.21	13.36	1.38	5.47
35	−14.03	0.156	1.52	26.73	99.00	0.81	29.80	1.13	15.95	1.29	9.03
40	0.00	0.184	2.00	13.74	99.96	1.14	17.52	1.60	0.00	1.82	0.00
40	−5.71	0.156	1.80	18.44	99.93	1.03	20.10	1.45	2.33	1.65	0.00
40	−11.31	0.130	1.58	27.09	99.33	0.90	22.35	1.27	8.76	1.45	1.97
40	−14.03	0.117	1.50	38.59	93.87	0.84	23.03	1.18	11.30	1.34	5.44

B. Physical Dimensions for a Technoeconomic Wall Section

ϕ Degree	α Degree	Top Width, mm	Bottom Width, mm	Horizontal Angle θ Degree	Depth of Shear Key D_{sk}, mm $\mu = 0.50$	$\mu = 0.70$	$\mu = 0.80$	Volume of Wall, m³
25	0.00	300	2910	90.00	1100	700	300	8.025
25	−5.71	300	2465	84.29	1100	700	500	6.913
25	−11.31	300	2030	78.69	1100	800	600	5.825
25	−14.03	300	1820	75.97	1200	900	800	5.300
30	0.00	300	2635	90.00	600	200	0	7.338
30	−5.71	300	2170	84.29	700	400	200	6.175
30	−11.31	300	1720	78.69	700	500	300	5.050
30	−14.03	300	1505	75.97	800	600	500	4.513
35	0.00	300	2385	90.00	300	0	0	6.713
35	−5.71	300	1910	84.29	500	200	0	5.525
35	−11.31	300	1445	78.69	500	300	100	4.363
35	−14.03	300	1225	75.97	600	400	300	3.813

(Continued)

TABLE 5.50 (Continued)

H = 5.0 m, $\gamma_{masonry}$ = 22.56 kN/m³, γ_{sat} = 17.27 kN/m³ and ι = 0°

ϕ Degree	α Degree	Top Width, mm	Bottom Width, mm	Horizontal Angle θ Degree	Depth of Shear Key D_{sk}, mm			Volume of Wall, m³
					μ = 0.50	μ = 0.70	μ = 0.80	
40	0.00	300	2170	90.00	100	0	0	6.175
40	−5.71	300	1670	84.29	300	0	0	4.925
40	−11.31	300	1195	78.69	400	200	100	3.738
40	−14.03	300	985	75.97	400	300	200	3.213

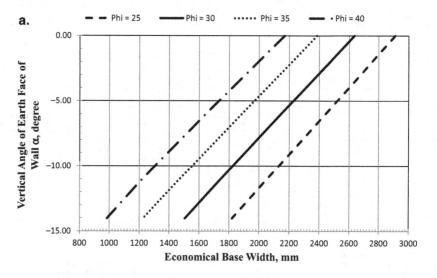

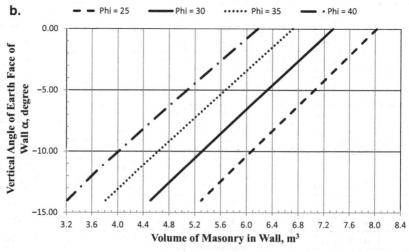

FIGURE 5.51 Design Charts for a Masonry Wall H = 5.0 m, $\gamma_{masonry}$ = 22.56 kN/m³, γ_{sat} = 17.27 kN/m³ and ι = 0°: (a) Economical Base Width for a Wall, (b) Volume of a Masonry in Wall

TABLE 5.51

$H = 5.0$ m, $\gamma_{masonry} = 22.56$ kN/m³, $\gamma_{sat} = 15.70$ kN/m³ and $\iota = \frac{1}{2}\phi$

A. Results of Stability Analysis

ϕ Degree	α Degree	K_{ah}	F_{OVT}	Pressure at Heel, kN/m²	Toe, kN/m²	$\mu = 0.50$ F_{SLD}	Net Force, kN	$\mu = 0.70$ F_{SLD}	Net Force, kN	$\mu = 0.80$ F_{SLD}	Net Force, kN
25	0.00	0.424	1.93	9.44	99.71	0.84	58.88	1.18	28.81	1.35	13.77
25	−5.71	0.385	1.79	11.41	99.71	0.79	58.33	1.11	32.36	1.26	19.37
25	−11.31	0.346	1.66	13.94	99.90	0.73	57.42	1.03	35.41	1.17	24.41
25	−14.03	0.326	1.59	15.70	99.89	0.70	56.81	0.99	36.70	1.13	26.64
30	0.00	0.348	1.94	10.34	99.89	0.92	44.01	1.28	16.36	1.47	2.53
30	−5.71	0.309	1.80	13.31	99.44	0.86	43.72	1.20	20.30	1.37	8.59
30	−11.31	0.270	1.64	16.71	99.78	0.79	43.34	1.11	24.07	1.26	14.43
30	−14.03	0.251	1.57	19.77	99.22	0.75	42.96	1.05	25.63	1.21	16.96
35	0.00	0.284	1.96	11.49	99.96	1.00	31.85	1.40	6.45	1.60	0.00
35	−5.71	0.246	1.80	15.15	99.58	0.93	32.18	1.30	11.19	1.49	0.69
35	−11.31	0.208	1.62	20.17	99.76	0.84	32.52	1.18	15.80	1.35	7.45
35	−14.03	0.190	1.53	23.97	99.77	0.79	32.62	1.11	17.94	1.27	10.59
40	0.00	0.232	1.99	13.10	99.65	1.09	22.26	1.52	0.00	1.74	0.00
40	−5.71	0.194	1.80	17.38	99.68	1.00	23.37	1.40	4.58	1.60	0.00
40	−11.31	0.158	1.59	24.97	99.47	0.89	24.50	1.25	10.12	1.43	2.94
40	−14.03	0.142	1.50	32.85	97.33	0.83	24.93	1.16	12.58	1.33	6.41

B. Physical Dimensions for a Technoeconomic Wall Section

ϕ Degree	α Degree	Top Width, mm	Bottom Width, mm	Horizontal Angle θ Degree	Depth of Shear Key D_{sk}, mm $\mu = 0.50$	$\mu = 0.70$	$\mu = 0.80$	Volume of Wall, m³
25	0.00	300	3065	90.00	1100	700	400	8.413
25	−5.71	300	2600	84.29	1200	900	600	7.250
25	−11.31	300	2150	78.69	1200	900	700	6.125
25	−14.03	300	1935	75.97	1300	1000	900	5.588
30	0.00	300	2790	90.00	700	400	0	7.725
30	−5.71	300	2310	84.29	800	500	300	6.525
30	−11.31	300	1840	78.69	800	600	400	5.350
30	−14.03	300	1620	75.97	900	700	600	4.800
35	0.00	300	2535	90.00	400	100	0	7.088
35	−5.71	300	2035	84.29	500	300	0	5.838
35	−11.31	300	1550	78.69	500	300	200	4.625
35	−14.03	300	1320	75.97	600	500	300	4.050

(Continued)

TABLE 5.51 (Continued)

H = 5.0 m, $\gamma_{masonry}$ = 22.56 kN/m³, γ_{sat} = 15.70 kN/m³ and $\iota = \frac{1}{2}\phi$

ϕ Degree	α Degree	Top Width, mm	Bottom Width, mm	Horizontal Angle θ Degree	Depth of Shear Key D_{sk}, mm			Volume of Wall, m³
					$\mu = 0.50$	$\mu = 0.70$	$\mu = 0.80$	
40	0.00	300	2310	90.00	200	0	0	6.525
40	−5.71	300	1785	84.29	300	100	0	5.213
40	−11.31	300	1285	78.69	300	200	0	3.963
40	−14.03	300	1055	75.97	400	300	200	3.388

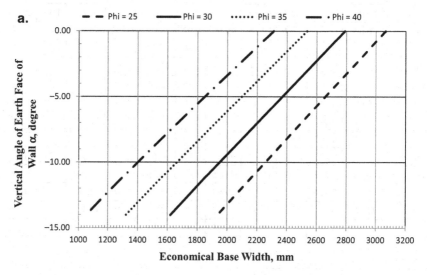

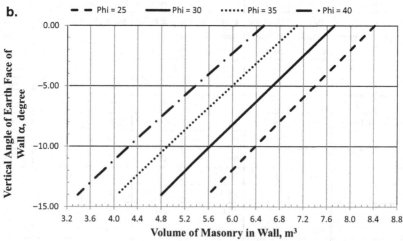

FIGURE 5.52 Design Charts for a Masonry Wall H = 5.0 m, $\gamma_{masonry}$ = 22.56 kN/m³, γ_{sat} = 15.70 kN/m³ and $\iota = \frac{1}{2}\phi$: (a) Economical Base Width for a Wall, (b) Volume of a Masonry in Wall

TABLE 5.52

$H = 5.0$ m, $\gamma_{masonry} = 22.56$ kN/m³, $\gamma_{sat} = 17.27$ kN/m³ and $\iota = \frac{1}{2}\phi$

A. Results of Stability Analysis

ϕ Degree	α Degree	K_{ah}	F_{OVT}	Pressure at		$\mu = 0.50$		$\mu = 0.70$		$\mu = 0.80$	
				Heel, kN/m²	Toe, kN/m²	F_{SLD}	Net Force, kN	F_{SLD}	Net Force, kN	F_{SLD}	Net Force, kN
25	0.00	0.424	1.92	8.90	99.74	0.80	68.14	1.12	36.78	1.28	21.11
25	−5.71	0.385	1.79	10.71	99.74	0.76	66.56	1.06	39.35	1.21	25.74
25	−11.31	0.346	1.66	13.14	99.77	0.71	64.60	0.99	41.41	1.14	29.81
25	−14.03	0.326	1.60	14.62	99.87	0.68	63.53	0.96	42.28	1.09	31.65
30	0.00	0.348	1.93	9.72	99.97	0.88	51.38	1.23	22.58	1.40	8.18
30	−5.71	0.309	1.79	12.06	99.97	0.82	50.15	1.15	25.68	1.32	13.44
30	−11.31	0.270	1.65	15.83	99.54	0.77	48.64	1.07	28.31	1.23	18.14
30	−14.03	0.251	1.57	17.78	99.88	0.73	47.92	1.03	29.62	1.17	20.47
35	0.00	0.284	1.95	11.06	99.79	0.96	37.55	1.34	11.09	1.53	0.00
35	−5.71	0.246	1.80	14.12	99.76	0.90	36.97	1.25	15.01	1.43	4.03
35	−11.31	0.208	1.63	19.03	99.52	0.82	36.32	1.15	18.68	1.32	9.86
35	−14.03	0.190	1.54	22.41	99.52	0.78	36.01	1.09	20.44	1.25	12.66
40	0.00	0.232	1.98	12.40	99.72	1.04	26.75	1.46	2.41	1.67	0.00
40	−5.71	0.194	1.80	16.53	99.53	0.97	26.86	1.36	7.19	1.55	0.00
40	−11.31	0.158	1.60	23.21	99.50	0.87	27.17	1.22	12.00	1.40	4.41
40	−14.03	0.142	1.51	28.95	98.96	0.82	27.33	1.14	14.27	1.31	7.75

B. Physical Dimensions for a Technoeconomic Wall Section

ϕ Degree	α Degree	Top Width, mm	Bottom Width, mm	Horizontal Angle θ Degree	Depth of Shear Key D_{sk}, mm			Volume of Wall, m³
					$\mu = 0.50$	$\mu = 0.70$	$\mu = 0.80$	
25	0.00	300	3210	90.00	1200	800	500	8.775
25	−5.71	300	2740	84.29	1300	900	700	7.600
25	−11.31	300	2285	78.69	1200	1000	800	6.463
25	−14.03	300	2065	75.97	1200	1000	800	5.913
30	0.00	300	2920	90.00	800	400	200	8.050
30	−5.71	300	2430	84.29	900	600	400	6.825
30	−11.31	300	1960	78.69	800	600	500	5.650
30	−14.03	300	1730	75.97	800	600	500	5.075
35	0.00	300	2655	90.00	500	100	0	7.388
35	−5.71	300	2145	84.29	400	200	0	6.113
35	−11.31	300	1655	78.69	500	300	200	4.888
35	−14.03	300	1420	75.97	500	400	300	4.300

(Continued)

TABLE 5.52 (Continued)

H = 5.0 m, $\gamma_{masonry}$ = 22.56 kN/m³, γ_{sat} = 17.27 kN/m³ and ι = ½ϕ

ϕ Degree	α Degree	Top Width, mm	Bottom Width, mm	Horizontal Angle θ Degree	Depth of Shear Key D_{sk}, mm			Volume of Wall, m³
					μ = 0.50	μ = 0.70	μ = 0.80	
40	0.00	300	2415	90.00	200	0	0	6.788
40	−5.71	300	1885	84.29	200	0	0	5.463
40	−11.31	300	1375	78.69	300	200	100	4.188
40	−14.03	300	1135	75.97	400	300	200	3.588

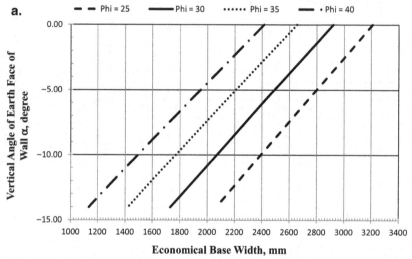

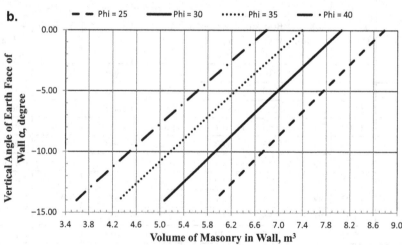

FIGURE 5.53 Design Charts for a Masonry Wall H = 5.0 m, $\gamma_{masonry}$ = 22.56 kN/m³, γ_{sat} = 17.27 kN/m³ and ι = ½ϕ: (a) Economical Base Width for a Wall, (b) Volume of a Masonry in Wall

TABLE 5.53

H = 5.0 m, $\gamma_{masonry}$ = 22.56 kN/m³, γ_{sat} = 15.70 kN/m³ and $\iota = \phi$

A. Results of Stability Analysis

ϕ Degree	α Degree	K_{ah}	F_{OVT}	Pressure at Heel, kN/m²	Toe, kN/m²	$\mu = 0.50$ F_{SLD}	Net Force, kN	$\mu = 0.70$ F_{SLD}	Net Force, kN	$\mu = 0.80$ F_{SLD}	Net Force, kN
25	0.00	0.821	1.87	6.20	99.93	0.62	143.86	0.86	103.64	0.99	83.53
25	−5.71	0.746	1.78	7.38	99.84	0.60	134.22	0.84	98.46	0.96	80.57
25	−11.31	0.675	1.69	8.70	99.91	0.58	125.24	0.81	93.80	0.93	78.08
25	−14.03	0.641	1.64	9.63	99.81	0.57	120.93	0.79	91.56	0.91	76.87
30	0.00	0.750	1.88	6.70	99.79	0.65	128.00	0.90	89.37	1.03	70.06
30	−5.71	0.666	1.79	7.93	99.87	0.63	116.67	0.88	82.85	1.01	65.94
30	−11.31	0.587	1.69	9.66	99.86	0.61	106.29	0.85	77.09	0.98	62.50
30	−14.03	0.549	1.63	10.62	99.97	0.60	101.47	0.84	74.53	0.96	61.06
35	0.00	0.671	1.89	7.05	99.91	0.68	110.82	0.95	74.08	1.09	55.71
35	−5.71	0.580	1.79	8.85	99.68	0.67	98.32	0.94	66.66	1.07	50.83
35	−11.31	0.496	1.68	11.06	99.65	0.65	87.32	0.91	60.60	1.04	47.23
35	−14.03	0.456	1.62	12.55	99.56	0.64	82.31	0.89	57.97	1.02	45.80
40	0.00	0.587	1.90	7.66	99.90	0.72	92.74	1.01	58.12	1.16	40.81
40	−5.71	0.492	1.79	9.75	99.74	0.72	80.00	1.00	50.76	1.15	36.14
40	−11.31	0.406	1.67	12.39	99.96	0.70	69.22	0.97	45.23	1.11	33.24
40	−14.03	0.366	1.60	14.43	99.86	0.68	64.44	0.95	42.97	1.09	32.23

B. Physical Dimensions for a Technoeconomic Wall Section

ϕ Degree	α Degree	Top Width, mm	Bottom Width, mm	Horizontal Angle θ Degree	Depth of Shear Key D_{sk}, mm $\mu = 0.50$	$\mu = 0.70$	$\mu = 0.80$	Volume of Wall, m³
25	0.00	300	4215	90.00	1800	1500	1300	11.288
25	−5.71	300	3710	84.29	1800	1500	1400	10.025
25	−11.31	300	3220	78.69	1800	1500	1300	8.800
25	−14.03	300	2985	75.97	1800	1600	1400	8.213
30	0.00	300	4035	90.00	1300	1100	900	10.838
30	−5.71	300	3490	84.29	1300	1000	900	9.475
30	−11.31	300	2965	78.69	1300	1100	900	8.163
30	−14.03	300	2710	75.97	1300	1000	900	7.525
35	0.00	300	3820	90.00	800	600	400	10.300
35	−5.71	300	3245	84.29	800	600	500	8.863
35	−11.31	300	2685	78.69	900	700	600	7.463
35	−14.03	300	2415	75.97	800	700	600	6.788

(Continued)

TABLE 5.53 (Continued)

H = 5.0 m, $\gamma_{masonry}$ = 22.56 kN/m³, γ_{sat} = 15.70 kN/m³ and $\iota = \phi$

ϕ Degree	α Degree	Top Width, mm	Bottom Width, mm	Horizontal Angle θ Degree	Depth of Shear Key D_{sk}, mm			Volume of Wall, m³
					$\mu = 0.50$	$\mu = 0.70$	$\mu = 0.80$	
40	0.00	300	3580	90.00	500	300	200	9.700
40	−5.71	300	2970	84.29	500	400	300	8.175
40	−11.31	300	2375	78.69	600	400	300	6.688
40	−14.03	300	2090	75.97	600	400	300	5.975

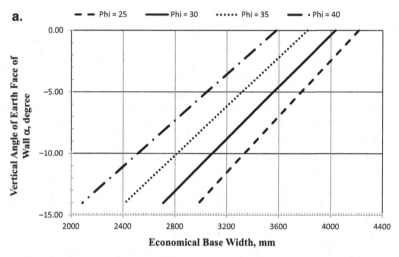

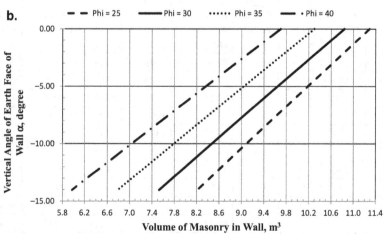

FIGURE 5.54 Design Charts for a Masonry Wall H = 5.0 m, $\gamma_{masonry}$ = 22.56 kN/m³, γ_{sat} = 15.70 kN/m³ and $\iota = \phi$: (a) Economical Base Width for a Wall, (b) Volume of a Masonry in Wall

TABLE 5.54
$H = 5.0m$, $\gamma_{masonry} = 22.56$ kN/m^3, $\gamma_{sat} = 17.27$ kN/m^3 and $\iota = \phi$

A. Results of Stability Analysis

ϕ Degree	α Degree	K_{ah}	F_{OVT}	Pressure at Heel, kN/m^2	Toe, kN/m^2	$\mu = 0.50$ F_{SLD}	Net Force, kN	$\mu = 0.70$ F_{SLD}	Net Force, kN	$\mu = 0.80$ F_{SLD}	Net Force, kN
25	0.00	0.821	1.87	5.84	99.91	0.59	163.48	0.82	121.46	0.94	100.44
25	−5.71	0.746	1.78	6.84	99.93	0.57	151.87	0.80	114.38	0.92	95.64
25	−11.31	0.675	1.70	8.29	99.73	0.56	140.90	0.78	107.78	0.89	91.22
25	−14.03	0.641	1.65	9.15	99.62	0.55	135.69	0.76	104.69	0.87	89.19
30	0.00	0.750	1.87	6.29	99.81	0.61	145.75	0.86	105.40	0.98	85.22
30	−5.71	0.666	1.78	7.42	99.89	0.60	132.17	0.84	96.71	0.96	78.98
30	−11.31	0.587	1.69	9.05	99.83	0.59	119.67	0.82	88.93	0.94	73.57
30	−14.03	0.549	1.64	10.00	99.86	0.58	113.86	0.81	85.42	0.92	71.20
35	0.00	0.671	1.88	6.66	99.89	0.65	126.46	0.91	88.09	1.04	68.91
35	−5.71	0.580	1.79	8.13	99.88	0.64	111.63	0.89	78.46	1.02	61.88
35	−11.31	0.496	1.68	10.29	99.70	0.63	98.37	0.88	70.24	1.00	56.17
35	−14.03	0.456	1.63	11.33	99.96	0.61	92.42	0.86	66.76	0.98	53.92
40	0.00	0.587	1.89	7.36	99.75	0.69	106.12	0.97	69.96	1.10	51.88
40	−5.71	0.492	1.79	9.22	99.69	0.69	90.94	0.96	60.29	1.10	44.97
40	−11.31	0.406	1.67	11.57	99.96	0.67	77.96	0.94	52.69	1.07	40.06
40	−14.03	0.366	1.61	13.69	99.58	0.66	72.12	0.93	49.41	1.06	38.06

B. Physical Dimensions for a Technoeconomic Wall Section

ϕ Degree	α Degree	Top Width, mm	Bottom Width, mm	Horizontal Angle θ Degree	Depth of Shear Key D_{sk}, mm $\mu = 0.50$	$\mu = 0.70$	$\mu = 0.80$	Volume of Wall, m^3
25	0.00	300	4420	90.00	1900	1500	1400	11.800
25	−5.71	300	3905	84.29	1900	1600	1400	10.513
25	−11.31	300	3410	78.69	1800	1500	1400	9.275
25	−14.03	300	3170	75.97	1800	1500	1400	8.675
30	0.00	300	4230	90.00	1300	1000	900	11.325
30	−5.71	300	3675	84.29	1300	1000	900	9.938
30	−11.31	300	3140	78.69	1300	1100	1000	8.600
30	−14.03	300	2880	75.97	1300	1100	900	7.950
35	0.00	300	4005	90.00	800	600	500	10.763
35	−5.71	300	3415	84.29	900	700	600	9.288
35	−11.31	300	2845	78.69	800	600	500	7.863
35	−14.03	300	2565	75.97	900	700	600	7.163

(Continued)

TABLE 5.54 (Continued)

H = 5.0m, $\gamma_{masonry}$ = 22.56 kN/m³, γ_{sat} = 17.27 kN/m³ and $\iota = \phi$

ϕ Degree	α Degree	Top Width, mm	Bottom Width, mm	Horizontal Angle θ Degree	Depth of Shear Key D_{sk}, mm			Volume of Wall, m³
					$\mu = 0.50$	$\mu = 0.70$	$\mu = 0.80$	
40	0.00	300	3755	90.00	400	200	100	10.138
40	−5.71	300	3130	84.29	500	300	200	8.575
40	−11.31	300	2520	78.69	500	300	300	7.050
40	−14.03	300	2230	75.97	600	400	300	6.325

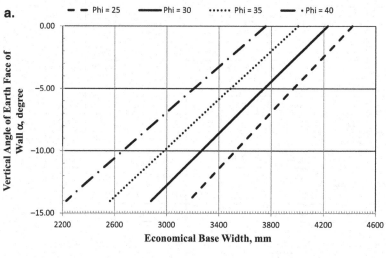

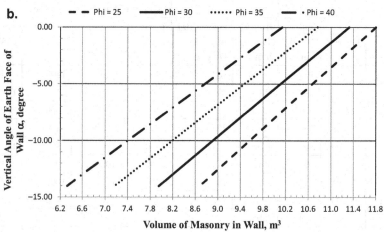

FIGURE 5.55 Design Charts for a Masonry Wall H = 5.0 m, $\gamma_{masonry}$ = 22.56 kN/m³, γ_{sat} = 17.27 kN/m³ and $\iota = \phi$: (a) Economical Base Width for a Wall, (b) Volume of a Masonry in Wall

6 Charts for Concrete Breast Walls

6.1 PARAMETERS FOR THE DEVELOPMENT OF CHARTS

The typical section adopted for design of concrete breast wall is shown in Figure 6.1. The following assumptions are made for carrying out the analysis of breast walls:

Top width of the wall	:	200 mm for the height of a wall <3.0 m
		300 mm for the height of a wall ≥3.0 m
Angle of friction between the concrete wall and the backfill, δ	:	⅓φ or 22.50°, whichever is less
Effect of water for uplift pressure	:	30% of the total height
Unit weight of water, γ_{water}	:	1.00 t/m³ or 9.81 kN/m³
Maximum allowable soil-bearing pressure	:	100 kN/m²
Width of the shear key at the bottom	:	150 mm
Location of the shear key	:	At the centre of the base

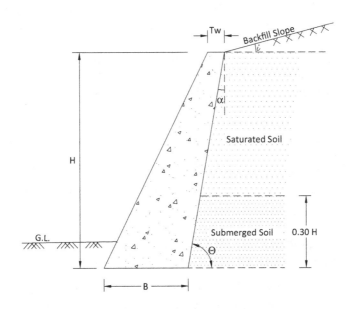

FIGURE 6.1 Typical Section of a Concrete Breast Wall

DOI: 10.1201/9781003162995-6

The following parameters have been varied for developing the design charts for breast walls:

Height of the breast wall, H	:	1.0 m
Angle of the surcharge of the backfill, ι	:	0°, ½φ, φ
Unit weight of the concrete, $\gamma_{concrete}$	:	2.50 t/m³ or 24.52 kN/m³
Unit weight of saturated earthfill, γ_{sat}	:	1.60 t/m³ or 15.70 kN/m³
		1.76 t/m³ or 17.27 kN/m³

For a 1.0-m- high concrete breast wall, astability analysis has been carried out for the previously discussed 6 combinations for 16 combinations of the following parameters:

Angle of repose of the backfill, φ	:	25°, 30°, 35°, 40°
Angle which the earth face of the wall makes with the vertical, α / Slope 1.0(H):x(V)	:	0° vertical face
		−5.71°1.0(H):10.0(V)
		−11.31°1.0(H):5.0(V)
		−14.03°1.0(H):4.0(V)
Coefficient of the friction between the concrete andthe soil/rock mass, μ	:	0.50, 0.70, 0.80

The base width of the concrete breast wall has been fixed to satisfy the following conditions:

1. The factor of safety against overturning is 1.50 or more.
2. There is no tension at the heel.
3. The maximum allowable soil-bearing pressure is 100 kN/m². In other words, pressure at the toe does not exceed 100 kN/m².

For all the preceding combinations, following wall heights have been considered for developing the charts/tables:

Height of the breast wall, H	:	1.0 m, 1.5 m, 2.0 m, 2.5 m, 3.0 m,
		3.5 m, 4.0 m, 4.5 m, 5.0 m

6.2 CHARTS FOR THE DESIGN OF A 1.0-M-HIGH CONCRETE BREAST WALL

The charts/tables developed for a 1.0-m-high concrete breast wall for the previously discussed six combinations are presented in the Tables 6.1 to 6.6 and Figures 6.2 to 6.7.

TABLE 6.1

H = 1.0 m, $\gamma_{concrete}$ = 24.52 kN/m³, γ_{sat} = 15.70 kN/m³ and ι = 0°

A. Results for Stability Analysis

				Pressure at		μ = 0.50		μ = 0.70		μ = 0.80	
ϕ Degree	α Degree	K_{ah}	F_{OVT}	Heel, kN/m²	Toe, kN/m²	F_{SLD}	Net Force, kN	F_{SLD}	Net Force, kN	F_{SLD}	Net Force, kN
25	0.00	0.373	1.96	0.22	31.37	1.09	1.32	1.52	0.00	1.74	0.00
25	−5.71	0.339	1.75	0.31	34.02	1.01	1.46	1.41	0.27	1.61	0.00
25	−11.31	0.305	1.52	0.08	39.02	0.91	1.59	1.28	0.60	1.46	0.11
25	−14.03	0.288	1.52	5.66	35.94	0.89	1.57	1.25	0.66	1.42	0.20
30	0.00	0.304	2.00	0.44	32.16	1.21	0.77	1.70	0.00	1.94	0.00
30	−5.71	0.270	1.75	0.55	35.67	1.12	0.93	1.56	0.00	1.79	0.00
30	−11.31	0.237	1.52	2.17	40.74	1.01	1.07	1.41	0.19	1.62	0.00
30	−14.03	0.221	1.57	13.77	32.34	1.00	1.04	1.40	0.21	1.60	0.00
35	0.00	0.246	2.02	0.17	33.68	1.35	0.34	1.89	0.00	2.16	0.00
35	−5.71	0.213	1.76	1.04	37.46	1.24	0.52	1.74	0.00	1.99	0.00
35	−11.31	0.182	1.65	11.97	34.14	1.16	0.60	1.63	0.00	1.86	0.00
35	−14.03	0.166	2.00	36.75	9.36	1.24	0.44	1.74	0.00	1.98	0.00
40	0.00	0.196	2.05	0.03	35.18	1.50	0.00	2.11	0.00	2.41	0.00
40	−5.71	0.166	1.74	0.62	40.98	1.37	0.21	1.92	0.00	2.20	0.00
40	−11.31	0.136	2.10	31.05	15.06	1.43	0.10	2.00	0.00	2.29	0.00
40	−14.03	0.122	2.00	43.18	0.04	1.23	0.37	1.72	0.00	1.96	0.00

B. Physical Dimensions for a Technoeconomic Wall Section

					Depth of Shear Key D_{sk}, mm			
ϕ Degree	α Degree	Top Width, mm	Bottom Width, mm	Horizontal Angle θ Degree	μ = 0.50	μ = 0.70	μ = 0.80	Volume of Wall, m³
25	0.00	200	490	90.00	300	0	0	0.345
25	−5.71	200	385	84.29	300	200	0	0.293
25	−11.31	200	280	78.69	300	200	100	0.240
25	−14.03	200	245	75.97	300	200	100	0.223
30	0.00	200	445	90.00	200	0	0	0.323
30	−5.71	200	335	84.29	200	0	0	0.268
30	−11.31	200	230	78.69	200	100	0	0.215
30	−14.03	200	200	75.97	200	100	0	0.200
35	0.00	200	400	90.00	100	0	0	0.300
35	−5.71	200	290	84.29	200	0	0	0.245
35	−11.31	200	200	78.69	200	0	0	0.200

(Continued)

TABLE 6.1 (Continued)

H = 1.0 m, $\gamma_{concrete}$ = 24.52 kN/m³, γ_{sat} = 15.70 kN/m³ and ι = 0°

ϕ Degree	α Degree	Top Width, mm	Bottom Width, mm	Horizontal Angle θ Degree	μ = 0.50	Depth of Shear Key D_{sk}, mm μ = 0.70	μ = 0.80	Volume of Wall, m³
35	−14.03	200	200	75.97	100	0	0	0.200
40	0.00	200	360	90.00	0	0	0	0.280
40	−5.71	200	245	84.29	100	0	0	0.223
40	−11.31	200	200	78.69	100	0	0	0.200
40	−14.03	150	170	75.97	100	0	0	0.160

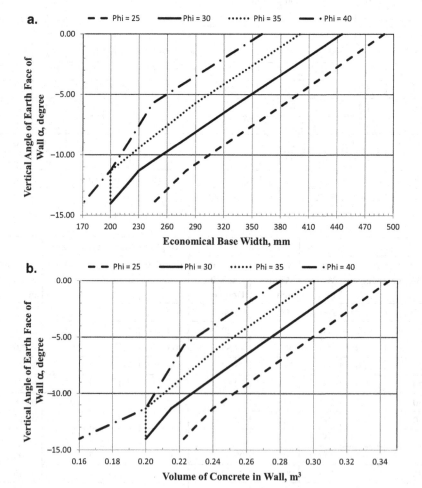

FIGURE 6.2 Design Charts for a Concrete Wall H = 1.0 m, $\gamma_{concrete}$ = 24.52 kN/m³, γ_{sat} = 15.70 kN/m³ and ι = 0°: (a) Economical Base Width for a Wall, (b) Volume of a Concrete Wall

TABLE 6.2

$H = 1.0$ m, $\gamma_{concrete} = 24.52$ kN/m^3, $\gamma_{sat} = 17.27$ kN/m^3 and $\iota = 0°$

A. Results for Stability Analysis

ϕ Degree	α Degree	K_{ah}	F_{OVT}	Pressure at Heel, kN/m^2	Toe, kN/m^2	$\mu = 0.50$ F_{SLD}	Net Force kN	$\mu = 0.70$ F_{SLD}	Net Force, kN	$\mu = 0.80$ F_{SLD}	Net Force, kN
25	0.00	0.373	1.95	0.29	30.82	1.03	1.63	1.44	0.19	1.65	0.00
25	−5.71	0.339	1.76	0.44	33.10	0.96	1.73	1.35	0.50	1.54	0.00
25	−11.31	0.305	1.55	0.45	37.21	0.88	1.83	1.23	0.79	1.41	0.28
25	−14.03	0.288	1.51	4.05	36.04	0.85	1.81	1.19	0.86	1.36	0.38
30	0.00	0.304	1.98	0.26	31.87	1.15	1.03	1.61	0.00	1.84	0.00
30	−5.71	0.270	1.75	0.33	35.07	1.06	1.15	1.49	0.02	1.70	0.00
30	−11.31	0.237	1.50	0.13	41.47	0.96	1.28	1.35	0.36	1.54	0.00
30	−14.03	0.221	1.51	8.38	36.56	0.94	1.25	1.32	0.40	1.51	0.00
35	0.00	0.246	2.01	0.32	32.94	1.28	0.54	1.79	0.00	2.05	0.00
35	−5.71	0.213	1.74	0.25	37.41	1.18	0.70	1.65	0.00	1.89	0.00
35	−11.31	0.182	1.52	4.15	41.95	1.08	0.82	1.51	0.00	1.72	0.00
35	−14.03	0.166	1.84	29.59	16.52	1.15	0.63	1.61	0.00	1.84	0.00
40	0.00	0.196	2.06	0.57	33.92	1.44	0.13	2.01	0.00	2.30	0.00
40	−5.71	0.166	1.74	0.39	40.06	1.32	0.33	1.84	0.00	2.10	0.00
40	−11.31	0.136	1.94	25.18	20.92	1.33	0.26	1.86	0.00	2.13	0.00
40	−14.03	0.122	2.14	39.28	0.70	1.25	0.37	1.74	0.00	1.99	0.00

B. Physical Dimensions for a Technoeconomic Wall Section

ϕ Degree	α Degree	Top Width, mm	Bottom Width, mm	Horizontal Angle θ Degree	Depth of Shear Key D_{sk}, mm $\mu = 0.50$	$\mu = 0.70$	$\mu = 0.80$	Volume of Wall, m^3
25	0.00	200	515	90.00	300	100	0	0.358
25	−5.71	200	410	84.29	300	200	0	0.305
25	−11.31	200	305	78.69	300	200	200	0.253
25	−14.03	200	265	75.97	300	200	200	0.233
30	0.00	200	465	90.00	200	0	0	0.333
30	−5.71	200	355	84.29	200	100	0	0.278
30	−11.31	200	245	78.69	200	200	0	0.223
30	−14.03	200	210	75.97	200	200	0	0.205
35	0.00	200	420	90.00	200	0	0	0.310
35	−5.71	200	305	84.29	200	0	0	0.253
35	−11.31	200	200	78.69	200	0	0	0.200

(Continued)

TABLE 6.2 (Continued)

H = 1.0 m, $\gamma_{concrete}$ = 24.52 kN/m³, γ_{sat} = 17.27 kN/m³ and ι = 0°

ϕ Degree	α Degree	Top Width, mm	Bottom Width, mm	Horizontal Angle θ Degree	Depth of Shear Key D_{sk}, mm $\mu = 0.50$	$\mu = 0.70$	$\mu = 0.80$	Volume of Wall, m³
35	−14.03	200	200	75.97	200	0	0	0.200
40	0.00	200	380	90.00	100	0	0	0.290
40	−5.71	200	260	84.29	100	0	0	0.230
40	−11.31	200	200	78.69	100	0	0	0.200
40	−14.03	150	200	75.97	100	0	0	0.175

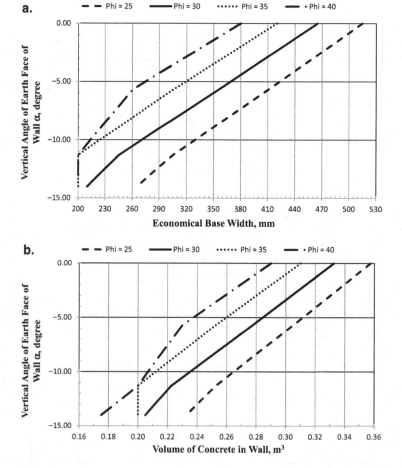

FIGURE 6.3 Design Charts for Concrete Wall H = 1.0 m, $\gamma_{concrete}$ = 24.52 kN/m³, γ_{sat} = 17.27 kN/m³ and ι = 0°: (a) Economical Base Width for a Wall, (b) Volume of a Concrete Wall

TABLE 6.3

$H = 1.0$ m, $\gamma_{concrete} = 24.52$ kN/m^3, $\gamma_{sat} = 15.70$ kN/m^3 and $\iota = \frac{1}{2}\phi$

A. Results for Stability Analysis

				Pressure at		$\mu = 0.50$		$\mu = 0.70$		$\mu = 0.80$	
ϕ Degree	α Degree	K_{ah}	F_{OVT}	Heel, kN/ m^2	Toe, kN/ m^2	F_{SLD}	Net Force, kN	F_{SLD}	Net Force, kN	F_{SLD}	Net Force, kN
25	0.00	0.450	1.95	0.36	30.31	0.99	1.93	1.38	0.44	1.58	0.00
25	−5.71	0.407	1.74	0.12	32.87	0.92	1.99	1.29	0.72	1.48	0.08
25	−11.31	0.364	1.56	0.34	36.34	0.85	2.03	1.20	0.96	1.37	0.42
25	−14.03	0.344	1.50	2.43	36.66	0.82	2.01	1.15	1.03	1.32	0.54
30	0.00	0.373	1.96	0.21	31.38	1.09	1.32	1.52	0.00	1.74	0.00
30	−5.71	0.329	1.76	0.68	33.81	1.02	1.37	1.43	0.19	1.64	0.00
30	−11.31	0.286	1.51	0.01	40.08	0.93	1.45	1.31	0.50	1.49	0.02
30	−14.03	0.266	1.53	7.58	35.33	0.92	1.39	1.29	0.51	1.47	0.06
35	0.00	0.305	1.99	0.36	32.24	1.21	0.78	1.69	0.00	1.94	0.00
35	−5.71	0.262	1.77	0.93	35.51	1.14	0.86	1.59	0.00	1.82	0.00
35	−11.31	0.220	1.51	2.26	42.13	1.04	0.96	1.45	0.10	1.66	0.00
35	−14.03	0.201	1.70	22.16	23.95	1.08	0.82	1.51	0.00	1.72	0.00
40	0.00	0.245	2.02	0.24	33.60	1.35	0.34	1.89	0.00	2.16	0.00
40	−5.71	0.204	1.75	0.65	38.45	1.26	0.46	1.77	0.00	2.02	0.00
40	−11.31	0.165	1.79	18.82	27.28	1.25	0.42	1.74	0.00	1.99	0.00
40	−14.03	0.147	2.22	44.71	1.40	1.35	0.23	1.89	0.00	2.16	0.00

B. Physical Dimensions for a Technoeconomic Wall Section

					Depth of Shear Key D_{sk}, mm			
ϕ Degree	α Degree	Top Width, mm	Bottom Width, mm	Horizontal Angle θDegree	$\mu = 0.50$	$\mu = 0.70$	$\mu = 0.80$	Volume of Wall, m^3
25	0.00	200	540	90.00	300	200	0	0.370
25	−5.71	200	430	84.29	300	200	100	0.315
25	−11.31	200	325	78.69	300	200	200	0.263
25	−14.03	200	280	75.97	300	300	200	0.240
30	0.00	200	490	90.00	300	0	0	0.345
30	−5.71	200	380	84.29	300	100	0	0.290
30	−11.31	200	265	78.69	300	200	100	0.233
30	−14.03	200	230	75.97	300	200	100	0.215
35	0.00	200	445	90.00	200	0	0	0.323
35	−5.71	200	330	84.29	200	0	0	0.265
35	−11.31	200	215	78.69	200	100	0	0.208

(Continued)

TABLE 6.3 (Continued)

$H = 1.0$ m, $\gamma_{concrete} = 24.52$ kN/m³, $\gamma_{sat} = 15.70$ kN/m³ and $\iota = \frac{1}{2}\phi$

ϕ Degree	α Degree	Top Width, mm	Bottom Width, mm	Horizontal Angle θDegree	Depth of Shear Key D_{sk}, mm			Volume of Wall, m³
					$\mu = 0.50$	$\mu = 0.70$	$\mu = 0.80$	
35	−14.03	200	200	75.97	200	0	0	0.200
40	0.00	200	400	90.00	100	0	0	0.300
40	−5.71	200	280	84.29	100	0	0	0.240
40	−11.31	200	200	78.69	100	0	0	0.200
40	−14.03	200	200	75.97	100	0	0	0.200

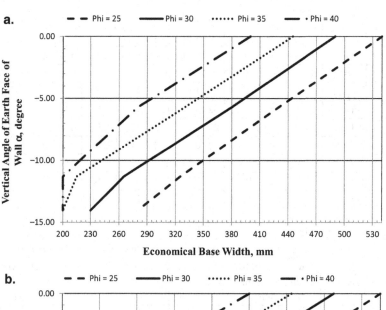

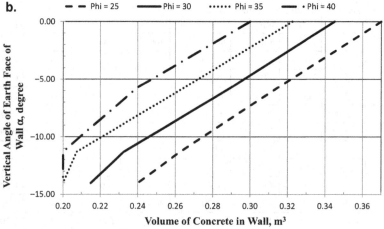

FIGURE 6.4 Design Charts for a Concrete Wall $H = 1.0$ m, $\gamma_{concrete} = 24.52$ kN/m³, $\gamma_{sat} = 15.70$ kN/m³ and $\iota = \frac{1}{2}\phi$: (a) Economical Base Width for a Wall, (b) Volume of a Concrete Wall

TABLE 6.4

$H = 1.0$ m, $\gamma_{concrete} = 24.52$ kN/m^3, $\gamma_{sat} = 17.27$ kN/m^3 and $\iota = \frac{1}{2}\phi$

A. Results for Stability Analysis

ϕ Degree	α Degree	K_{ah}	F_{OVT}	Pressure at Heel, kN/m²	Pressure at Toe, kN/m²	$\mu = 0.50$ F_{SLD}	$\mu = 0.50$ Net Force, kN	$\mu = 0.70$ F_{SLD}	$\mu = 0.70$ Net Force, kN	$\mu = 0.80$ F_{SLD}	$\mu = 0.80$ Net Force, kN
25	0.00	0.450	1.92	0.15	30.11	0.93	2.34	1.31	0.80	1.49	0.03
25	−5.71	0.407	1.76	0.56	31.68	0.88	2.32	1.24	0.99	1.41	0.32
25	−11.31	0.364	1.56	0.15	35.45	0.82	2.34	1.15	1.21	1.31	0.65
25	−14.03	0.344	1.51	2.06	35.61	0.79	2.30	1.11	1.26	1.27	0.75
30	0.00	0.373	1.95	0.28	30.83	1.03	1.64	1.44	0.20	1.65	0.00
30	−5.71	0.329	1.74	0.10	33.75	0.97	1.66	1.36	0.44	1.55	0.00
30	−11.31	0.286	1.54	0.62	37.87	0.90	1.67	1.26	0.66	1.44	0.16
30	−14.03	0.266	1.50	4.38	37.22	0.88	1.63	1.23	0.72	1.40	0.26
35	0.00	0.305	1.97	0.18	31.95	1.14	1.04	1.60	0.00	1.83	0.00
35	−5.71	0.262	1.76	0.76	34.84	1.09	1.07	1.52	0.00	1.74	0.00
35	−11.31	0.220	1.52	2.10	40.36	1.00	1.12	1.40	0.23	1.60	0.00
35	−14.03	0.201	1.56	13.51	32.59	0.99	1.06	1.39	0.23	1.59	0.00
40	0.00	0.245	2.01	0.40	32.86	1.28	0.53	1.80	0.00	2.06	0.00
40	−5.71	0.204	1.77	1.09	36.84	1.21	0.60	1.70	0.00	1.94	0.00
40	−11.31	0.165	1.65	11.71	34.40	1.16	0.62	1.62	0.00	1.85	0.00
40	−14.03	0.147	2.04	38.37	7.74	1.26	0.40	1.76	0.00	2.01	0.00

B. Physical Dimensions for a Technoeconomic Wall Section

ϕ Degree	α Degree	Top Width, mm	Bottom Width, mm	Horizontal Angle θ Degree	Depth of Shear Key D_{sk}, mm $\mu = 0.50$	Depth of Shear Key D_{sk}, mm $\mu = 0.70$	Depth of Shear Key D_{sk}, mm $\mu = 0.80$	Volume of Wall, m³
25	0.00	200	565	90.00	300	200	100	0.383
25	−5.71	200	460	84.29	300	200	200	0.330
25	−11.31	200	350	78.69	300	300	200	0.275
25	−14.03	200	305	75.97	300	300	200	0.253
30	0.00	200	515	90.00	300	100	0	0.358
30	−5.71	200	400	84.29	300	200	0	0.300
30	−11.31	200	290	78.69	300	200	100	0.245
30	−14.03	200	245	75.97	300	200	100	0.223
35	0.00	200	465	90.00	200	0	0	0.333
35	−5.71	200	350	84.29	200	0	0	0.275
35	−11.31	200	235	78.69	200	100	0	0.218

(Continued)

TABLE 6.4 (Continued)

H = 1.0 m, $\gamma_{concrete}$ = 24.52 kN/m³, γ_{sat} = 17.27 kN/m³ and ι = ½ϕ

ϕ Degree	α Degree	Top Width, mm	Bottom Width, mm	Horizontal Angle θ Degree	Depth of Shear Key D_{sk}, mm			Volume of Wall, m³
					$\mu = 0.50$	$\mu = 0.70$	$\mu = 0.80$	
35	−14.03	200	200	75.97	200	100	0	0.200
40	0.00	200	420	90.00	100	0	0	0.310
40	−5.71	200	300	84.29	100	0	0	0.250
40	−11.31	200	200	78.69	100	0	0	0.200
40	−14.03	200	200	75.97	100	0	0	0.200

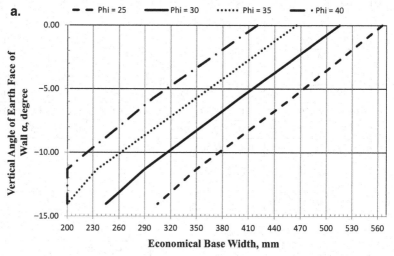

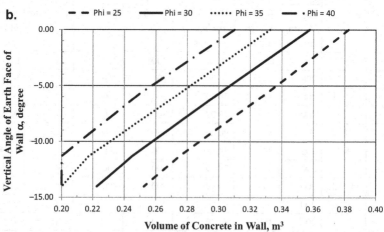

FIGURE 6.5 Design Charts for a Concrete Wall H = 1.0 m, $\gamma_{concrete}$ = 24.52 kN/m³, γ_{sat} = 17.27 kN/m³ and ι = ½ϕ: (a) Economical Base Width for a Wall, (b) Volume of a Concrete Wall

TABLE 6.5

H = 1.0 m, $\gamma_{concrete}$ = 24.52 kN/m³, γ_{sat} = 15.70 kN/m³ and $\iota = \phi$

A. Results for Stability Analysis

ϕ Degree	α Degree	K_{ah}	F_{OVT}	Pressure at		$\mu = 0.50$		$\mu = 0.70$		$\mu = 0.80$	
				Heel, kN/ m²	Toe, kN/ m²	F_{SLD}	Net Force, kN	F_{SLD}	Net Force, kN	F_{SLD}	Net Force, kN
25	0.00	0.821	1.89	0.30	27.96	0.72	5.11	1.00	3.24	1.15	2.31
25	−5.71	0.746	1.77	0.44	28.99	0.69	4.81	0.97	3.16	1.11	2.33
25	−11.31	0.675	1.63	0.39	30.71	0.66	4.55	0.93	3.11	1.06	2.39
25	−14.03	0.641	1.56	0.17	32.07	0.64	4.44	0.90	3.11	1.03	2.44
30	0.00	0.750	1.89	0.23	28.36	0.75	4.49	1.05	2.69	1.20	1.79
30	−5.71	0.666	1.75	0.10	29.94	0.73	4.13	1.02	2.57	1.17	1.78
30	−11.31	0.587	1.61	0.23	31.90	0.70	3.81	0.98	2.47	1.12	1.80
30	−14.03	0.549	1.55	0.58	32.96	0.69	3.66	0.96	2.43	1.10	1.81
35	0.00	0.671	1.90	0.17	28.84	0.80	3.80	1.12	2.08	1.27	1.22
35	−5.71	0.580	1.76	0.22	30.53	0.78	3.40	1.09	1.92	1.25	1.18
35	−11.31	0.496	1.59	0.00	33.54	0.75	3.07	1.05	1.84	1.20	1.22
35	−14.03	0.456	1.50	0.10	35.49	0.73	2.93	1.03	1.81	1.17	1.25
40	0.00	0.587	1.91	0.10	29.46	0.85	3.09	1.20	1.45	1.37	0.63
40	−5.71	0.492	1.76	0.33	31.37	0.85	2.67	1.18	1.29	1.35	0.60
40	−11.31	0.406	1.58	0.55	34.85	0.82	2.34	1.15	1.21	1.31	0.65
40	−14.03	0.366	1.50	1.86	36.35	0.80	2.19	1.13	1.18	1.29	0.67

B. Physical Dimensions for a Technoeconomic Wall Section

ϕ Degree	α Degree	Top Width, mm	Bottom Width, mm	Horizontal Angle θ Degree	Depth of Shear Key Dsk, mm			Volume of Wall, m³
					$\mu = 0.50$	$\mu = 0.70$	$\mu = 0.80$	
25	0.00	200	735	90.00	500	400	400	0.468
25	−5.71	200	625	84.29	500	400	400	0.413
25	−11.31	200	515	78.69	500	400	400	0.358
25	−14.03	200	460	75.97	500	400	400	0.330
30	0.00	200	700	90.00	400	300	300	0.450
30	−5.71	200	580	84.29	400	300	300	0.390
30	−11.31	200	465	78.69	400	300	300	0.333
30	−14.03	200	410	75.97	400	300	300	0.305
35	0.00	200	660	90.00	300	300	200	0.430
35	−5.71	200	535	84.29	300	300	200	0.368
35	−11.31	200	410	78.69	300	300	200	0.305

(Continued)

TABLE 6.5 (Continued)

H = 1.0 m, $\gamma_{concrete}$ = 24.52 kN/m³, γ_{sat} = 15.70 kN/m³ and $\iota = \phi$

ϕ Degree	α Degree	Top Width, mm	Bottom Width, mm	Horizontal Angle θ Degree	Depth of Shear Key Dsk, mm			Volume of Wall, m³
					μ = 0.50	μ = 0.70	μ = 0.80	
35	−14.03	200	350	75.97	300	300	200	0.275
40	0.00	200	615	90.00	200	100	100	0.408
40	−5.71	200	485	84.29	300	200	100	0.343
40	−11.31	200	355	78.69	200	200	200	0.278
40	−14.03	200	295	75.97	200	200	200	0.248

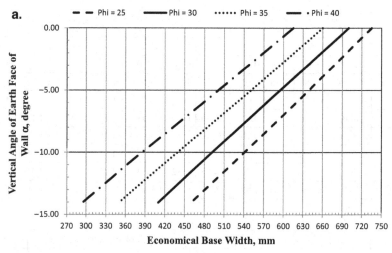

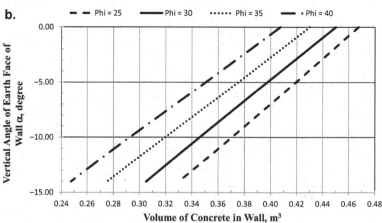

FIGURE 6.6 Design Charts for a Concrete Wall H = 1.0 m, $\gamma_{concrete}$ = 24.52 kN/m³, γ_{sat} = 15.70 kN/m³ and $\iota = \phi$: (a) Economical Base Width for a Wall, (b) Volume of a Concrete Wall

TABLE 6.6

$H = 1.0$ m, $\gamma_{concrete} = 24.52$ kN/m^3, $\gamma_{sat} = 17.27$ kN/m^3 and $\iota = \phi$

A. Results for Stability Analysis

ϕ Degree	α Degree	K_{ah}	F_{OVT}	Pressure at		$\mu = 0.50$		$\mu = 0.70$		$\mu = 0.80$	
				Heel, kN/m^2	Toe, kN/m^2	F_{SLD}	Net Force, kN	F_{SLD}	Net Force, kN	F_{SLD}	Net Force, kN
25	0.00	0.821	1.87	0.10	27.86	0.68	5.90	0.95	3.97	1.08	3.00
25	−5.71	0.746	1.76	0.26	28.75	0.66	5.52	0.92	3.80	1.05	2.94
25	−11.31	0.675	1.64	0.30	30.20	0.63	5.18	0.88	3.67	1.01	2.92
25	−14.03	0.641	1.57	0.17	31.32	0.62	5.02	0.86	3.62	0.99	2.92
30	0.00	0.750	1.88	0.14	28.11	0.71	5.20	0.99	3.33	1.14	2.40
30	−5.71	0.666	1.76	0.12	29.43	0.69	4.75	0.97	3.11	1.11	2.29
30	−11.31	0.587	1.63	0.45	30.94	0.67	4.33	0.94	2.92	1.08	2.22
30	−14.03	0.549	1.55	0.18	32.55	0.66	4.16	0.92	2.87	1.05	2.22
35	0.00	0.671	1.90	0.24	28.40	0.75	4.42	1.06	2.63	1.21	1.74
35	−5.71	0.580	1.77	0.48	29.71	0.75	3.91	1.04	2.37	1.19	1.59
35	−11.31	0.496	1.62	0.69	31.91	0.72	3.49	1.01	2.18	1.16	1.53
35	−14.03	0.456	1.52	0.21	34.28	0.71	3.32	0.99	2.14	1.13	1.55
40	0.00	0.587	1.91	0.37	28.76	0.81	3.61	1.14	1.90	1.30	1.05
40	−5.71	0.492	1.76	0.37	30.74	0.81	3.10	1.13	1.66	1.29	0.94
40	−11.31	0.406	1.58	0.09	34.40	0.78	2.70	1.10	1.52	1.25	0.93
40	−14.03	0.366	1.51	1.37	35.54	0.77	2.50	1.08	1.44	1.24	0.91

B. Physical Dimensions for a Technoeconomic Wall Section

ϕ Degree	α Degree	Top Width, mm	Bottom Width, mm	Horizontal Angle θ Degree	Depth of Shear Key D_{sk}, mm			Volume of Wall, m^3
					$\mu = 0.50$	$\mu = 0.70$	$\mu = 0.80$	
25	0.00	200	770	90.00	500	400	400	0.485
25	−5.71	200	660	84.29	500	400	400	0.430
25	−11.31	200	550	78.69	500	400	400	0.375
25	−14.03	200	495	75.97	500	400	400	0.348
30	0.00	200	735	90.00	400	400	300	0.468
30	−5.71	200	615	84.29	400	300	300	0.408
30	−11.31	200	500	78.69	400	300	300	0.350
30	−14.03	200	440	75.97	400	300	300	0.320
35	0.00	200	695	90.00	400	300	200	0.448
35	−5.71	200	570	84.29	300	300	200	0.385
35	−11.31	200	445	78.69	300	300	200	0.323

(Continued)

TABLE 6.6 (Continued)

H = 1.0 m, $\gamma_{concrete}$ = 24.52 kN/m³, γ_{sat} = 17.27 kN/m³ and $\iota = \phi$

ϕ Degree	α Degree	Top Width, mm	Bottom Width, mm	Horizontal Angle θ Degree	Depth of Shear Key D_{sk}, mm			Volume of Wall, m³
					$\mu = 0.50$	$\mu = 0.70$	$\mu = 0.80$	
35	−14.03	200	380	75.97	300	300	200	0.290
40	0.00	200	650	90.00	200	100	100	0.425
40	−5.71	200	515	84.29	300	200	200	0.358
40	−11.31	200	380	78.69	200	200	200	0.290
40	−14.03	200	320	75.97	200	200	200	0.260

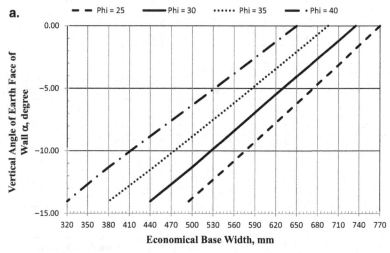

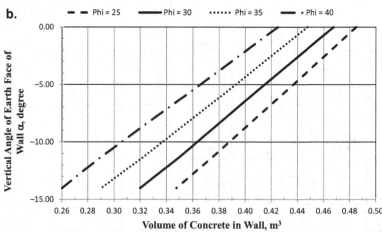

FIGURE 6.7 Design Charts for a Concrete Wall H = 1.0 m, $\gamma_{concrete}$ = 24.52 kN/m³, γ_{sat} = 17.27 kN/m³ and $\iota = \phi$: (a) Economical Base Width for a Wall, (b) Volume of a Concrete Wall

6.3 CHARTS FOR THE DESIGN OF A 1.5-M-HIGH CONCRETE BREAST WALL

The charts/tables developed for a 1.5-m-high concrete breast wall for the previously discussed six combinations are presented in the Tables 6.7 to 6.12 and Figures 6.8 to 6.13.

TABLE 6.7
H = 1.5 m, $\gamma_{concrete}$ = 24.52 kN/m³, γ_{sat} = 15.70 kN/m³ and ι = 0°

A. Results for Stability Analysis

				Pressure at		$\mu = 0.50$		$\mu = 0.70$		$\mu = 0.80$	
ϕ Degree	α Degree	K_{ah}	F_{OVT}	Heel, kN/m²	Toe, kN/m²	F_{SLD}	Net Force, kN	F_{SLD}	Net Force, kN	F_{SLD}	Net Force, kN
25	0.00	0.373	1.88	0.26	41.92	0.99	3.68	1.38	0.84	1.58	0.00
25	−5.71	0.339	1.70	0.47	44.07	0.91	3.90	1.28	1.48	1.46	0.27
25	−11.31	0.305	1.53	1.04	47.16	0.83	4.07	1.16	2.06	1.33	1.05
25	−14.03	0.288	1.52	6.29	43.82	0.81	4.02	1.13	2.15	1.29	1.21
30	0.00	0.304	1.89	0.13	43.14	1.09	2.51	1.52	0.00	1.74	0.00
30	−5.71	0.270	1.69	0.49	45.90	1.00	2.76	1.40	0.57	1.59	0.00
30	−11.31	0.237	1.51	2.91	48.57	0.90	2.95	1.26	1.17	1.44	0.28
30	−14.03	0.221	1.51	11.54	42.80	0.88	2.91	1.23	1.28	1.40	0.46
35	0.00	0.246	1.92	0.51	43.93	1.20	1.53	1.68	0.00	1.92	0.00
35	−5.71	0.213	1.67	0.26	48.46	1.09	1.88	1.52	0.00	1.74	0.00
35	−11.31	0.182	1.52	7.23	48.50	0.98	2.08	1.38	0.50	1.57	0.00
35	−14.03	0.166	1.50	20.22	40.46	0.94	2.10	1.32	0.68	1.51	0.00
40	0.00	0.196	1.93	0.26	45.62	1.32	0.77	1.85	0.00	2.11	0.00
40	−5.71	0.166	1.67	1.09	50.39	1.19	1.17	1.66	0.00	1.90	0.00
40	−11.31	0.136	1.52	14.54	47.26	1.06	1.42	1.49	0.03	1.70	0.00
40	−14.03	0.122	1.54	42.63	26.53	1.03	1.44	1.44	0.19	1.64	0.00

B. Physical Dimensions for a Technoeconomic Wall Section

					Depth of Shear Key D_{sk}, mm			
ϕ Degree	α Degree	Top Width, mm	Bottom Width, mm	Horizontal Angle θ Degree	$\mu = 0.50$	$\mu = 0.70$	$\mu = 0.80$	Volume of Wall, m³
25	0.00	200	750	90.00	400	200	0	0.713
25	−5.71	200	605	84.29	500	300	200	0.604
25	−11.31	200	465	78.69	500	300	300	0.499
25	−14.03	200	415	75.97	500	300	300	0.461
30	0.00	200	675	90.00	300	0	0	0.656
30	−5.71	200	525	84.29	300	200	0	0.544
30	−11.31	200	385	78.69	400	200	100	0.439

(Continued)

TABLE 6.7 (Continued)

H = 1.5 m, $\gamma_{concrete}$ = 24.52 kN/m³, γ_{sat} = 15.70 kN/m³ and ι = 0°

φ Degree	α Degree	Top Width, mm	Bottom Width, mm	Horizontal Angle θ Degree	Depth of Shear Key D_{sk}, mm			Volume of Wall, m³
					μ = 0.50	μ = 0.70	μ = 0.80	
30	−14.03	200	335	75.97	400	200	200	0.401
35	0.00	200	610	90.00	200	0	0	0.608
35	−5.71	200	450	84.29	300	0	0	0.488
35	−11.31	200	315	78.69	300	200	0	0.386
35	−14.03	200	260	75.97	300	200	0	0.345
40	0.00	200	545	90.00	200	0	0	0.559
40	−5.71	200	385	84.29	200	0	0	0.439
40	−11.31	200	250	78.69	200	100	0	0.338
40	−14.03	200	200	75.97	200	100	0	0.300

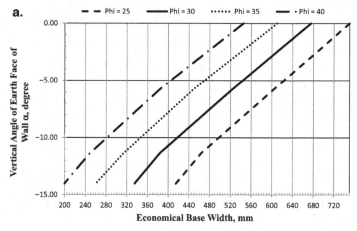

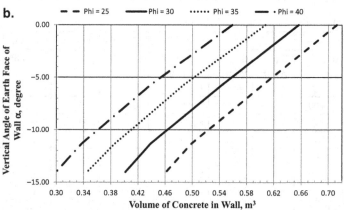

FIGURE 6.8 Design Charts for a Concrete Wall H = 1.5 m, $\gamma_{concrete}$ = 24.52 kN/m³, γ_{sat} = 15.70 kN/m³ and ι = 0°: (a) Economical Base Width for a Wall, (b) Volume of a Concrete Wall

TABLE 6.8

H = 1.5 m, $\gamma_{concrete}$ = 24.52 kN/m³, γ_{sat} = 17.27 kN/m³ and ι = 0°

A. Results for Stability Analysis

ϕ Degree	α Degree	K_{ah}	F_{OVT}	Pressure at Heel, kN/ m²	Toe, kN/ m²	$\mu = 0.50$ F_{SLD}	Net Force, kN	$\mu = 0.70$ F_{SLD}	Net Force, kN	$\mu = 0.80$ F_{SLD}	Net Force, kN
25	0.00	0.373	1.88	0.43	41.25	0.94	4.38	1.32	1.42	1.51	0.00
25	−5.71	0.339	1.70	0.15	43.72	0.87	4.54	1.22	2.02	1.40	0.76
25	−11.31	0.305	1.54	0.76	46.33	0.80	4.63	1.12	2.51	1.28	1.45
25	−14.03	0.288	1.51	4.43	44.47	0.78	4.56	1.09	2.61	1.24	1.63
30	0.00	0.304	1.89	0.25	42.49	1.04	3.05	1.45	0.33	1.66	0.00
30	−5.71	0.270	1.68	0.00	45.63	0.95	3.26	1.34	0.98	1.53	0.00
30	−11.31	0.237	1.50	0.94	49.38	0.87	3.40	1.21	1.54	1.38	0.62
30	−14.03	0.221	1.50	8.52	44.29	0.84	3.31	1.18	1.61	1.35	0.75
35	0.00	0.246	1.91	0.48	43.39	1.14	1.96	1.60	0.00	1.83	0.00
35	−5.71	0.213	1.68	0.45	47.25	1.05	2.23	1.47	0.17	1.67	0.00
35	−11.31	0.182	1.52	5.65	48.36	0.95	2.38	1.33	0.72	1.52	0.00
35	−14.03	0.166	1.51	16.87	41.32	0.92	2.36	1.29	0.86	1.47	0.12
40	0.00	0.196	1.92	0.07	45.21	1.26	1.11	1.76	0.00	2.01	0.00
40	−5.71	0.166	1.67	0.98	49.34	1.15	1.43	1.61	0.00	1.83	0.00
40	−11.31	0.136	1.51	11.27	48.35	1.03	1.64	1.45	0.19	1.65	0.00
40	−14.03	0.122	1.51	32.73	33.87	0.99	1.65	1.39	0.36	1.59	0.00

B. Physical Dimensions for a Technoeconomic Wall Section

ϕ Degree	α Degree	Top Width, mm	Bottom Width, mm	Horizontal Angle θ Degree	Depth of Shear Key D_{sk}, mm $\mu = 0.50$	$\mu = 0.70$	$\mu = 0.80$	Volume of Wall, m³
25	0.00	200	790	90.00	500	300	0	0.743
25	−5.71	200	640	84.29	500	300	200	0.630
25	−11.31	200	500	78.69	500	400	300	0.525
25	−14.03	200	445	75.97	500	400	300	0.484
30	0.00	200	710	90.00	300	100	0	0.683
30	−5.71	200	555	84.29	400	200	0	0.566
30	−11.31	200	410	78.69	400	300	200	0.458
30	−14.03	200	360	75.97	400	300	200	0.420
35	0.00	200	640	90.00	300	0	0	0.630
35	−5.71	200	480	84.29	300	100	0	0.510
35	−11.31	200	340	78.69	300	200	0	0.405

(Continued)

TABLE 6.8 (Continued)

H = 1.5 m, $\gamma_{concrete}$ = 24.52 kN/m³, γ_{sat} = 17.27 kN/m³ and ι = 0°

ϕ Degree	α Degree	Top Width, mm	Bottom Width, mm	Horizontal Angle θ Degree	Depth of Shear Key D_{sk}, mm			Volume of Wall, m³
					μ = 0.50	μ = 0.70	μ = 0.80	
35	−14.03	200	285	75.97	300	200	100	0.364
40	0.00	200	570	90.00	200	0	0	0.578
40	−5.71	200	410	84.29	200	0	0	0.458
40	−11.31	200	270	78.69	200	100	0	0.353
40	−14.03	200	215	75.97	200	100	0	0.311

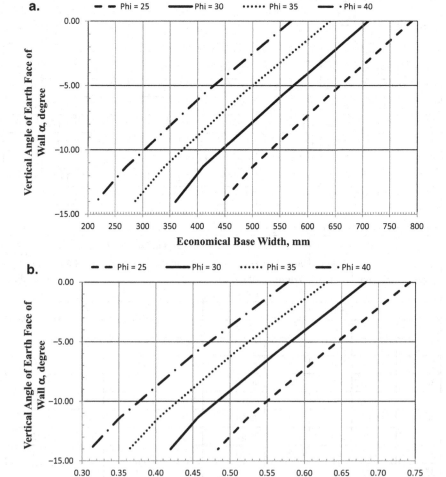

FIGURE 6.9 Design Charts for a Concrete Wall H = 1.5 m, $\gamma_{concrete}$ = 24.52 kN/m³, γ_{sat} = 17.27 kN/m³ and ι = 0°: (a) Economical Base Width for a Wall, (b) Volume of a Concrete Wall

TABLE 6.9

H = 1.5 m, $\gamma_{concrete}$ = 24.52 kN/m³, γ_{sat} = 15.70 kN/m³ and ι = ½ϕ

A. Results for Stability Analysis

				Pressure at		μ = 0.50		μ = 0.70		μ = 0.80	
ϕ Degree	α Degree	K_{ah}	F_{OVT}	Heel, kN/ m²	Toe, kN/ m²	F_{SLD}	Net Force, kN	F_{SLD}	Net Force, kN	F_{SLD}	Net Force, kN
25	0.00	0.450	1.87	0.17	41.12	0.90	5.07	1.26	2.00	1.44	0.47
25	−5.71	0.407	1.71	0.23	43.04	0.85	5.09	1.18	2.46	1.35	1.15
25	−11.31	0.364	1.54	0.50	45.76	0.78	5.08	1.09	2.88	1.25	1.77
25	−14.03	0.344	1.50	3.07	44.96	0.76	5.00	1.06	2.97	1.21	1.95
30	0.00	0.373	1.88	0.25	41.94	0.99	3.68	1.38	0.84	1.58	0.00
30	−5.71	0.329	1.71	0.66	44.08	0.92	3.72	1.29	1.33	1.48	0.13
30	−11.31	0.286	1.51	0.54	48.55	0.84	3.78	1.18	1.84	1.35	0.87
30	−14.03	0.266	1.50	6.45	45.03	0.82	3.67	1.15	1.89	1.32	1.00
35	0.00	0.305	1.89	0.02	43.25	1.08	2.53	1.52	0.00	1.73	0.00
35	−5.71	0.262	1.69	0.64	46.02	1.01	2.62	1.41	0.46	1.62	0.00
35	−11.31	0.220	1.51	3.95	48.58	0.92	2.68	1.29	0.96	1.48	0.10
35	−14.03	0.201	1.51	14.33	41.78	0.90	2.60	1.26	1.03	1.44	0.25
40	0.00	0.245	1.93	0.62	43.82	1.20	1.51	1.68	0.00	1.92	0.00
40	−5.71	0.204	1.68	0.92	48.18	1.11	1.72	1.55	0.00	1.77	0.00
40	−11.31	0.165	1.50	8.25	49.50	1.01	1.85	1.41	0.35	1.61	0.00
40	−14.03	0.147	1.51	27.38	36.30	0.98	1.81	1.37	0.46	1.56	0.00

B. Physical Dimensions for a Technoeconomic Wall Section

					Depth of Shear Key D_{sk}, mm			
ϕ Degree	α Degree	Top Width, mm	Bottom Width, mm	Horizontal Angle θ Degree	μ = 0.50	μ = 0.70	μ = 0.80	Volume of Wall, m³
25	0.00	200	825	90.00	500	300	200	0.769
25	−5.71	200	675	84.29	500	400	300	0.656
25	−11.31	200	530	78.69	500	400	300	0.548
25	−14.03	200	470	75.97	500	400	300	0.503
30	0.00	200	750	90.00	400	200	0	0.713
30	−5.71	200	595	84.29	400	300	100	0.596
30	−11.31	200	440	78.69	400	300	200	0.480
30	−14.03	200	385	75.97	400	300	200	0.439
35	0.00	200	675	90.00	300	0	0	0.656
35	−5.71	200	515	84.29	300	200	0	0.536
35	−11.31	200	365	78.69	300	200	100	0.424
35	−14.03	200	310	75.97	300	200	100	0.383

(Continued)

TABLE 6.9 (Continued)

H = 1.5 m, $\gamma_{concrete}$ = 24.52 kN/m³, γ_{sat} = 15.70 kN/m³ and ι = ½ϕ

ϕ Degree	α Degree	Top Width, mm	Bottom Width, mm	Horizontal Angle θ Degree	Depth of Shear Key D_{sk}, mm			Volume of Wall, m³
					μ = 0.50	μ = 0.70	μ = 0.80	
40	0.00	200	610	90.00	200	0	0	0.608
40	−5.71	200	440	84.29	200	0	0	0.480
40	−11.31	200	290	78.69	200	100	0	0.368
40	−14.03	200	235	75.97	200	100	0	0.326

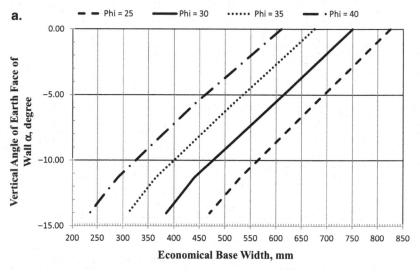

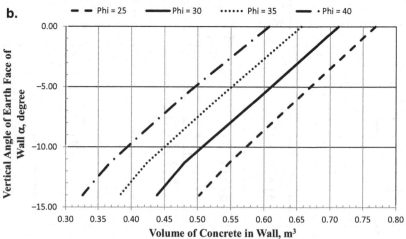

FIGURE 6.10 Design Charts for a Concrete Wall H = 1.5 m, $\gamma_{concrete}$ = 24.52 kN/m³, γ_{sat} = 15.70 kN/m³ and ι = ½ϕ: (a) Economical Base Width for a Wall, (b) Volume of a Concrete Wall

TABLE 6.10

$H = 1.5$ m, $\gamma_{concrete} = 24.52$ kN/m^3, $\gamma_{sat} = 17.27$ kN/m^3 and $\iota = \frac{1}{2}\phi$

A. Results for Stability Analysis

				Pressure at		$\mu = 0.50$		$\mu = 0.70$		$\mu = 0.80$	
ϕ Degree	α Degree	K_{ah}	F_{OVT}	Heel, kN/m^2	Toe, kN/m^2	F_{SLD}	Net Force, kN	F_{SLD}	Net Force, kN	F_{SLD}	Net Force, kN
25	0.00	0.450	1.87	0.40	40.43	0.86	5.93	1.20	2.74	1.38	1.14
25	−5.71	0.407	1.71	0.11	42.55	0.81	5.87	1.13	3.13	1.29	1.76
25	−11.31	0.364	1.56	0.57	44.71	0.75	5.75	1.05	3.43	1.20	2.27
25	−14.03	0.344	1.51	2.07	44.87	0.73	5.65	1.02	3.52	1.16	2.45
30	0.00	0.373	1.88	0.42	41.27	0.94	4.38	1.32	1.42	1.51	0.00
30	−5.71	0.329	1.70	0.40	43.65	0.88	4.34	1.24	1.84	1.42	0.59
30	−11.31	0.286	1.52	0.54	47.32	0.82	4.29	1.14	2.24	1.30	1.22
30	−14.03	0.266	1.50	4.78	45.32	0.79	4.16	1.11	2.29	1.27	1.35
35	0.00	0.305	1.88	0.13	42.60	1.03	3.08	1.45	0.35	1.65	0.00
35	−5.71	0.262	1.69	0.22	45.65	0.97	3.09	1.35	0.84	1.55	0.00
35	−11.31	0.220	1.50	2.08	49.16	0.89	3.09	1.24	1.29	1.42	0.39
35	−14.03	0.201	1.51	11.23	43.10	0.87	2.95	1.22	1.31	1.40	0.49
40	0.00	0.245	1.92	0.59	43.28	1.15	1.94	1.61	0.00	1.84	0.00
40	−5.71	0.204	1.67	0.23	47.97	1.06	2.08	1.49	0.06	1.70	0.00
40	−11.31	0.165	1.51	6.88	48.85	0.98	2.11	1.37	0.53	1.56	0.00
40	−14.03	0.147	1.50	21.46	39.76	0.95	2.05	1.32	0.65	1.51	0.00

B. Physical Dimensions for a Technoeconomic Wall Section

					Depth of Shear Key D_{sk}, mm			
ϕ Degree	α Degree	Top Width, mm	Bottom Width, mm	Horizontal Angle θ Degree	$\mu = 0.50$	$\mu = 0.70$	$\mu = 0.80$	Volume of Wall, m^3
25	0.00	200	870	90.00	500	400	300	0.803
25	−5.71	200	715	84.29	500	400	300	0.686
25	−11.31	200	570	78.69	500	400	300	0.578
25	−14.03	200	505	75.97	500	400	400	0.529
30	0.00	200	790	90.00	400	300	0	0.743
30	−5.71	200	630	84.29	400	300	200	0.623
30	−11.31	200	475	78.69	400	300	200	0.506
30	−14.03	200	415	75.97	400	300	200	0.461
35	0.00	200	710	90.00	200	0	0	0.683
35	−5.71	200	545	84.29	300	200	0	0.559
35	−11.31	200	390	78.69	300	200	100	0.443

(Continued)

TABLE 6.10 (Continued)

H = 1.5 m, $\gamma_{concrete}$ = 24.52 kN/m³, γ_{sat} = 17.27 kN/m³ and ι = ½φ

φ Degree	α Degree	Top Width, mm	Bottom Width, mm	Horizontal Angle θ Degree	Depth of Shear Key D_{sk}, mm			Volume of Wall, m³
					μ = 0.50	μ = 0.70	μ = 0.80	
35	−14.03	200	335	75.97	300	200	200	0.401
40	0.00	200	640	90.00	100	0	0	0.630
40	−5.71	200	465	84.29	200	100	0	0.499
40	−11.31	200	315	78.69	200	100	0	0.386
40	−14.03	200	255	75.97	200	100	0	0.341

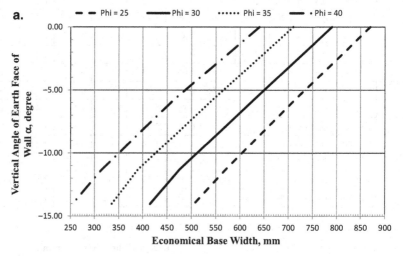

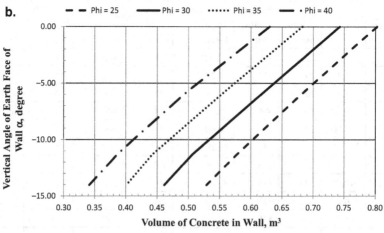

FIGURE 6.11 Design Charts for a Concrete Wall H = 1.5 m, $\gamma_{concrete}$ = 24.52 kN/m³, γ_{sat} = 17.27 kN/m³ and ι = ½φ: (a) Economical Base Width for a Wall, (b) Volume of a Concrete Wall

TABLE 6.11

H = 1.5 m, $\gamma_{concrete}$ = 24.52 kN/m³, γ_{sat} = 15.70 kN/m³ and ι = ϕ

A. Results for Stability Analysis

ϕ Degree	α Degree	K_{ah}	F_{OVT}	Pressure at Heel, kN/m²	Toe, kN/m²	$\mu = 0.50$ F_{SLD}	Net Force, kN	$\mu = 0.70$ F_{SLD}	Net Force, kN	$\mu = 0.80$ F_{SLD}	Net Force, kN
25	0.00	0.821	1.84	0.09	38.82	0.67	12.16	0.94	8.22	1.07	6.25
25	−5.71	0.746	1.73	0.31	39.65	0.65	11.42	0.91	7.93	1.04	6.19
25	−11.31	0.675	1.61	0.16	41.24	0.62	10.76	0.87	7.73	0.99	6.21
25	−14.03	0.641	1.55	0.07	42.24	0.60	10.45	0.84	7.64	0.97	6.23
30	0.00	0.750	1.85	0.22	38.99	0.70	10.74	0.98	6.95	1.13	5.05
30	−5.71	0.666	1.72	0.13	40.37	0.68	9.87	0.96	6.58	1.09	4.93
30	−11.31	0.587	1.61	0.59	41.66	0.66	9.06	0.92	6.23	1.05	4.81
30	−14.03	0.549	1.54	0.63	42.81	0.64	8.70	0.90	6.11	1.03	4.81
35	0.00	0.671	1.85	0.25	39.37	0.74	9.20	1.04	5.58	1.19	3.78
35	−5.71	0.580	1.72	0.19	40.99	0.73	8.24	1.02	5.15	1.16	3.61
35	−11.31	0.496	1.58	0.13	43.39	0.70	7.42	0.98	4.83	1.12	3.54
35	−14.03	0.456	1.50	0.09	45.08	0.68	7.05	0.95	4.71	1.09	3.55
40	0.00	0.587	1.85	0.11	40.05	0.79	7.61	1.11	4.19	1.27	2.49
40	−5.71	0.492	1.72	0.52	41.53	0.78	6.59	1.09	3.72	1.25	2.28
40	−11.31	0.406	1.56	0.64	44.53	0.75	5.79	1.05	3.45	1.20	2.29
40	−14.03	0.366	1.50	1.83	45.56	0.74	5.41	1.03	3.33	1.18	2.28

B. Physical Dimensions for a Technoeconomic Wall Section

ϕ Degree	α Degree	Top Width, mm	Bottom Width, mm	Horizontal Angle θ Degree	Depth of Shear Key D_{sk}, mm $\mu = 0.50$	$\mu = 0.70$	$\mu = 0.80$	Volume of Wall, m³
25	0.00	200	1125	90.00	700	500	500	0.994
25	−5.71	200	970	84.29	700	600	600	0.878
25	−11.31	200	815	78.69	700	600	600	0.761
25	−14.03	200	740	75.97	700	600	600	0.705
30	0.00	200	1075	90.00	500	400	300	0.956
30	−5.71	200	905	84.29	500	400	300	0.829
30	−11.31	200	745	78.69	600	500	400	0.709
30	−14.03	200	665	75.97	600	500	400	0.649
35	0.00	200	1015	90.00	400	300	200	0.911
35	−5.71	200	835	84.29	400	300	200	0.776
35	−11.31	200	660	78.69	500	400	300	0.645

(Continued)

TABLE 6.11 (Continued)

H = 1.5 m, $\gamma_{concrete}$ = 24.52 kN/m³, γ_{sat} = 15.70 kN/m³ and $\iota = \phi$

ϕ Degree	α Degree	Top Width, mm	Bottom Width, mm	Horizontal Angle θ Degree	Depth of Shear Key D_{sk}, mm			Volume of Wall, m³
					μ = 0.50	μ = 0.70	μ = 0.80	
35	−14.03	200	575	75.97	500	400	300	0.581
40	0.00	200	945	90.00	300	200	100	0.859
40	−5.71	200	760	84.29	300	200	100	0.720
40	−11.31	200	575	78.69	400	300	200	0.581
40	−14.03	200	490	75.97	300	300	200	0.518

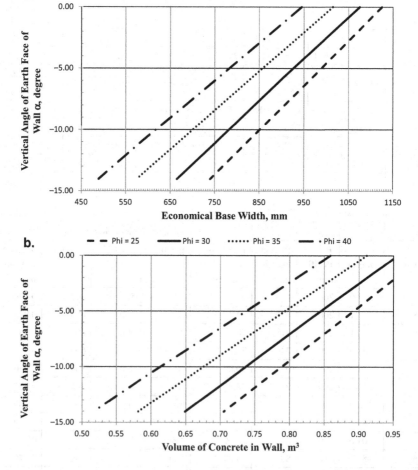

FIGURE 6.12 Design Charts for a Concrete Wall H = 1.5 m, $\gamma_{concrete}$ = 24.52 kN/m³, γ_{sat} = 15.70 kN/m³ and $\iota = \phi$: (a) Economical Base Width for a Wall and (b) Volume of a Concrete Wall

TABLE 6.12

$H = 1.5$ m, $\gamma_{concrete} = 24.52$ kN/m³, $\gamma_{sat} = 17.27$ kN/m³ and $\iota = \phi$

A. *Results for Stability Analysis*

ϕ Degree	α Degree	K_{ah}	F_{OVT}	Pressure at Heel, kN/m²	Toe, kN/m²	$\mu = 0.50$ F_{SLD}	Net Force, kN	$\mu = 0.70$ F_{SLD}	Net Force, kN	$\mu = 0.80$ F_{SLD}	Net Force, kN
25	0.00	0.821	1.84	0.20	38.38	0.64	13.89	0.89	9.78	1.02	7.73
25	−5.71	0.746	1.73	0.21	39.34	0.62	12.99	0.87	9.35	0.99	7.52
25	−11.31	0.675	1.62	0.26	40.57	0.59	12.15	0.83	8.95	0.95	7.36
25	−14.03	0.641	1.57	0.34	41.29	0.58	11.75	0.81	8.77	0.93	7.29
30	0.00	0.750	1.84	0.18	38.70	0.67	12.32	0.93	8.37	1.07	6.39
30	−5.71	0.666	1.73	0.29	39.75	0.65	11.23	0.91	7.78	1.04	6.05
30	−11.31	0.587	1.62	0.53	41.10	0.63	10.25	0.88	7.27	1.01	5.78
30	−14.03	0.549	1.54	0.09	42.65	0.61	9.83	0.86	7.10	0.98	5.74
35	0.00	0.671	1.84	0.09	39.20	0.70	10.61	0.99	6.85	1.13	4.97
35	−5.71	0.580	1.73	0.25	40.44	0.69	9.41	0.97	6.18	1.11	4.56
35	−11.31	0.496	1.60	0.62	42.11	0.67	8.36	0.94	5.64	1.08	4.27
35	−14.03	0.456	1.52	0.19	44.05	0.66	7.93	0.92	5.46	1.05	4.23
40	0.00	0.587	1.85	0.20	39.57	0.75	8.79	1.06	5.24	1.21	3.46
40	−5.71	0.492	1.73	0.47	41.04	0.75	7.57	1.05	4.56	1.20	3.06
40	−11.31	0.406	1.57	0.35	43.99	0.72	6.57	1.01	4.12	1.16	2.90
40	−14.03	0.366	1.51	1.71	44.55	0.71	6.09	1.00	3.89	1.14	2.78

B. *Physical Dimensions for a Technoeconomic Wall Section*

ϕ Degree	α Degree	Top Width, mm	Bottom Width, mm	Horizontal Angle θ Degree	Depth of Shear Key D_{sk}, mm $\mu = 0.50$	$\mu = 0.70$	$\mu = 0.80$	Volume of Wall, m³
25	0.00	200	1185	90.00	700	600	500	1.039
25	−5.71	200	1025	84.29	600	500	500	0.919
25	−11.31	200	870	78.69	700	600	600	0.803
25	−14.03	200	795	75.97	700	600	600	0.746
30	0.00	200	1130	90.00	500	400	400	0.998
30	−5.71	200	960	84.29	500	400	400	0.870
30	−11.31	200	795	78.69	600	500	500	0.746
30	−14.03	200	710	75.97	600	500	500	0.683
35	0.00	200	1065	90.00	400	300	300	0.949
35	−5.71	200	885	84.29	400	300	300	0.814
35	−11.31	200	710	78.69	400	300	200	0.683
35	−14.03	200	620	75.97	500	400	300	0.615

(Continued)

TABLE 6.12 (Continued)
$H = 1.5$ m, $\gamma_{concrete} = 24.52$ kN/m³, $\gamma_{sat} = 17.27$ kN/m³ and $\iota = \phi$

ϕ Degree	α Degree	Top Width, mm	Bottom Width, mm	Horizontal Angle θ Degree	Depth of Shear Key D_{sk}, mm			Volume of Wall, m³
					$\mu = 0.50$	$\mu = 0.70$	$\mu = 0.80$	
40	0.00	200	995	90.00	300	200	200	0.896
40	−5.71	200	805	84.29	300	200	200	0.754
40	−11.31	200	615	78.69	300	200	200	0.611
40	−14.03	200	530	75.97	300	300	300	0.548

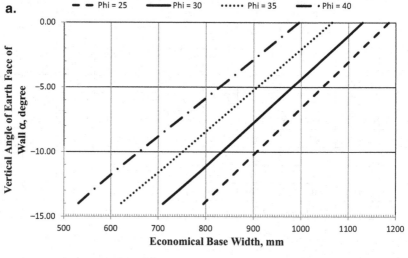

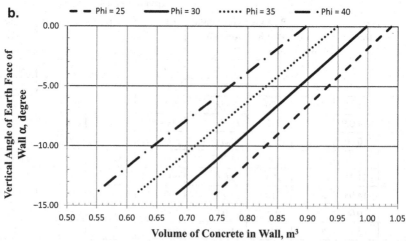

FIGURE 6.13 Design Charts for a Concrete Wall $H = 1.5$ m, $\gamma_{concrete} = 24.52$ kN/m³, $\gamma_{sat} = 17.27$ kN/m³ and $\iota = \phi$: (a) Economical Base Width for a Wall, (b) Volume of a Concrete Wall

6.4 CHARTS FOR THE DESIGN OF A 2.0-M-HIGH CONCRETE BREAST WALL

The charts/tables developed for a 2.0-m-high concrete breast wall for the previously discussed six combinations are presented in Tables 6.13 to 6.18 and Figures 6.14 to 6.19.

TABLE 6.13

$H = 2.0$ m, $\gamma_{concrete} = 24.52$ kN/m³, $\gamma_{sat} = 15.70$ kN/m³ and $\iota = 0°$

A. Results for Stability Analysis

ϕ Degree	α Degree	K_{ah}	F_{OVT}	Pressure at Heel, kN/m²	Toe, kN/m²	$\mu = 0.50$ F_{SLD}	Net Force, kN	$\mu = 0.70$ F_{SLD}	Net Force, kN	$\mu = 0.80$ F_{SLD}	Net Force, kN
25	0.00	0.373	1.84	0.05	52.78	0.94	7.13	1.32	2.31	1.51	0.00
25	−5.71	0.339	1.68	0.13	54.85	0.87	7.44	1.22	3.34	1.39	1.29
25	−11.31	0.305	1.52	0.86	57.28	0.79	7.64	1.11	4.21	1.27	2.50
25	−14.03	0.288	1.51	5.98	53.81	0.77	7.52	1.08	4.35	1.23	2.76
30	0.00	0.304	1.85	0.06	53.83	1.03	5.04	1.44	0.61	1.65	0.00
30	−5.71	0.270	1.67	0.52	56.17	0.95	5.41	1.32	1.71	1.51	0.00
30	−11.31	0.237	1.50	2.12	59.04	0.85	5.68	1.19	2.68	1.37	1.18
30	−14.03	0.221	1.51	11.58	51.82	0.83	5.54	1.17	2.78	1.33	1.39
35	0.00	0.246	1.87	0.37	54.69	1.13	3.34	1.58	0.00	1.81	0.00
35	−5.71	0.213	1.65	0.49	58.37	1.02	3.84	1.43	0.53	1.64	0.00
35	−11.31	0.182	1.50	6.15	58.81	0.92	4.14	1.29	1.51	1.47	0.19
35	−14.03	0.166	1.50	19.49	49.16	0.89	4.09	1.24	1.72	1.42	0.53
40	0.00	0.196	1.89	0.56	55.86	1.24	1.98	1.73	0.00	1.98	0.00
40	−5.71	0.166	1.64	1.14	60.36	1.11	2.61	1.55	0.00	1.77	0.00
40	−11.31	0.136	1.51	13.96	56.08	0.99	2.95	1.39	0.65	1.59	0.00
40	−14.03	0.122	1.52	37.62	38.25	0.95	2.97	1.33	0.92	1.52	0.00

B. Physical Dimensions for a Technoeconomic Wall Section

ϕ Degree	α Degree	Top Width, mm	Bottom Width, mm	Horizontal Angle θ Degree	Depth of Shear Key D_{sk}, mm $\mu = 0.50$	$\mu = 0.70$	$\mu = 0.80$	Volume of Wall, m³
25	0.00	200	1015	90.00	500	300	0	1.215
25	−5.71	200	830	84.29	600	400	300	1.030
25	−11.31	200	655	78.69	600	500	400	0.855
25	−14.03	200	590	75.97	600	500	400	0.790
30	0.00	200	915	90.00	300	100	0	1.115
30	−5.71	200	725	84.29	500	300	0	0.925
30	−11.31	200	545	78.69	500	300	200	0.745

(Continued)

TABLE 6.13 (Continued)

H = 2.0 m, $\gamma_{concrete}$ = 24.52 kN/m³, γ_{sat} = 15.70 kN/m³ and ι = 0°

ϕ Degree	α Degree	Top Width, mm	Bottom Width, mm	Horizontal Angle θ Degree	Depth of Shear Key D_{sk}, mm μ = 0.50	μ = 0.70	μ = 0.80	Volume of Wall, m³
30	−14.03	200	485	75.97	500	300	300	0.685
35	0.00	200	825	90.00	200	0	0	1.025
35	−5.71	200	625	84.29	300	200	0	0.825
35	−11.31	200	450	78.69	400	200	100	0.650
35	−14.03	200	385	75.97	400	200	200	0.585
40	0.00	200	740	90.00	100	0	0	0.940
40	−5.71	200	535	84.29	300	0	0	0.735
40	−11.31	200	365	78.69	300	200	0	0.565
40	−14.03	200	300	75.97	300	200	0	0.500

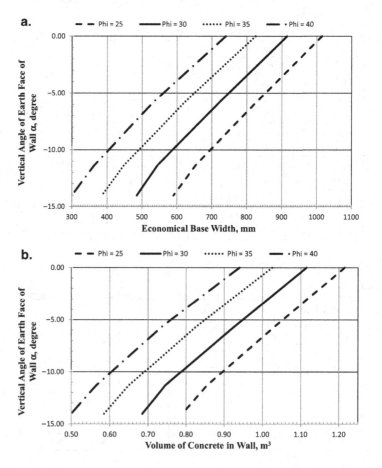

FIGURE 6.14 Design Charts for a Concrete Wall H = 2.0 m, $\gamma_{concrete}$ = 24.52 kN/m³, γ_{sat} = 15.70 kN/m³ and ι = 0°: (a) Economical Base Width for a Wall, (b) Volume of a Concrete Wall

TABLE 6.14
$H = 2.0$ m, $\gamma_{concrete} = 24.52$ kN/m^3, $\gamma_{sat} = 17.27$ kN/m^3 and $\iota = 0°$

A. Results for Stability Analysis

ϕ Degree	α Degree	K_{ah}	F_{OVT}	Pressure at Heel, kN/m^2	Toe, kN/m^2	$\mu = 0.50$ F_{SLD}	Net Force, kN	$\mu = 0.70$ F_{SLD}	Net Force, kN	$\mu = 0.80$ F_{SLD}	Net Force, kN
25	0.00	0.373	1.85	0.36	51.97	0.90	8.35	1.26	3.32	1.44	0.80
25	−5.71	0.339	1.68	0.12	54.19	0.84	8.55	1.17	4.26	1.34	2.11
25	−11.31	0.305	1.53	0.25	56.92	0.77	8.64	1.07	5.04	1.22	3.24
25	−14.03	0.288	1.50	4.01	54.73	0.74	8.49	1.04	5.16	1.19	3.50
30	0.00	0.304	1.86	0.45	52.88	0.99	5.99	1.38	1.36	1.58	0.00
30	−5.71	0.270	1.68	0.58	55.33	0.91	6.24	1.28	2.37	1.46	0.44
30	−11.31	0.237	1.51	1.45	58.48	0.83	6.41	1.16	3.25	1.32	1.68
30	−14.03	0.221	1.50	8.89	53.14	0.81	6.24	1.13	3.34	1.29	1.89
35	0.00	0.246	1.86	0.24	54.26	1.08	4.10	1.51	0.00	1.73	0.00
35	−5.71	0.213	1.66	0.61	57.30	0.99	4.46	1.39	0.99	1.58	0.00
35	−11.31	0.182	1.50	5.06	58.33	0.90	4.65	1.26	1.89	1.43	0.51
35	−14.03	0.166	1.51	17.10	49.42	0.87	4.53	1.22	2.02	1.39	0.77
40	0.00	0.196	1.88	0.39	55.43	1.19	2.56	1.66	0.00	1.90	0.00
40	−5.71	0.166	1.63	0.18	60.34	1.07	3.10	1.50	0.02	1.71	0.00
40	−11.31	0.136	1.51	11.86	56.14	0.97	3.30	1.36	0.89	1.55	0.00
40	−14.03	0.122	1.51	30.61	42.74	0.93	3.30	1.30	1.16	1.49	0.09

B. Physical Dimensions for a Technoeconomic Wall Section

ϕ Degree	α Degree	Top Width, mm	Bottom Width, mm	Horizontal Angle θ Degree	Depth of Shear Key D_{sk}, mm $\mu = 0.50$	$\mu = 0.70$	$\mu = 0.80$	Volume of Wall, m^3
25	0.00	200	1070	90.00	500	300	100	1.270
25	−5.71	200	880	84.29	600	400	300	1.080
25	−11.31	200	700	78.69	600	500	400	0.900
25	−14.03	200	630	75.97	600	500	400	0.830
30	0.00	200	965	90.00	400	100	0	1.165
30	−5.71	200	770	84.29	500	300	200	0.970
30	−11.31	200	585	78.69	500	400	300	0.785
30	−14.03	200	520	75.97	500	400	300	0.720
35	0.00	200	865	90.00	200	0	0	1.065
35	−5.71	200	665	84.29	400	200	0	0.865
35	−11.31	200	485	78.69	400	200	200	0.685

(Continued)

TABLE 6.14 (Continued)

H = 2.0 m, $\gamma_{concrete}$ = 24.52 kN/m³, γ_{sat} = 17.27 kN/m³ and ι = 0°

					Depth of Shear Key D_{sk}, mm			
ϕ Degree	α Degree	Top Width, mm	Bottom Width, mm	Horizontal Angle θ Degree	$\mu = 0.50$	$\mu = 0.70$	$\mu = 0.80$	Volume of Wall, m³
35	−14.03	200	420	75.97	400	300	200	0.620
40	0.00	200	775	90.00	100	0	0	0.975
40	−5.71	200	565	84.29	300	100	0	0.765
40	−11.31	200	395	78.69	300	200	0	0.595
40	−14.03	200	325	75.97	300	200	100	0.525

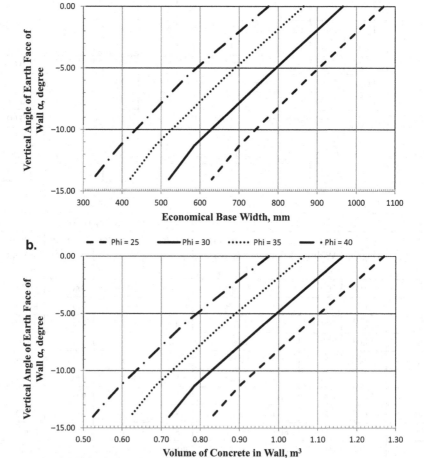

FIGURE 6.15 Design Charts for a Concrete Wall H = 2.0 m, $\gamma_{concrete}$ = 24.52 kN/m³, γ_{sat} = 17.27 kN/m³ and ι = 0°: (a) Economical Base Width for a Wall, (b) Volume of a Concrete Wall

TABLE 6.15
$H = 2.0$ m, $\gamma_{concrete} = 24.52$ kN/m^3, $\gamma_{sat} = 15.70$ kN/m^3 and $\iota = \frac{1}{2}\phi$

A. Results for Stability Analysis

				Pressure at		$\mu = 0.50$		$\mu = 0.70$		$\mu = 0.80$	
ϕ Degree	α Degree	K_{ah}	F_{OVT}	Heel, kN/m^2	Toe, kN/m^2	F_{SLD}	Net Force, kN	F_{SLD}	Net Force, kN	F_{SLD}	Net Force, kN
25	0.00	0.450	1.84	0.25	51.67	0.87	9.55	1.21	4.32	1.39	1.71
25	−5.71	0.407	1.70	0.56	53.15	0.81	9.49	1.14	5.00	1.30	2.75
25	−11.31	0.364	1.55	0.83	55.50	0.75	9.39	1.05	5.62	1.20	3.73
25	−14.03	0.344	1.50	2.87	55.05	0.73	9.24	1.02	5.78	1.16	4.05
30	0.00	0.373	1.84	0.03	52.79	0.94	7.14	1.32	2.32	1.51	0.00
30	−5.71	0.329	1.68	0.16	55.05	0.88	7.14	1.23	3.09	1.41	1.07
30	−11.31	0.286	1.51	0.95	57.91	0.81	7.09	1.13	3.78	1.29	2.13
30	−14.03	0.266	1.50	7.13	53.71	0.79	6.86	1.10	3.82	1.26	2.30
35	0.00	0.305	1.86	0.44	53.39	1.03	5.04	1.45	0.58	1.65	0.00
35	−5.71	0.262	1.67	0.47	56.51	0.96	5.16	1.34	1.52	1.53	0.00
35	−11.31	0.220	1.51	3.67	58.36	0.87	5.18	1.22	2.28	1.40	0.83
35	−14.03	0.201	1.50	13.80	51.17	0.85	5.00	1.19	2.37	1.36	1.05
40	0.00	0.245	1.88	0.51	54.54	1.13	3.30	1.59	0.00	1.81	0.00
40	−5.71	0.204	1.64	0.06	59.32	1.04	3.61	1.45	0.38	1.66	0.00
40	−11.31	0.165	1.50	8.11	58.41	0.94	3.71	1.32	1.19	1.51	0.00
40	−14.03	0.147	1.50	25.50	45.69	0.91	3.59	1.28	1.35	1.46	0.23

B. Physical Dimensions for a Technoeconomic Wall Section

					Depth of Shear Key D_{sk}, mm			
ϕ Degree	α Degree	Top Width, mm	Bottom Width, mm	Horizontal Angle θ Degree	$\mu = 0.50$	$\mu = 0.70$	$\mu = 0.80$	Volume of Wall, m^3
25	0.00	200	1120	90.00	600	400	200	1.320
25	−5.71	200	930	84.29	700	500	400	1.130
25	−11.31	200	745	78.69	700	500	400	0.945
25	−14.03	200	665	75.97	700	500	500	0.865
30	0.00	200	1015	90.00	400	200	0	1.215
30	−5.71	200	815	84.29	500	400	200	1.015
30	−11.31	200	625	78.69	500	400	300	0.825
30	−14.03	200	555	75.97	500	400	300	0.755
35	0.00	200	920	90.00	300	100	0	1.120
35	−5.71	200	710	84.29	300	100	0	0.910
35	−11.31	200	520	78.69	400	300	200	0.720

(Continued)

TABLE 6.15 (Continued)

$H = 2.0 \text{ m}, \gamma_{concrete} = 24.52 \text{ kN/m}^3, \gamma_{sat} = 15.70 \text{ kN/m}^3 \text{ and } \iota = \frac{1}{2}\phi$

| | | | | | Depth of Shear Key D_{sk}, mm | | | |
ϕ Degree	α Degree	Top Width, mm	Bottom Width, mm	Horizontal Angle θ Degree	$\mu = 0.50$	$\mu = 0.70$	$\mu = 0.80$	Volume of Wall, m^3
35	−14.03	200	450	75.97	400	300	200	0.650
40	0.00	200	825	90.00	200	0	0	1.025
40	−5.71	200	605	84.29	300	100	0	0.805
40	−11.31	200	420	78.69	300	200	0	0.620
40	−14.03	200	350	75.97	300	200	100	0.550

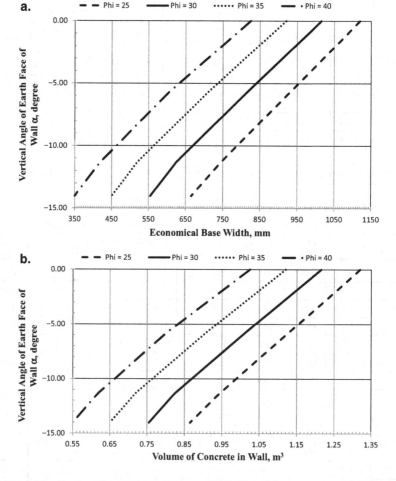

FIGURE 6.16 Design Charts for a Concrete Wall $H = 2.0$ m, $\gamma_{concrete} = 24.52$ kN/m³, $\gamma_{sat} = 15.70$ kN/m³ and $\iota = \frac{1}{2}\phi$: (a) Economical Base Width for a Wall, (b) Volume of a Concrete Wall

TABLE 6.16

$H = 2.0$ m, $\gamma_{concrete} = 24.52$ kN/m^3, $\gamma_{sat} = 17.27$ kN/m^3 and $\iota = \frac{1}{2}\phi$

A. Results for Stability Analysis

				Pressure at		$\mu = 0.50$		$\mu = 0.70$		$\mu = 0.80$	
ϕ Degree	α Degree	K_{ah}	F_{OVT}	Heel, kN/ m^2	Toe, kN/ m^2	F_{SLD}	Net Force, kN	F_{SLD}	Net Force, kN	F_{SLD}	Net Force, kN
25	0.00	0.450	1.84	0.08	51.43	0.82	11.13	1.15	5.69	1.32	2.97
25	−5.71	0.407	1.70	0.53	52.60	0.78	10.87	1.09	6.16	1.25	3.81
25	−11.31	0.364	1.55	0.38	55.12	0.72	10.62	1.01	6.66	1.16	4.67
25	−14.03	0.344	1.50	1.49	55.49	0.70	10.42	0.98	6.79	1.12	4.97
30	0.00	0.373	1.85	0.34	51.99	0.90	8.36	1.26	3.33	1.44	0.81
30	−5.71	0.329	1.68	0.24	54.26	0.85	8.20	1.18	3.96	1.35	1.84
30	−11.31	0.286	1.52	0.60	57.20	0.78	8.00	1.09	4.52	1.25	2.78
30	−14.03	0.266	1.50	5.36	54.30	0.76	7.72	1.07	4.53	1.22	2.93
35	0.00	0.305	1.86	0.30	53.03	0.99	6.03	1.38	1.40	1.58	0.00
35	−5.71	0.262	1.68	0.64	55.51	0.92	5.96	1.29	2.14	1.48	0.24
35	−11.31	0.220	1.50	1.99	58.85	0.85	5.88	1.18	2.84	1.35	1.33
35	−14.03	0.201	1.50	11.23	52.16	0.83	5.60	1.16	2.84	1.33	1.46
40	0.00	0.245	1.87	0.39	54.12	1.08	4.07	1.52	0.00	1.73	0.00
40	−5.71	0.204	1.66	0.39	57.98	1.00	4.18	1.41	0.80	1.61	0.00
40	−11.31	0.165	1.51	7.28	57.45	0.92	4.14	1.29	1.50	1.48	0.17
40	−14.03	0.147	1.51	21.26	47.72	0.89	4.00	1.25	1.64	1.43	0.46

B. Physical Dimensions for a Technoeconomic Wall Section

					Depth of Shear Key D_{sk}, mm			
ϕ Degree	α Degree	Top Width, mm	Bottom Width, mm	Horizontal Angle θ Degree	$\mu = 0.50$	$\mu = 0.70$	$\mu = 0.80$	Volume of Wall, m^3
25	0.00	200	1175	90.00	600	400	300	1.375
25	−5.71	200	985	84.29	700	500	400	1.185
25	−11.31	200	795	78.69	700	500	500	0.995
25	−14.03	200	710	75.97	700	600	500	0.910
30	0.00	200	1070	90.00	400	300	100	1.270
30	−5.71	200	865	84.29	400	300	200	1.065
30	−11.31	200	670	78.69	500	400	300	0.870
30	−14.03	200	595	75.97	500	400	300	0.795
35	0.00	200	965	90.00	300	100	0	1.165
35	−5.71	200	755	84.29	300	200	0	0.955
35	−11.31	200	555	78.69	400	300	200	0.755

(Continued)

TABLE 6.16 (Continued)

H = 2.0 m, $\gamma_{concrete}$ = 24.52 kN/m³, γ_{sat} = 17.27 kN/m³ and ι = ½ϕ

ϕ Degree	α Degree	Top Width, mm	Bottom Width, mm	Horizontal Angle θ Degree	Depth of Shear Key D_{sk}, mm			Volume of Wall, m³
					μ = 0.50	μ = 0.70	μ = 0.80	
35	−14.03	200	485	75.97	400	300	200	0.685
40	0.00	200	865	90.00	200	0	0	1.065
40	−5.71	200	645	84.29	200	100	0	0.845
40	−11.31	200	455	78.69	300	200	100	0.655
40	−14.03	200	380	75.97	300	200	100	0.580

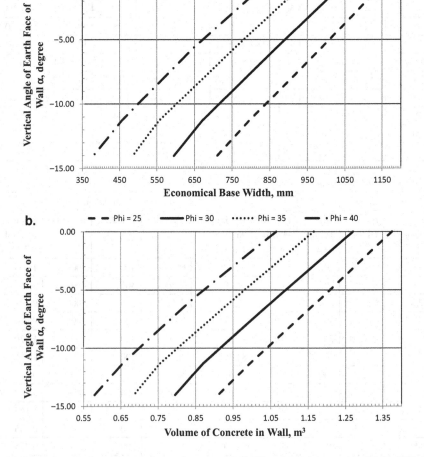

FIGURE 6.17 Design Charts for a Concrete Wall H = 2.0 m, $\gamma_{concrete}$ = 24.52 kN/m³, γ_{sat} = 17.27 kN/m³ and ι = ½ϕ: (a) Economical Base Width for a Wall, (b) Volume of a Concrete Wall

TABLE 6.17

H = 2.0 m, $\gamma_{concrete}$ = 24.52 kN/m³, γ_{sat} = 15.70 kN/m³ and $\iota = \phi$

A. Results for Stability Analysis

				Pressure at		$\mu = 0.50$		$\mu = 0.70$		$\mu = 0.80$	
ϕ Degree	α Degree	K_{ah}	F_{OVT}	Heel, kN/ m²	Toe, kN/ m²	F_{SLD}	Net Force, kN	F_{SLD}	Net Force, kN	F_{SLD}	Net Force, kN
25	0.00	0.821	1.82	0.18	49.42	0.65	22.10	0.91	15.30	1.04	11.90
25	−5.71	0.746	1.72	0.12	50.48	0.63	20.77	0.88	14.77	1.01	11.76
25	−11.31	0.675	1.61	0.31	51.57	0.60	19.50	0.84	14.25	0.97	11.62
25	−14.03	0.641	1.56	0.36	52.33	0.59	18.90	0.82	14.02	0.94	11.58
30	0.00	0.750	1.83	0.17	49.73	0.68	19.61	0.95	13.08	1.09	9.82
30	−5.71	0.666	1.71	0.04	51.07	0.66	18.01	0.93	12.34	1.06	9.50
30	−11.31	0.587	1.59	0.12	52.61	0.64	16.53	0.89	11.67	1.02	9.24
30	−14.03	0.549	1.54	0.39	53.38	0.62	15.83	0.87	11.36	0.99	9.13
35	0.00	0.671	1.83	0.25	50.05	0.72	16.88	1.01	10.66	1.15	7.55
35	−5.71	0.580	1.70	0.01	51.76	0.70	15.13	0.98	9.83	1.12	7.17
35	−11.31	0.496	1.58	0.53	53.29	0.68	13.53	0.95	9.08	1.08	6.85
35	−14.03	0.456	1.51	0.39	54.89	0.66	12.84	0.92	8.82	1.05	6.81
40	0.00	0.587	1.83	0.02	50.81	0.76	14.06	1.07	8.21	1.22	5.29
40	−5.71	0.492	1.71	0.47	52.13	0.75	12.20	1.05	7.28	1.20	4.82
40	−11.31	0.406	1.55	0.22	55.21	0.72	10.70	1.01	6.72	1.16	4.72
40	−14.03	0.366	1.50	2.10	55.18	0.71	9.95	0.99	6.37	1.14	4.59

B. Physical Dimensions for a Technoeconomic Wall Section

					Depth of Shear Key D_{sk}, mm			
ϕ Degree	α Degree	Top Width, mm	Bottom Width, mm	Horizontal Angle θ Degree	$\mu = 0.50$	$\mu = 0.70$	$\mu = 0.80$	Volume of Wall, m³
25	0.00	200	1525	90.00	900	700	700	1.725
25	−5.71	200	1320	84.29	900	700	600	1.520
25	−11.31	200	1125	78.69	800	700	600	1.325
25	−14.03	200	1030	75.97	800	700	600	1.230
30	0.00	200	1455	90.00	700	600	500	1.655
30	−5.71	200	1235	84.29	700	600	500	1.435
30	−11.31	200	1025	78.69	700	600	500	1.225
30	−14.03	200	925	75.97	700	500	500	1.125
35	0.00	200	1375	90.00	600	400	400	1.575
35	−5.71	200	1140	84.29	500	400	400	1.340
35	−11.31	200	920	78.69	500	400	300	1.120

(Continued)

TABLE 6.17 (Continued)

$H = 2.0$ m, $\gamma_{concrete} = 24.52$ kN/m³, $\gamma_{sat} = 15.70$ kN/m³ and $\iota = \phi$

| | | | | | Depth of Shear Key D_{sk}, mm | | | |
ϕ Degree	α Degree	Top Width, mm	Bottom Width, mm	Horizontal Angle θ Degree	$\mu = 0.50$	$\mu = 0.70$	$\mu = 0.80$	Volume of Wall, m³
35	−14.03	200	810	75.97	500	400	300	1.010
40	0.00	200	1280	90.00	300	200	100	1.480
40	−5.71	200	1040	84.29	400	300	200	1.240
40	−11.31	200	800	78.69	400	300	200	1.000
40	−14.03	200	695	75.97	300	300	200	0.895

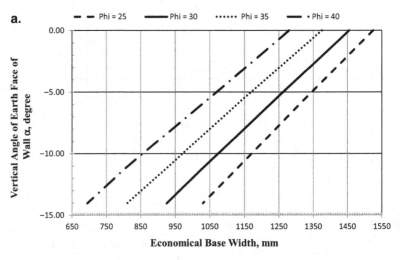

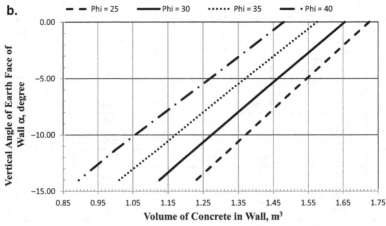

FIGURE 6.18 Design Charts for a Concrete Wall $H = 2.0$ m, $\gamma_{concrete} = 24.52$ kN/m³, $\gamma_{sat} = 15.70$ kN/m³ and $\iota = \phi$: (a) Economical Base Width of a Wall, (b) Volume of a Concrete Wall

TABLE 6.18

$H = 2.0$ m, $\gamma_{concrete} = 24.52$ kN/m^3, $\gamma_{sat} = 17.27$ kN/m^3 and $\iota = \phi$

A. Results for Stability Analysis

ϕ Degree	α Degree	K_{ah}	F_{OVT}	Pressure at Heel, kN/ m^2	Toe, kN/ m^2	$\mu = 0.50$ F_{SLD}	Net Force, kN	$\mu = 0.70$ F_{SLD}	Net Force, kN	$\mu = 0.80$ F_{SLD}	Net Force, kN
25	0.00	0.821	1.83	0.27	49.00	0.62	25.19	0.87	18.08	0.99	14.53
25	−5.71	0.746	1.72	0.10	50.10	0.60	23.55	0.84	17.26	0.96	14.11
25	−11.31	0.675	1.62	0.12	51.25	0.58	21.99	0.81	16.47	0.93	13.71
25	−14.03	0.641	1.57	0.33	51.75	0.57	21.23	0.79	16.08	0.91	13.51
30	0.00	0.750	1.82	0.19	49.39	0.65	22.41	0.91	15.59	1.04	12.18
30	−5.71	0.666	1.72	0.36	50.30	0.63	20.41	0.89	14.45	1.01	11.47
30	−11.31	0.587	1.61	0.43	51.69	0.61	18.61	0.86	13.48	0.98	10.92
30	−14.03	0.549	1.55	0.39	52.69	0.60	17.79	0.84	13.06	0.96	10.70
35	0.00	0.671	1.83	0.21	49.74	0.68	19.36	0.96	12.87	1.09	9.62
35	−5.71	0.580	1.72	0.34	50.93	0.67	17.18	0.94	11.60	1.08	8.81
35	−11.31	0.496	1.59	0.32	52.85	0.65	15.28	0.91	10.60	1.04	8.25
35	−14.03	0.456	1.53	0.51	53.93	0.64	14.41	0.89	10.15	1.02	8.02
40	0.00	0.587	1.83	0.32	50.11	0.73	16.14	1.02	10.02	1.17	6.96
40	−5.71	0.492	1.71	0.37	51.71	0.72	13.93	1.01	8.78	1.15	6.21
40	−11.31	0.406	1.56	0.14	54.50	0.70	12.08	0.98	7.88	1.12	5.78
40	−14.03	0.366	1.50	0.56	55.86	0.68	11.24	0.96	7.49	1.09	5.61

B. Physical Dimensions for a Technoeconomic Wall Section

ϕ Degree	α Degree	Top Width, mm	Bottom Width, mm	Horizontal Angle θ Degree	Depth of Shear Key D_{sk}, mm m = 0.50	m = 0.70	m = 0.80	Volume of Wall, m^3
25	0.00	200	1605	90.00	900	800	700	1.805
25	−5.71	200	1395	84.29	900	800	700	1.595
25	−11.31	200	1195	78.69	900	700	700	1.395
25	−14.03	200	1100	75.97	800	700	700	1.300
30	0.00	200	1530	90.00	700	600	500	1.730
30	−5.71	200	1310	84.29	700	600	500	1.510
30	−11.31	200	1095	78.69	700	600	500	1.295
30	−14.03	200	990	75.97	700	600	500	1.190
35	0.00	200	1445	90.00	500	400	300	1.645
35	−5.71	200	1210	84.29	500	400	400	1.410
35	−11.31	200	980	78.69	500	400	400	1.180

(Continued)

TABLE 6.18 (Continued)
H = 2.0 m, $\gamma_{concrete}$ = 24.52 kN/m^3, γ_{sat} = 17.27 kN/m^3 and ι = ϕ

ϕ Degree	α Degree	Top Width, mm	Bottom Width, mm	Horizontal Angle θ Degree	Depth of Shear Key D_{sk}, mm			Volume of Wall, m^3
					m = 0.50	m = 0.70	m = 0.80	
35	−14.03	200	870	75.97	500	400	400	1.070
40	0.00	200	1350	90.00	300	200	200	1.550
40	−5.71	200	1100	84.29	400	300	300	1.300
40	−11.31	200	855	78.69	400	300	200	1.055
40	−14.03	200	740	75.97	400	300	200	0.940

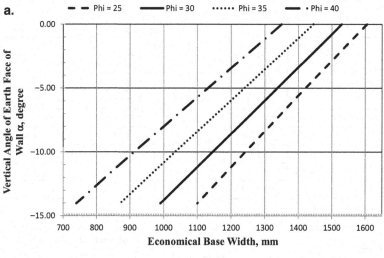

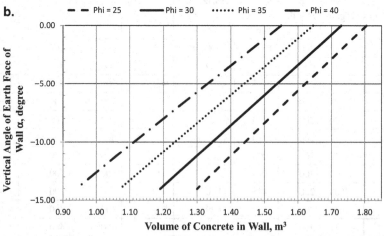

FIGURE 6.19 Design Charts for a Concrete Wall H = 2.0 m, $\gamma_{concrete}$ = 24.52 kN/m^3, γ_{sat} = 17.27 kN/m^3 and ι = ϕ: (a) Economical Base Width for a Wall and (b) Volume of a Concrete Wall

6.5 CHARTS FOR THE DESIGN OF A 2.5-M-HIGH CONCRETE BREAST WALL

The charts/tables developed for a 2.5-m-high concrete breast wall for the previously discussedsix combinations are presented in Tables 6.19 to 6.24 and Figures 6.20 to 6.25.

TABLE 6.19
H = 2.5 m, $\gamma_{concrete}$ = 24.52 kN/m³, γ_{sat} = 15.70 kN/m³ and ι = 0°

A. Results for Stability Analysis

				Pressure at		$\mu = 0.50$		$\mu = 0.70$		$\mu = 0.80$	
ϕ Degree	α Degree	K_{ah}	F_{OVT}	Heel, kN/m²	Toe, kN/m²	F_{SLD}	Net Force, kN	F_{SLD}	Net Force, kN	F_{SLD}	Net Force, kN
25	0.00	0.373	1.83	0.02	63.48	0.92	11.64	1.28	4.30	1.47	0.64
25	−5.71	0.339	1.67	0.01	65.51	0.85	12.05	1.19	5.80	1.35	2.68
25	−11.31	0.305	1.51	0.01	68.46	0.77	12.30	1.08	7.10	1.23	4.50
25	−14.03	0.288	1.51	6.14	63.74	0.75	12.05	1.05	7.21	1.20	4.79
30	0.00	0.304	1.84	0.14	64.38	1.00	8.37	1.40	1.64	1.60	0.00
30	−5.71	0.270	1.67	0.71	66.43	0.92	8.85	1.29	3.23	1.47	0.43
30	−11.31	0.237	1.50	1.68	69.54	0.83	9.21	1.16	4.67	1.33	2.39
30	−14.03	0.221	1.50	10.88	62.39	0.81	9.00	1.13	4.81	1.29	2.72
35	0.00	0.246	1.85	0.38	65.31	1.09	5.74	1.53	0.00	1.75	0.00
35	−5.71	0.213	1.65	0.80	68.39	0.99	6.40	1.39	1.40	1.59	0.00
35	−11.31	0.182	1.50	6.91	67.65	0.89	6.76	1.25	2.77	1.43	0.78
35	−14.03	0.166	1.50	19.74	58.03	0.86	6.67	1.21	3.07	1.38	1.27
40	0.00	0.196	1.86	0.31	66.76	1.19	3.66	1.67	0.00	1.91	0.00
40	−5.71	0.166	1.62	0.15	71.71	1.06	4.57	1.49	0.14	1.70	0.00
40	−11.31	0.136	1.50	13.78	65.46	0.95	4.96	1.33	1.50	1.53	0.00
40	−14.03	0.122	1.50	35.30	48.93	0.91	4.96	1.28	1.89	1.46	0.36

B. Physical Dimensions for a Technoeconomic Wall Section

					Depth of Shear Key D_{sk}, mm			
ϕ Degree	α Degree	Top Width, mm	Bottom Width, mm	Horizontal Angle θ Degree	$\mu = 0.50$	$\mu = 0.70$	$\mu = 0.80$	Volume of Wall, m³
25	0.00	200	1285	90.00	600	400	100	1.856
25	−5.71	200	1060	84.29	700	400	300	1.575
25	−11.31	200	845	78.69	800	600	500	1.306
25	−14.03	200	770	75.97	800	600	500	1.213
30	0.00	200	1160	90.00	500	200	0	1.700
30	−5.71	200	930	84.29	500	300	100	1.413
30	−11.31	200	710	78.69	600	400	300	1.138
30	−14.03	200	635	75.97	600	400	300	1.044

(Continued)

TABLE 6.19 (Continued)

H = 2.5 m, $\gamma_{concrete}$ = 24.52 kN/m³, γ_{sat} = 15.70 kN/m³ and ι = 0°

ϕ Degree	α Degree	Top Width, mm	Bottom Width, mm	Horizontal Angle θ Degree	Depth of Shear Key D_{sk}, mm			Volume of Wall, m³
					$\mu = 0.50$	$\mu = 0.70$	$\mu = 0.80$	
35	0.00	200	1045	90.00	300	0	0	1.556
35	−5.71	200	805	84.29	300	100	0	1.256
35	−11.31	200	595	78.69	400	300	200	0.994
35	−14.03	200	515	75.97	400	300	200	0.894
40	0.00	200	935	90.00	200	0	0	1.419
40	−5.71	200	685	84.29	200	0	0	1.106
40	−11.31	200	485	78.69	300	200	0	0.856
40	−14.03	200	405	75.97	300	200	100	0.756

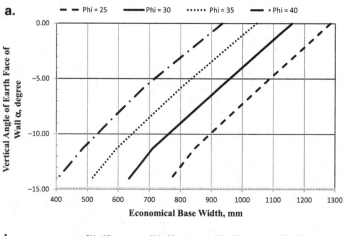

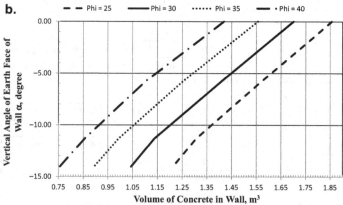

FIGURE 6.20 Design Charts for a Concrete Wall H = 2.5 m, $\gamma_{concrete}$ = 24.52 kN/m³, γ_{sat} = 15.70 kN/m³ and ι = 0°: (a) Economical Base Width for a Wall, (b) Volume of a Concrete Wall

TABLE 6.20

$H = 2.5$ m, $\gamma_{concrete} = 24.52$ kN/m³, $\gamma_{sat} = 17.27$ kN/m³ and $\iota = 0°$

A. Results for Stability Analysis

				Pressure at		$\mu = 0.50$		$\mu = 0.70$		$\mu = 0.80$	
ϕ Degree	α Degree	K_{ah}	F_{OVT}	Heel, kN/m²	Toe, kN/m²	F_{SLD}	Net Force, kN	F_{SLD}	Net Force, kN	F_{SLD}	Net Force, kN
25	0.00	0.373	1.83	0.03	63.01	0.88	13.59	1.23	5.94	1.40	2.11
25	−5.71	0.339	1.68	0.28	64.58	0.82	13.75	1.14	7.19	1.31	3.91
25	−11.31	0.305	1.53	0.05	67.45	0.75	13.82	1.05	8.33	1.20	5.58
25	−14.03	0.288	1.50	4.02	64.89	0.73	13.56	1.02	8.48	1.16	5.94
30	0.00	0.304	1.84	0.31	63.69	0.96	9.88	1.34	2.85	1.54	0.00
30	−5.71	0.270	1.67	0.57	65.84	0.89	10.17	1.24	4.29	1.42	1.35
30	−11.31	0.237	1.50	1.02	69.07	0.80	10.35	1.13	5.56	1.29	3.17
30	−14.03	0.221	1.50	8.47	63.51	0.78	10.07	1.10	5.67	1.25	3.47
35	0.00	0.246	1.84	0.17	64.98	1.05	6.93	1.47	0.52	1.68	0.00
35	−5.71	0.213	1.64	0.06	68.32	0.96	7.42	1.34	2.20	1.53	0.00
35	−11.31	0.182	1.50	4.80	68.46	0.87	7.61	1.22	3.43	1.39	1.33
35	−14.03	0.166	1.50	16.54	59.51	0.84	7.41	1.18	3.62	1.35	1.72
40	0.00	0.196	1.85	0.20	66.27	1.14	4.56	1.60	0.00	1.83	0.00
40	−5.71	0.166	1.63	0.55	70.21	1.03	5.24	1.45	0.60	1.65	0.00
40	−11.31	0.136	1.50	11.00	66.53	0.93	5.53	1.30	1.91	1.49	0.10
40	−14.03	0.122	1.51	30.29	51.53	0.90	5.44	1.26	2.20	1.44	0.58

B. Physical Dimensions for a Technoeconomic Wall Section

					Depth of Shear Key D_{sk}, mm			
ϕ Degree	α Degree	Top Width, mm	Bottom Width, mm	Horizontal Angle θ Degree	$\mu = 0.50$	$\mu = 0.70$	$\mu = 0.80$	Volume of Wall, m³
25	0.00	200	1350	90.00	700	400	200	1.938
25	−5.71	200	1125	84.29	700	500	300	1.656
25	−11.31	200	905	78.69	800	600	500	1.381
25	−14.03	200	820	75.97	800	600	500	1.275
30	0.00	200	1220	90.00	500	200	0	1.775
30	−5.71	200	985	84.29	500	300	100	1.481
30	−11.31	200	760	78.69	600	400	300	1.200
30	−14.03	200	680	75.97	600	500	400	1.100
35	0.00	200	1095	90.00	300	100	0	1.619
35	−5.71	200	850	84.29	300	200	0	1.313
35	−11.31	200	635	78.69	400	300	200	1.044

(Continued)

TABLE 6.20 (Continued)

H = 2.5 m, $\gamma_{concrete}$ = 24.52 kN/m³, γ_{sat} = 17.27 kN/m³ and ι = 0°

| ϕ Degree | α Degree | Top Width, mm | Bottom Width, mm | Horizontal Angle θ Degree | Depth of Shear Key D_{sk}, mm | | | Volume of Wall, m³ |
					$\mu = 0.50$	$\mu = 0.70$	$\mu = 0.80$	
35	−14.03	200	555	75.97	400	300	200	0.944
40	0.00	200	980	90.00	200	0	0	1.475
40	−5.71	200	730	84.29	200	0	0	1.163
40	−11.31	200	520	78.69	300	200	100	0.900
40	−14.03	200	440	75.97	300	200	100	0.800

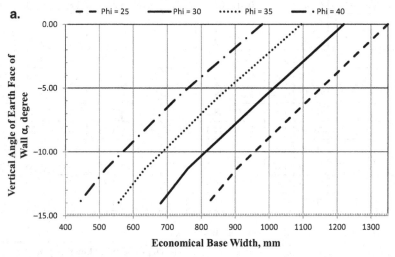

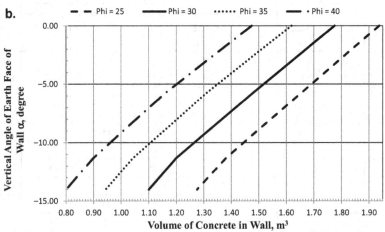

FIGURE 6.21 Design Charts for a Concrete Wall H = 2.5 m, $\gamma_{concrete}$ = 24.52 kN/m³, γ_{sat} = 17.27 kN/m³ and ι = 0°: (a) Economical Base Width for a Wall, (b) Volume of a Concrete Wall

TABLE 6.21

$H = 2.5$ m, $\gamma_{concrete} = 24.52$ kN/m^3, $\gamma_{sat} = 15.70$ kN/m^3 and $\iota = \frac{1}{2}\phi$

A. Results for Stability Analysis

ϕ Degree	α Degree	K_{ah}	F_{OVT}	Pressure at Heel, kN/m^2	Toe, kN/m^2	$\mu = 0.50$ F_{SLD}	Net Force, kN	$\mu = 0.70$ F_{SLD}	Net Force, kN	$\mu = 0.80$ F_{SLD}	Net Force, kN
25	0.00	0.450	1.82	0.09	62.53	0.85	15.43	1.18	7.46	1.35	3.48
25	−5.71	0.407	1.69	0.50	63.80	0.79	15.24	1.11	8.39	1.27	4.96
25	−11.31	0.364	1.55	0.57	66.16	0.73	15.02	1.03	9.26	1.17	6.38
25	−14.03	0.344	1.50	3.03	65.10	0.71	14.72	0.99	9.43	1.14	6.78
30	0.00	0.373	1.84	0.45	63.01	0.92	11.59	1.29	4.23	1.47	0.55
30	−5.71	0.329	1.68	0.52	65.17	0.86	11.52	1.20	5.35	1.37	2.27
30	−11.31	0.286	1.51	0.53	68.56	0.79	11.42	1.10	6.39	1.26	3.87
30	−14.03	0.266	1.50	7.01	63.86	0.77	11.03	1.07	6.41	1.23	4.10
35	0.00	0.305	1.84	0.47	64.01	1.00	8.37	1.40	1.62	1.60	0.00
35	−5.71	0.262	1.66	0.48	66.95	0.93	8.48	1.30	2.96	1.49	0.21
35	−11.31	0.220	1.51	3.68	68.31	0.85	8.42	1.19	4.02	1.36	1.82
35	−14.03	0.201	1.50	13.99	60.57	0.83	8.10	1.16	4.11	1.32	2.12
40	0.00	0.245	1.85	0.56	65.13	1.10	5.68	1.53	0.00	1.75	0.00
40	−5.71	0.204	1.64	0.32	69.35	1.00	6.05	1.40	1.16	1.61	0.00
40	−11.31	0.165	1.50	8.23	67.82	0.91	6.12	1.28	2.32	1.46	0.42
40	−14.03	0.147	1.50	25.10	54.95	0.88	5.91	1.24	2.53	1.41	0.84

B. Physical Dimensions for a Technoeconomic Wall Section

ϕ Degree	α Degree	Top Width, mm	Bottom Width, mm	Horizontal Angle θ Degree	Depth of Shear Key D_{sk}, mm $\mu = 0.50$	$\mu = 0.70$	$\mu = 0.80$	Volume of Wall, m^3
25	0.00	200	1415	90.00	700	500	300	2.019
25	−5.71	200	1185	84.29	700	500	400	1.731
25	−11.31	200	960	78.69	800	700	600	1.450
25	−14.03	200	865	75.97	800	700	600	1.331
30	0.00	200	1290	90.00	600	300	100	1.863
30	−5.71	200	1045	84.29	500	400	200	1.556
30	−11.31	200	810	78.69	600	500	400	1.263
30	−14.03	200	725	75.97	600	500	400	1.156
35	0.00	200	1165	90.00	400	100	0	1.706
35	−5.71	200	910	84.29	400	200	0	1.388
35	−11.31	200	680	78.69	500	400	300	1.100

(Continued)

TABLE 6.21 (Continued)

H = 2.5 m, $\gamma_{concrete}$ = 24.52 kN/m³, γ_{sat} = 15.70 kN/m³ and ι = ½ϕ

ϕ Degree	α Degree	Top Width, mm	Bottom Width, mm	Horizontal Angle θ Degree	Depth of Shear Key D_{sk}, mm			Volume of Wall, m³
					μ = 0.50	μ = 0.70	μ = 0.80	
35	−14.03	200	595	75.97	500	400	300	0.994
40	0.00	200	1045	90.00	300	0	0	1.556
40	−5.71	200	780	84.29	300	100	0	1.225
40	−11.31	200	555	78.69	400	200	100	0.944
40	−14.03	200	470	75.97	400	300	200	0.838

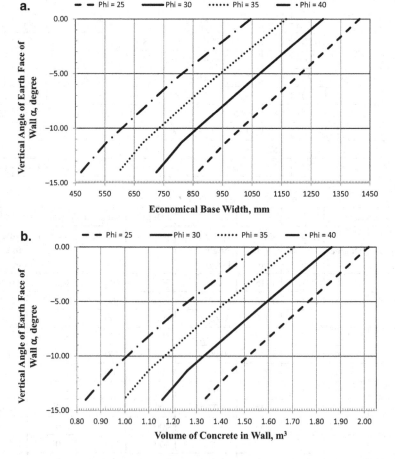

FIGURE 6.22 Design Charts for a Concrete Wall H = 2.5 m, $\gamma_{concrete}$ = 24.52 kN/m³, γ_{sat} = 15.70 kN/m³ and ι = ½ϕ: (a) Economical Base Width for a Wall, (b) Volume of a Concrete Wall

TABLE 6.22

$H = 2.5$ m, $\gamma_{concrete} = 24.52$ kN/m^3, $\gamma_{sat} = 17.27$ kN/m^3 and $\iota = \frac{1}{2}\phi$

A. Results for Stability Analysis

ϕ Degree	α Degree	K_{ah}	F_{OVT}	Pressure at Heel, kN/m^2	Toe, kN/m^2	$\mu = 0.50$ F_{SLD}	Net Force, kN	$\mu = 0.70$ F_{SLD}	Net Force, kN	$\mu = 0.80$ F_{SLD}	Net Force, kN
25	0.00	0.450	1.83	0.34	61.85	0.81	17.83	1.13	9.49	1.29	5.33
25	−5.71	0.407	1.69	0.06	63.70	0.76	17.44	1.06	10.28	1.22	6.69
25	−11.31	0.364	1.56	0.46	65.46	0.71	16.91	0.99	10.83	1.14	7.80
25	−14.03	0.344	1.50	1.31	65.97	0.68	16.58	0.96	11.02	1.09	8.23
30	0.00	0.373	1.83	0.01	63.03	0.88	13.60	1.23	5.95	1.40	2.12
30	−5.71	0.329	1.68	0.28	64.77	0.83	13.21	1.16	6.75	1.32	3.52
30	−11.31	0.286	1.53	0.87	67.18	0.76	12.80	1.07	7.48	1.22	4.81
30	−14.03	0.266	1.50	5.18	64.60	0.74	12.37	1.04	7.51	1.19	5.08
35	0.00	0.305	1.83	0.14	63.87	0.96	9.95	1.34	2.93	1.53	0.00
35	−5.71	0.262	1.67	0.48	66.18	0.90	9.74	1.25	3.95	1.43	1.06
35	−11.31	0.220	1.51	2.13	68.74	0.82	9.50	1.15	4.88	1.32	2.57
35	−14.03	0.201	1.51	11.78	61.33	0.81	9.03	1.13	4.83	1.29	2.72
40	0.00	0.245	1.85	0.35	64.81	1.05	6.88	1.47	0.46	1.68	0.00
40	−5.71	0.204	1.65	0.64	68.08	0.97	6.94	1.36	1.81	1.56	0.00
40	−11.31	0.165	1.50	6.46	68.10	0.89	6.85	1.25	2.86	1.42	0.86
40	−14.03	0.147	1.50	20.35	57.89	0.86	6.57	1.21	3.02	1.38	1.24

B. Physical Dimensions for a Technoeconomic Wall Section

ϕ Degree	α Degree	Top Width, mm	Bottom Width, mm	Horizontal Angle θ Degree	Depth of Shear Key D_{sk}, mm $\mu = 0.50$	$\mu = 0.70$	$\mu = 0.80$	Volume of Wall, m^3
25	0.00	200	1490	90.00	800	500	400	2.113
25	−5.71	200	1250	84.29	800	600	500	1.813
25	−11.31	200	1025	78.69	700	600	500	1.531
25	−14.03	200	920	75.97	800	700	600	1.400
30	0.00	200	1350	90.00	600	400	200	1.938
30	−5.71	200	1105	84.29	600	400	300	1.631
30	−11.31	200	870	78.69	600	400	300	1.338
30	−14.03	200	775	75.97	600	500	400	1.219
35	0.00	200	1220	90.00	400	200	0	1.775
35	−5.71	200	965	84.29	400	200	100	1.456
35	−11.31	200	725	78.69	400	300	200	1.156

(Continued)

TABLE 6.22 (Continued)

H = 2.5 m, $\gamma_{concrete}$ = 24.52 kN/m³, γ_{sat} = 17.27 kN/m³ and ι = ½ϕ

ϕ Degree	α Degree	Top Width, mm	Bottom Width, mm	Horizontal Angle θ Degree	Depth of Shear Key D_{sk}, mm			Volume of Wall, m³
					μ = 0.50	μ = 0.70	μ = 0.80	
35	−14.03	200	640	75.97	500	400	300	1.050
40	0.00	200	1095	90.00	300	0	0	1.619
40	−5.71	200	830	84.29	300	100	0	1.288
40	−11.31	200	595	78.69	400	300	200	0.994
40	−14.03	200	505	75.97	400	300	200	0.881

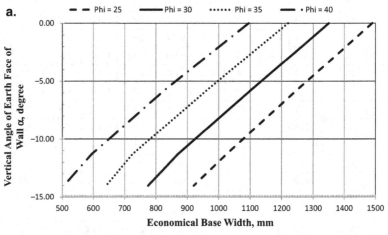

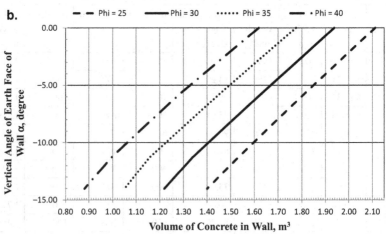

FIGURE 6.23 Design Charts for a Concrete Wall H = 2.5 m, $\gamma_{concrete}$ = 24.52 kN/m³, γ_{sat} = 17.27 kN/m³ and ι = ½ϕ: (a) Economical Base Width for a Wall, (b) Volume of Concrete Wall

TABLE 6.23

$H = 2.5$ m, $\gamma_{concrete} = 24.52$ kN/m^3, $\gamma_{sat} = 15.70$ kN/m^3 and $\iota = \phi$

A. Results for Stability Analysis

				Pressure at		$\mu = 0.50$		$\mu = 0.70$		$\mu = 0.80$	
ϕ Degree	α Degree	K_{ah}	F_{OVT}	Heel, kN/ m^2	Toe, kN/ m^2	F_{SLD}	Net Force, kN	F_{SLD}	Net Force, kN	F_{SLD}	Net Force, kN
25	0.00	0.821	1.82	0.41	59.89	0.64	34.94	0.90	24.47	1.03	19.24
25	−5.71	0.746	1.71	0.13	61.14	0.62	32.84	0.87	23.61	0.99	19.00
25	−11.31	0.675	1.61	0.20	62.30	0.59	30.80	0.83	22.74	0.95	18.71
25	−14.03	0.641	1.56	0.32	62.92	0.58	29.82	0.81	22.32	0.93	18.57
30	0.00	0.750	1.82	0.29	60.33	0.67	31.08	0.94	21.05	1.07	16.03
30	−5.71	0.666	1.71	0.15	61.62	0.65	28.51	0.91	19.79	1.04	15.43
30	−11.31	0.587	1.60	0.47	62.81	0.63	26.11	0.88	18.63	1.00	14.89
30	−14.03	0.549	1.54	0.44	63.82	0.61	25.02	0.86	18.15	0.98	14.71
35	0.00	0.671	1.82	0.09	60.93	0.70	26.87	0.99	17.35	1.13	12.59
35	−5.71	0.580	1.70	0.04	62.38	0.69	24.03	0.96	15.89	1.10	11.82
35	−11.31	0.496	1.58	0.52	63.82	0.66	21.47	0.93	14.64	1.06	11.23
35	−14.03	0.456	1.51	0.18	65.51	0.65	20.36	0.91	14.19	1.03	11.11
40	0.00	0.587	1.82	0.09	61.43	0.75	22.42	1.05	13.46	1.20	8.98
40	−5.71	0.492	1.69	0.13	63.11	0.74	19.51	1.03	12.00	1.18	8.25
40	−11.31	0.406	1.55	0.09	65.77	0.71	17.05	0.99	10.95	1.13	7.90
40	−14.03	0.366	1.50	1.77	65.81	0.69	15.86	0.97	10.39	1.11	7.66

B. Physical Dimensions for a Technoeconomic Wall Section

					Depth of Shear Key D_{sk}, mm			
ϕ Degree	α Degree	Top Width, mm	Bottom Width, mm	Horizontal Angle θ Degree	$\mu = 0.50$	$\mu = 0.70$	$\mu = 0.80$	Volume of Wall, m^3
25	0.00	200	1930	90.00	1200	1000	800	2.663
25	−5.71	200	1675	84.29	1100	900	800	2.344
25	−11.31	200	1435	78.69	1100	900	800	2.044
25	−14.03	200	1320	75.97	1100	900	800	1.900
30	0.00	200	1840	90.00	800	700	600	2.550
30	−5.71	200	1570	84.29	900	700	600	2.213
30	−11.31	200	1315	78.69	900	700	600	1.894
30	−14.03	200	1190	75.97	800	700	600	1.738
35	0.00	200	1735	90.00	600	500	400	2.419
35	−5.71	200	1450	84.29	600	500	400	2.063
35	−11.31	200	1180	78.69	700	500	500	1.725

(Continued)

TABLE 6.23 (Continued)

H = 2.5 m, $\gamma_{concrete}$ = 24.52 kN/m³, γ_{sat} = 15.70 kN/m³ and ι = ϕ

ϕ Degree	α Degree	Top Width, mm	Bottom Width, mm	Horizontal Angle θ Degree	Depth of Shear Key D_{sk}, mm			Volume of Wall, m³
					μ = 0.50	μ = 0.70	μ = 0.80	
35	−14.03	200	1045	75.97	600	500	500	1.556
40	0.00	200	1620	90.00	500	300	200	2.275
40	−5.71	200	1320	84.29	400	300	200	1.900
40	−11.31	200	1030	78.69	500	400	300	1.538
40	−14.03	200	900	75.97	500	400	300	1.375

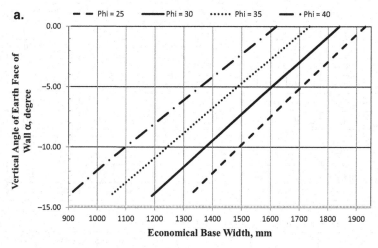

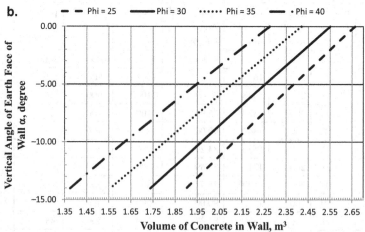

FIGURE 6.24 Design Charts for a Concrete Wall H = 2.5 m, $\gamma_{concrete}$ = 24.52 kN/m³, γ_{sat} = 15.70 kN/m³ and ι = ϕ: (a) Economical Base Width for a Wall, (b) Volume of a Concrete Wall

TABLE 6.24

$H = 2.5$ m, $\gamma_{concrete} = 24.52$ kN/m³, $\gamma_{sat} = 17.27$ kN/m³ and $\iota = \phi$

A. Results for Stability Analysis

ϕ Degree	α Degree	K_{ah}	F_{OVT}	Pressure at Heel, kN/m²	Toe, kN/m²	$\mu = 0.50$ F_{SLD}	Net Force, kN	$\mu = 0.70$ F_{SLD}	Net Force, kN	$\mu = 0.80$ F_{SLD}	Net Force, kN
25	0.00	0.821	1.81	0.22	59.79	0.61	39.82	0.85	28.90	0.98	23.43
25	−5.71	0.746	1.72	0.19	60.70	0.59	37.17	0.83	27.48	0.95	22.64
25	−11.31	0.675	1.62	0.22	61.78	0.57	34.67	0.80	26.17	0.91	21.92
25	−14.03	0.641	1.56	0.07	62.62	0.56	33.50	0.78	25.59	0.89	21.63
30	0.00	0.750	1.81	0.06	60.25	0.64	35.49	0.89	25.03	1.02	19.80
30	−5.71	0.666	1.71	0.23	61.11	0.62	32.31	0.87	23.16	1.00	18.58
30	−11.31	0.587	1.60	0.09	62.66	0.60	29.45	0.84	21.59	0.96	17.65
30	−14.03	0.549	1.55	0.33	63.28	0.59	28.09	0.82	20.82	0.94	17.19
35	0.00	0.671	1.82	0.19	60.49	0.67	30.70	0.94	20.75	1.07	15.77
35	−5.71	0.580	1.71	0.22	61.72	0.66	27.26	0.92	18.71	1.05	14.44
35	−11.31	0.496	1.59	0.30	63.43	0.64	24.20	0.90	17.01	1.02	13.42
35	−14.03	0.456	1.53	0.38	64.52	0.63	22.81	0.88	16.27	1.00	13.00
40	0.00	0.587	1.82	0.26	60.88	0.72	25.70	1.00	16.32	1.14	11.64
40	−5.71	0.492	1.70	0.03	62.72	0.70	22.22	0.99	14.35	1.13	10.41
40	−11.31	0.406	1.56	0.19	64.91	0.68	19.19	0.96	12.75	1.09	9.53
40	−14.03	0.366	1.50	0.70	66.03	0.67	17.83	0.94	12.07	1.07	9.19

B. Physical Dimensions for a Technoeconomic Wall Section

ϕ Degree	α Degree	Top Width, mm	Bottom Width, mm	Horizontal Angle θ Degree	Depth of Shear Key D_{sk}, mm $\mu = 0.50$	$\mu = 0.70$	$\mu = 0.80$	Volume of Wall, m³
25	0.00	200	2025	90.00	1100	900	800	2.781
25	−5.71	200	1770	84.29	1100	1000	900	2.463
25	−11.31	200	1525	78.69	1100	900	900	2.156
25	−14.03	200	1405	75.97	1100	900	900	2.006
30	0.00	200	1930	90.00	900	700	600	2.663
30	−5.71	200	1660	84.29	800	700	600	2.325
30	−11.31	200	1395	78.69	900	700	700	1.994
30	−14.03	200	1270	75.97	800	700	600	1.838
35	0.00	200	1825	90.00	600	500	400	2.531
35	−5.71	200	1535	84.29	600	500	400	2.169
35	−11.31	200	1255	78.69	700	500	500	1.819

(Continued)

TABLE 6.24 (Continued)

$H = 2.5$ m, $\gamma_{concrete} = 24.52$ kN/m³, $\gamma_{sat} = 17.27$ kN/m³ and $\iota = \phi$

ϕ Degree	α Degree	Top Width, mm	Bottom Width, mm	Horizontal Angle θ Degree	Depth of Shear Key D_{sk}, mm			Volume of Wall, m³
					$\mu = 0.50$	$\mu = 0.70$	$\mu = 0.80$	
35	−14.03	200	1120	75.97	600	500	500	1.650
40	0.00	200	1705	90.00	500	300	300	2.381
40	−5.71	200	1395	84.29	400	300	200	1.994
40	−11.31	200	1100	78.69	500	400	300	1.625
40	−14.03	200	960	75.97	500	400	300	1.450

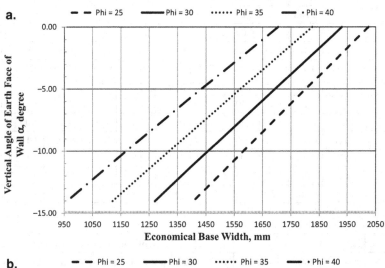

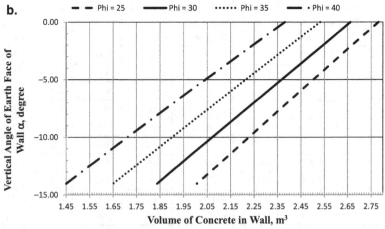

FIGURE 6.25 Design Charts for Concrete Wall $H = 2.5$ m, $\gamma_{concrete} = 24.52$ kN/m³, $\gamma_{sat} = 17.27$ kN/m³ and $\iota = \phi$: (a) Economical Base Width for a Wall, (b) Volume of a Concrete Wall

6.6 CHARTS FOR THE DESIGN OF A 3.0-M-HIGH CONCRETE BREAST WALL

The charts/tables developed for a 3.0-m-high concrete breast wall for the previously discussed six combinations are presented in Tables 6.25 to 6.30 and Figures 6.26 to 6.31.

TABLE 6.25

$H = 3.0$ m, $\gamma_{concrete} = 24.52$ kN/m³, $\gamma_{sat} = 15.70$ kN/m³ and $\iota = 0°$

A. Results for Stability Analysis

				Pressure at		$\mu = 0.50$		$\mu = 0.70$		$\mu = 0.80$	
ϕ Degree	α Degree	K_{ah}	F_{OVT}	Heel, kN/m²	Toe, kN/m²	F_{SLD}	Net Force, kN	F_{SLD}	Net Force, kN	F_{SLD}	Net Force, kN
25	0.00	0.373	1.85	0.31	78.91	0.94	16.02	1.32	5.15	1.51	0.00
25	−5.71	0.339	1.68	0.20	82.28	0.87	16.75	1.22	7.52	1.39	2.90
25	−11.31	0.305	1.52	0.80	86.47	0.79	17.22	1.11	9.53	1.27	5.69
25	−14.03	0.288	1.50	7.84	81.99	0.77	16.99	1.07	9.89	1.23	6.33
30	0.00	0.304	1.86	0.35	80.44	1.03	11.31	1.45	1.32	1.65	0.00
30	−5.71	0.270	1.67	0.40	84.69	0.94	12.20	1.32	3.90	1.51	0.00
30	−11.31	0.237	1.51	3.81	87.86	0.86	12.74	1.20	5.98	1.37	2.61
30	−14.03	0.221	1.50	16.67	78.52	0.83	12.50	1.16	6.30	1.33	3.19
35	0.00	0.246	1.87	0.26	82.36	1.13	7.54	1.58	0.00	1.81	0.00
35	−5.71	0.213	1.65	0.27	88.08	1.02	8.68	1.43	1.25	1.64	0.00
35	−11.31	0.182	1.50	9.23	88.21	0.92	9.31	1.29	3.39	1.47	0.44
35	−14.03	0.166	1.50	30.11	72.69	0.89	9.17	1.25	3.81	1.42	1.13
40	0.00	0.196	1.88	0.17	84.55	1.23	4.53	1.73	0.00	1.97	0.00
40	−5.71	0.166	1.64	1.15	91.19	1.11	5.90	1.55	0.00	1.77	0.00
40	−11.31	0.136	1.50	19.98	85.27	0.99	6.68	1.38	1.52	1.58	0.00
40	−14.03	0.122	1.50	54.29	60.06	0.94	6.75	1.32	2.17	1.51	0.00

B. Physical Dimensions for a Technoeconomic Wall Section

					Depth of Shear Key D_{sk}, mm			
ϕ Degree	α Degree	Top Width, mm	Bottom Width, mm	Horizontal Angle θ Degree	$\mu = 0.50$	$\mu = 0.70$	$\mu = 0.80$	Volume of Wall, m³
25	0.00	300	1525	90.00	800	400	0	2.738
25	−5.71	300	1245	84.29	800	500	300	2.318
25	−11.31	300	980	78.69	900	700	500	1.920
25	−14.03	300	880	75.97	900	700	600	1.770
30	0.00	300	1375	90.00	500	200	0	2.513
30	−5.71	300	1085	84.29	600	300	0	2.078
30	−11.31	300	820	78.69	700	500	300	1.680
30	−14.03	300	725	75.97	700	500	400	1.538
35	0.00	300	1235	90.00	400	0	0	2.303
35	−5.71	300	935	84.29	400	100	0	1.853

(Continued)

TABLE 6.25 (Continued)

$H = 3.0$ m, $\gamma_{concrete} = 24.52$ kN/m³, $\gamma_{sat} = 15.70$ kN/m³ and $\iota = 0°$

| ϕ Degree | α Degree | Top Width, mm | Bottom Width, mm | Horizontal Angle θ Degree | Depth of Shear Key D_{sk}, mm | | | Volume of Wall, m³ |
					$\mu = 0.50$	$\mu = 0.70$	$\mu = 0.80$	
35	−11.31	300	675	78.69	500	300	200	1.463
35	−14.03	300	580	75.97	500	300	200	1.320
40	0.00	300	1105	90.00	200	0	0	2.108
40	−5.71	300	800	84.29	300	0	0	1.650
40	−11.31	300	545	78.69	400	200	0	1.268
40	−14.03	300	445	75.97	400	200	0	1.118

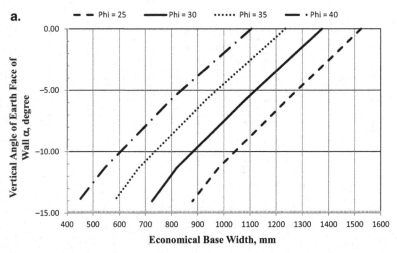

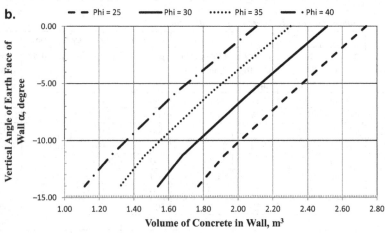

FIGURE 6.26 Design Charts for a Concrete Wall $H = 3.0$ m, $\gamma_{concrete} = 24.52$ kN/m³, $\gamma_{sat} = 15.70$ kN/m³ and $\iota = 0°$: (a) Economical Base Width for a Wall, (b) Volume of a Concrete Wall

TABLE 6.26

$H = 3.0$ m, $\gamma_{concrete} = 24.52$ kN/m^3, $\gamma_{sat} = 17.27$ kN/m^3 and $\iota = 0°$

A. Results for Stability Analysis

ϕ Degree	α Degree	K_{ah}	F_{OVT}	Pressure at Heel, kN/m^2	Toe, kN/m^2	$\mu = 0.50$ F_{SLD}	Net Force, kN	$\mu = 0.70$ F_{SLD}	Net Force, kN	$\mu = 0.80$ F_{SLD}	Net Force, kN
25	0.00	0.373	1.84	0.10	78.44	0.90	18.87	1.26	7.57	1.44	1.92
25	−5.71	0.339	1.68	0.18	81.28	0.84	19.25	1.17	9.58	1.34	4.74
25	−11.31	0.305	1.53	0.38	85.39	0.77	19.43	1.07	11.34	1.22	7.29
25	−14.03	0.288	1.50	6.01	82.09	0.74	19.10	1.04	11.62	1.19	7.87
30	0.00	0.304	1.86	0.43	79.59	0.99	13.50	1.38	3.11	1.58	0.00
30	−5.71	0.270	1.67	0.14	83.80	0.91	14.12	1.27	5.44	1.45	1.10
30	−11.31	0.237	1.50	1.60	88.37	0.83	14.45	1.16	7.37	1.32	3.83
30	−14.03	0.221	1.50	13.33	79.72	0.81	14.04	1.13	7.51	1.29	4.25
35	0.00	0.246	1.86	0.08	81.71	1.08	9.27	1.51	0.00	1.73	0.00
35	−5.71	0.213	1.66	0.49	86.44	0.99	10.06	1.38	2.29	1.58	0.00
35	−11.31	0.182	1.50	6.88	88.32	0.89	10.50	1.25	4.30	1.43	1.20
35	−14.03	0.166	1.50	24.03	76.03	0.87	10.28	1.21	4.65	1.39	1.84
40	0.00	0.196	1.87	0.27	83.51	1.18	5.81	1.66	0.00	1.90	0.00
40	−5.71	0.166	1.64	0.81	89.91	1.07	6.93	1.50	0.00	1.71	0.00
40	−11.31	0.136	1.51	16.91	85.24	0.97	7.47	1.35	2.05	1.55	0.00
40	−14.03	0.122	1.50	44.89	65.36	0.93	7.46	1.30	2.65	1.48	0.25

B. Physical Dimensions for a Technoeconomic Wall Section

ϕ Degree	α Degree	Top Width, mm	Bottom Width, mm	Horizontal Angle θ Degree	Depth of Shear Key D_{sk}, mm $\mu = 0.50$	$\mu = 0.70$	$\mu = 0.80$	Volume of Wall, m^3
25	0.00	300	1600	90.00	800	500	200	2.850
25	−5.71	300	1320	84.29	800	500	400	2.430
25	−11.31	300	1050	78.69	800	600	500	2.025
25	−14.03	300	945	75.97	900	700	600	1.868
30	0.00	300	1445	90.00	600	200	0	2.618
30	−5.71	300	1150	84.29	600	300	100	2.175
30	−11.31	300	875	78.69	600	400	300	1.763
30	−14.03	300	780	75.97	700	500	400	1.620
35	0.00	300	1295	90.00	400	0	0	2.393
35	−5.71	300	995	84.29	400	200	0	1.943
35	−11.31	300	725	78.69	400	200	100	1.538

(Continued)

TABLE 6.26 (Continued)

$H = 3.0$ m, $\gamma_{concrete} = 24.52$ kN/m³, $\gamma_{sat} = 17.27$ kN/m³ and $\iota = 0°$

ϕ Degree	α Degree	Top Width, mm	Bottom Width, mm	Horizontal Angle θ Degree	Depth of Shear Key D_{sk}, mm			Volume of Wall, m³
					$\mu = 0.50$	$\mu = 0.70$	$\mu = 0.80$	
35	−14.03	300	625	75.97	500	400	200	1.388
40	0.00	300	1160	90.00	200	0	0	2.190
40	−5.71	300	850	84.29	300	0	0	1.725
40	−11.31	300	590	78.69	400	200	0	1.335
40	−14.03	300	485	75.97	400	200	100	1.178

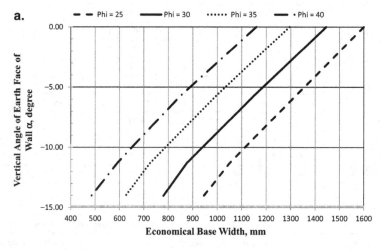

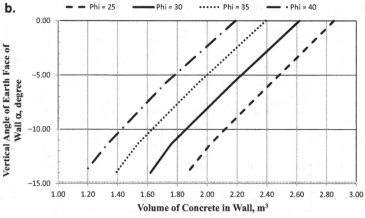

FIGURE 6.27 Design Charts for a Concrete Wall $H = 3.0$ m, $\gamma_{concrete} = 24.52$ kN/m³, $\gamma_{sat} = 17.27$ kN/m³ and $\iota = 0°$: (a) Economical Base Width for a Wall, (b) Volume of a Concrete Wall

TABLE 6.27

$H = 3.0$ m, $\gamma_{concrete} = 24.52$ kN/m³, $\gamma_{sat} = 15.70$ kN/m³ and $\iota = \frac{1}{2}\phi$

A. Results for Stability Analysis

				Pressure at		$\mu = 0.50$		$\mu = 0.70$		$\mu = 0.80$	
ϕ Degree	α Degree	K_{ah}	F_{OVT}	Heel, kN/m²	Toe, kN/m²	F_{SLD}	Net Force, kN	F_{SLD}	Net Force, kN	F_{SLD}	Net Force, kN
25	0.00	0.450	1.84	0.38	77.50	0.87	21.49	1.21	9.72	1.39	3.84
25	−5.71	0.407	1.69	0.27	80.36	0.81	21.42	1.13	11.34	1.30	6.31
25	−11.31	0.364	1.54	0.83	83.71	0.75	21.17	1.05	12.70	1.20	8.46
25	−14.03	0.344	1.50	3.80	83.13	0.72	20.84	1.01	13.06	1.16	9.17
30	0.00	0.373	1.85	0.29	78.93	0.94	16.03	1.32	5.17	1.51	0.00
30	−5.71	0.329	1.68	0.57	82.19	0.88	16.02	1.23	6.91	1.41	2.35
30	−11.31	0.286	1.51	0.90	87.45	0.81	15.99	1.13	8.57	1.29	4.85
30	−14.03	0.266	1.50	10.09	81.25	0.79	15.47	1.10	8.65	1.26	5.24
35	0.00	0.305	1.85	0.13	80.67	1.03	11.40	1.44	1.41	1.65	0.00
35	−5.71	0.262	1.67	0.71	84.76	0.96	11.61	1.34	3.43	1.53	0.00
35	−11.31	0.220	1.50	4.19	89.03	0.87	11.73	1.22	5.24	1.39	1.99
35	−14.03	0.201	1.50	20.69	76.75	0.85	11.24	1.19	5.33	1.36	2.37
40	0.00	0.245	1.87	0.48	82.14	1.13	7.47	1.58	0.00	1.81	0.00
40	−5.71	0.204	1.65	0.58	88.42	1.04	8.09	1.45	0.81	1.66	0.00
40	−11.31	0.165	1.50	12.17	87.62	0.94	8.34	1.32	2.69	1.51	0.00
40	−14.03	0.147	1.50	38.26	68.53	0.91	8.08	1.28	3.04	1.46	0.52

B. Physical Dimensions for a Technoeconomic Wall Section

					Depth of Shear Key D_{sk}, mm			
ϕ Degree	α Degree	Top Width, mm	Bottom Width, mm	Horizontal Angle θ Degree	$\mu = 0.50$	$\mu = 0.70$	$\mu = 0.80$	Volume of Wall, m³
25	0.00	300	1680	90.00	900	600	300	2.970
25	−5.71	300	1390	84.29	900	600	500	2.535
25	−11.31	300	1115	78.69	900	700	500	2.123
25	−14.03	300	995	75.97	1000	800	700	1.943
30	0.00	300	1525	90.00	700	400	0	2.738
30	−5.71	300	1225	84.29	700	400	200	2.288
30	−11.31	300	935	78.69	700	500	300	1.853
30	−14.03	300	830	75.97	600	500	400	1.695
35	0.00	300	1375	90.00	500	100	0	2.513
35	−5.71	300	1065	84.29	500	200	0	2.048
35	−11.31	300	775	78.69	500	300	200	1.613

(Continued)

TABLE 6.27 (Continued)
H = 3.0 m, $\gamma_{concrete}$ = 24.52 kN/m³, γ_{sat} = 15.70 kN/m³ and ι = ½ϕ

| | | | | | Depth of Shear Key D_{sk}, mm | | | |
ϕ Degree	α Degree	Top Width, mm	Bottom Width, mm	Horizontal Angle θ Degree	μ = 0.50	μ = 0.70	μ = 0.80	Volume of Wall, m³
35	−14.03	300	675	75.97	600	400	300	1.463
40	0.00	300	1235	90.00	200	0	0	2.303
40	−5.71	300	910	84.29	300	100	0	1.815
40	−11.31	300	630	78.69	300	200	0	1.395
40	−14.03	300	525	75.97	400	300	100	1.238

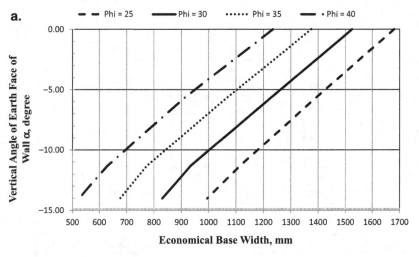

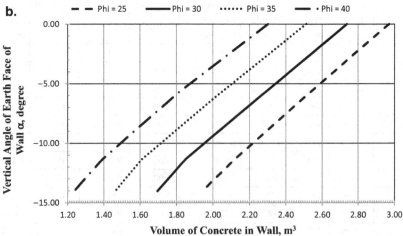

FIGURE 6.28 Design Charts for a Concrete Wall H = 3.0 m, $\gamma_{concrete}$ = 24.52 kN/m³, γ_{sat} = 15.70 kN/m³ and ι = ½ϕ: (a) Economical Base Width for a Wall, (b) Volume of a Concrete Wall

TABLE 6.28
$H = 3.0$ m, $\gamma_{concrete} = 24.52$ kN/m³, $\gamma_{sat} = 17.27$ kN/m³ and $\iota = \frac{1}{2}\phi$

A. Results for Stability Analysis

ϕ Degree	α Degree	K_{ah}	F_{OVT}	Pressure at Heel, kN/m²	Toe, kN/m²	$\mu = 0.50$ F_{SLD}	Net Force, kN	$\mu = 0.70$ F_{SLD}	Net Force, kN	$\mu = 0.80$ F_{SLD}	Net Force, kN
25	0.00	0.450	1.84	0.32	76.93	0.83	25.01	1.16	12.75	1.32	6.62
25	−5.71	0.407	1.70	0.52	79.19	0.78	24.49	1.09	13.92	1.25	8.63
25	−11.31	0.364	1.55	0.18	83.11	0.72	23.94	1.01	15.03	1.16	10.57
25	−14.03	0.344	1.50	2.24	83.23	0.70	23.45	0.98	15.27	1.12	11.18
30	0.00	0.373	1.84	0.07	78.47	0.90	18.88	1.26	7.58	1.44	1.93
30	−5.71	0.329	1.68	0.05	81.74	0.84	18.49	1.18	8.97	1.35	4.20
30	−11.31	0.286	1.52	0.90	85.81	0.78	18.01	1.09	10.17	1.25	6.25
30	−14.03	0.266	1.50	7.47	82.08	0.76	17.41	1.07	10.24	1.22	6.66
35	0.00	0.305	1.85	0.21	79.82	0.98	13.61	1.38	3.21	1.58	0.00
35	−5.71	0.262	1.67	0.60	83.68	0.92	13.44	1.29	4.87	1.47	0.59
35	−11.31	0.220	1.50	2.37	88.97	0.84	13.27	1.18	6.45	1.35	3.04
35	−14.03	0.201	1.50	16.15	79.04	0.83	12.64	1.16	6.44	1.32	3.33
40	0.00	0.245	1.86	0.30	81.49	1.08	9.19	1.52	0.00	1.73	0.00
40	−5.71	0.204	1.65	0.13	87.48	1.00	9.45	1.40	1.85	1.60	0.00
40	−11.31	0.165	1.50	10.16	87.04	0.92	9.36	1.29	3.42	1.47	0.45
40	−14.03	0.147	1.51	31.88	71.59	0.89	8.99	1.25	3.69	1.43	1.04

B. Physical Dimensions for a Technoeconomic Wall Section

ϕ Degree	α Degree	Top Width, mm	Bottom Width, mm	Horizontal Angle θ Degree	Depth of Shear Key D_{sk}, mm $\mu = 0.50$	$\mu = 0.70$	$\mu = 0.80$	Volume of Wall, m³
25	0.00	300	1765	90.00	900	600	400	3.098
25	−5.71	300	1475	84.29	900	700	500	2.663
25	−11.31	300	1190	78.69	900	700	600	2.235
25	−14.03	300	1065	75.97	900	700	600	2.048
30	0.00	300	1600	90.00	700	400	200	2.850
30	−5.71	300	1295	84.29	700	500	300	2.393
30	−11.31	300	1005	78.69	700	500	400	1.958
30	−14.03	300	890	75.97	700	500	400	1.785
35	0.00	300	1445	90.00	400	100	0	2.618
35	−5.71	300	1130	84.29	500	300	100	2.145
35	−11.31	300	830	78.69	500	300	200	1.695

(Continued)

TABLE 6.28 (Continued)

H = 3.0 m, $\gamma_{concrete}$ = 24.52 kN/m³, γ_{sat} = 17.27 kN/m³ and ι = ½ϕ

ϕ Degree	α Degree	Top Width, mm	Bottom Width, mm	Horizontal Angle θ Degree	Depth of Shear Key D_{sk}, mm			Volume of Wall, m³
					$\mu = 0.50$	$\mu = 0.70$	$\mu = 0.80$	
35	−14.03	300	725	75.97	500	300	200	1.538
40	0.00	300	1295	90.00	200	0	0	2.393
40	−5.71	300	965	84.29	300	100	0	1.898
40	−11.31	300	680	78.69	300	200	0	1.470
40	−14.03	300	570	75.97	400	300	200	1.305

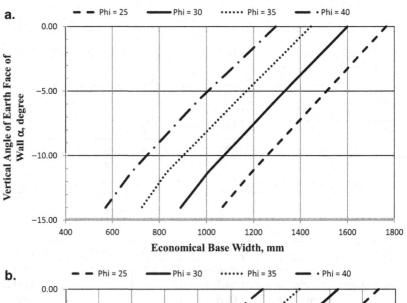

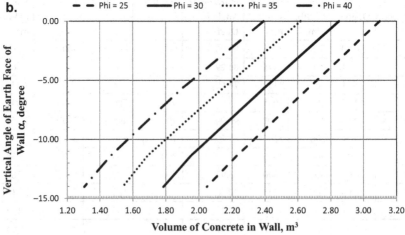

FIGURE 6.29 Design Charts for a Concrete Wall H = 3.0 m, $\gamma_{concrete}$ = 24.52 kN/m³, γ_{sat} = 17.27 kN/m³ and ι = ½ϕ: (a) Economical Base Width for a Wall, (b) Volume of a Concrete Wall

TABLE 6.29

$H = 3.0$ m, $\gamma_{concrete} = 24.52$ kN/m³, $\gamma_{sat} = 15.70$ kN/m³ and $\iota = \phi$

A. Results for Stability Analysis

ϕ Degree	α Degree	K_{ah}	F_{OVT}	Pressure at Heel, kN/m²	Toe, kN/m²	$\mu = 0.50$ F_{SLD}	Net Force, kN	$\mu = 0.70$ F_{SLD}	Net Force, kN	$\mu = 0.80$ F_{SLD}	Net Force, kN
25	0.00	0.821	1.82	0.11	74.29	0.65	49.77	0.91	34.49	1.04	26.84
25	−5.71	0.746	1.72	0.18	75.71	0.63	46.73	0.88	33.22	1.01	26.47
25	−11.31	0.675	1.61	0.22	77.63	0.60	43.90	0.84	32.11	0.96	26.21
25	−14.03	0.641	1.56	0.54	78.49	0.59	42.52	0.82	31.54	0.94	26.05
30	0.00	0.750	1.82	0.10	74.77	0.68	44.16	0.95	29.49	1.09	22.15
30	−5.71	0.666	1.71	0.27	76.38	0.66	40.49	0.93	27.71	1.06	21.31
30	−11.31	0.587	1.60	0.46	78.62	0.64	37.17	0.89	26.22	1.02	20.74
30	−14.03	0.549	1.53	0.25	80.44	0.62	35.66	0.87	25.62	0.99	20.59
35	0.00	0.671	1.83	0.21	75.25	0.72	38.01	1.01	24.04	1.15	17.05
35	−5.71	0.580	1.70	0.02	77.64	0.70	34.05	0.98	22.11	1.12	16.14
35	−11.31	0.496	1.57	0.16	80.64	0.68	30.52	0.95	20.53	1.08	15.53
35	−14.03	0.456	1.51	0.58	82.33	0.66	28.90	0.92	19.84	1.05	15.31
40	0.00	0.587	1.83	0.02	76.22	0.76	31.64	1.07	18.48	1.22	11.90
40	−5.71	0.492	1.70	0.20	78.74	0.75	27.52	1.05	16.49	1.20	10.97
40	−11.31	0.406	1.55	0.33	82.81	0.72	24.09	1.01	15.12	1.16	10.63
40	−14.03	0.366	1.50	2.68	83.29	0.71	22.43	0.99	14.39	1.13	10.37

B. Physical Dimensions for a Technoeconomic Wall Section

ϕ Degree	α Degree	Top Width, mm	Bottom Width, mm	Horizontal Angle θ Degree	Depth of Shear Key D_{sk}, mm $\mu = 0.50$	$\mu = 0.70$	$\mu = 0.80$	Volume of Wall, m³
25	0.00	300	2285	90.00	1300	1000	900	3.878
25	−5.71	300	1980	84.29	1300	1100	1000	3.420
25	−11.31	300	1685	78.69	1300	1100	1000	2.978
25	−14.03	300	1545	75.97	1300	1100	1000	2.768
30	0.00	300	2180	90.00	1000	800	700	3.720
30	−5.71	300	1855	84.29	1000	800	700	3.233
30	−11.31	300	1540	78.69	1000	900	800	2.760
30	−14.03	300	1385	75.97	1000	800	800	2.528
35	0.00	300	2060	90.00	800	600	500	3.540
35	−5.71	300	1710	84.29	700	600	500	3.015
35	−11.31	300	1375	78.69	800	600	500	2.513

(Continued)

TABLE 6.29 (Continued)

H = 3.0 m, $\gamma_{concrete}$ = 24.52 kN/m³, γ_{sat} = 15.70 kN/m³ and ι = ϕ

ϕ Degree	α Degree	Top Width, mm	Bottom Width, mm	Horizontal Angle θ Degree	Depth of Shear Key D_{sk}, mm			Volume of Wall, m³
					μ = 0.50	μ = 0.70	μ = 0.80	
35	−14.03	300	1215	75.97	800	600	500	2.273
40	0.00	300	1920	90.00	500	300	200	3.330
40	−5.71	300	1555	84.29	500	400	300	2.783
40	−11.31	300	1200	78.69	600	400	400	2.250
40	−14.03	300	1040	75.97	600	400	400	2.010

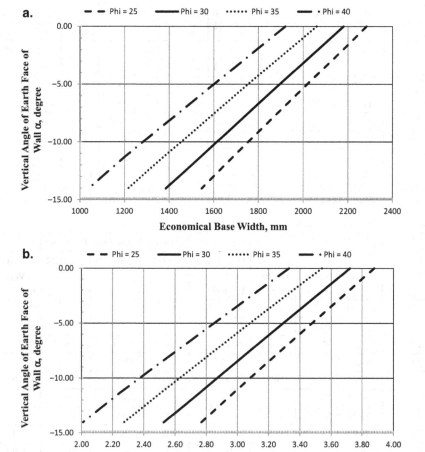

FIGURE 6.30 Design Charts for a Concrete Wall H = 3.0 m, $\gamma_{concrete}$ = 24.52 kN/m³, γ_{sat} = 15.70 kN/m³ and ι = ϕ: (a) Economical Base Width for a Wall and (b) Volume of a Concrete Wall

TABLE 6.30

$H = 3.0$ m, $\gamma_{concrete} = 24.52$ kN/m^3, $\gamma_{sat} = 17.27$ kN/m^3 and $\iota = \phi$

A. Results for Stability Analysis

ϕ Degree	α Degree	K_{ah}	F_{OVT}	Pressure at Heel, kN/ m^2	Toe, kN/ m^2	$\mu = 0.50$ F_{SLD}	Net Force, kN	$\mu = 0.70$ F_{SLD}	Net Force, kN	$\mu = 0.80$ F_{SLD}	Net Force, kN
25	0.00	0.821	1.82	0.26	73.66	0.62	56.72	0.87	40.73	0.99	32.74
25	−5.71	0.746	1.72	0.33	74.95	0.60	52.96	0.84	38.78	0.96	31.69
25	−11.31	0.675	1.62	0.41	76.63	0.58	49.45	0.81	37.01	0.93	30.80
25	−14.03	0.641	1.57	0.50	77.63	0.57	47.78	0.79	36.19	0.91	30.39
30	0.00	0.750	1.82	0.28	74.08	0.65	50.43	0.91	35.08	1.04	27.41
30	−5.71	0.666	1.72	0.15	75.86	0.63	46.01	0.88	32.61	1.01	25.91
30	−11.31	0.587	1.61	0.39	77.81	0.61	41.92	0.86	30.39	0.98	24.62
30	−14.03	0.549	1.55	0.58	79.03	0.60	40.02	0.84	29.39	0.96	24.08
35	0.00	0.671	1.83	0.16	74.78	0.68	43.59	0.96	29.00	1.09	21.71
35	−5.71	0.580	1.71	0.09	76.85	0.67	38.73	0.94	26.21	1.07	19.95
35	−11.31	0.496	1.59	0.49	79.27	0.65	34.38	0.91	23.84	1.04	18.57
35	−14.03	0.456	1.52	0.04	81.69	0.64	32.49	0.89	22.94	1.02	18.16
40	0.00	0.587	1.83	0.14	75.54	0.73	36.39	1.02	22.65	1.17	15.78
40	−5.71	0.492	1.70	0.09	78.08	0.72	31.42	1.01	19.86	1.15	14.08
40	−11.31	0.406	1.57	0.55	81.37	0.70	27.15	0.98	17.68	1.12	12.95
40	−14.03	0.366	1.50	0.84	83.79	0.68	25.29	0.96	16.84	1.09	12.62

B. Physical Dimensions for a Technoeconomic Wall Section

ϕ Degree	α Degree	Top Width, mm	Bottom Width, mm	Horizontal Angle θ Degree	Depth of Shear Key D_{sk}, mm $\mu = 0.50$	$\mu = 0.70$	$\mu = 0.80$	Volume of Wall, m^3
25	0.00	300	2405	90.00	1300	1100	1000	4.058
25	−5.71	300	2095	84.29	1300	1100	900	3.593
25	−11.31	300	1795	78.69	1300	1100	1000	3.143
25	−14.03	300	1650	75.97	1300	1100	1000	2.925
30	0.00	300	2295	90.00	1000	800	700	3.893
30	−5.71	300	1960	84.29	1000	800	700	3.390
30	−11.31	300	1640	78.69	1000	900	800	2.910
30	−14.03	300	1485	75.97	1000	900	800	2.678
35	0.00	300	2165	90.00	700	500	400	3.698
35	−5.71	300	1810	84.29	700	600	500	3.165
35	−11.31	300	1470	78.69	700	600	500	2.655

(Continued)

TABLE 6.30 (Continued)

H = 3.0 m, $\gamma_{concrete}$ = 24.52 kN/m³, γ_{sat} = 17.27 kN/m³ and $\iota = \phi$

ϕ Degree	α Degree	Top Width, mm	Bottom Width, mm	Horizontal Angle θ Degree	Depth of Shear Key D_{sk}, mm			Volume of Wall, m³
					$\mu = 0.50$	$\mu = 0.70$	$\mu = 0.80$	
35	−14.03	300	1300	75.97	800	600	600	2.400
40	0.00	300	2020	90.00	500	300	200	3.480
40	−5.71	300	1645	84.29	500	400	300	2.918
40	−11.31	300	1285	78.69	500	400	300	2.378
40	−14.03	300	1110	75.97	600	400	400	2.115

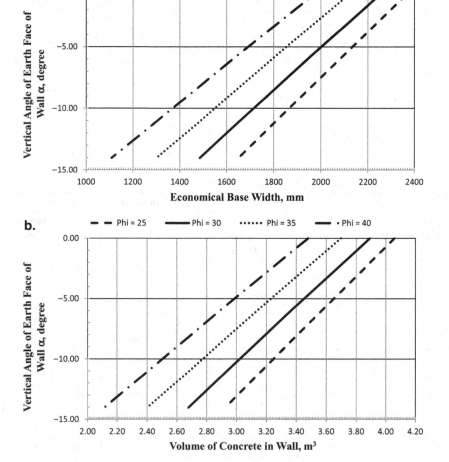

FIGURE 6.31 Design Charts for a Concrete Wall H = 3.0 m, $\gamma_{concrete}$ = 24.52 kN/m³, γ_{sat} = 17.27 kN/m³ and $\iota = \phi$: (a) Economical Base Width for a Wall, (b) Volume of a Concrete Wall

6.7 CHARTS FOR THE DESIGN OF A 3.5-M-HIGH CONCRETE BREAST WALL

The charts/tables developed for a 3.5-m-high concrete breast wall for previously discussed six combinations are presented in Tables 6.31 to 6.36 and Figures 6.32 to 6.37.

TABLE 6.31

$H = 3.5$ m, $\gamma_{concrete} = 24.52$ kN/m³, $\gamma_{sat} = 15.70$ kN/m³ and $\iota = 0°$

A. Results for Stability Analysis

ϕ Degree	α Degree	K_{ah}	F_{OVT}	Pressure at Heel, kN/m²	Pressure at Toe, kN/m²	$\mu = 0.50$ F_{SLD}	$\mu = 0.50$ Net Force, kN	$\mu = 0.70$ F_{SLD}	$\mu = 0.70$ Net Force, kN	$\mu = 0.80$ F_{SLD}	$\mu = 0.80$ Net Force, kN
25	0.00	0.373	1.84	0.31	89.57	0.93	22.50	1.30	7.99	1.48	0.74
25	−5.71	0.339	1.67	0.12	92.87	0.85	23.38	1.19	11.05	1.36	4.88
25	−11.31	0.305	1.51	0.03	97.52	0.78	23.95	1.09	13.69	1.24	8.56
25	−14.03	0.288	1.50	8.12	91.71	0.75	23.53	1.06	14.02	1.21	9.26
30	0.00	0.304	1.85	0.47	90.96	1.01	16.09	1.42	2.77	1.62	0.00
30	−5.71	0.270	1.66	0.67	94.83	0.93	17.16	1.30	6.08	1.48	0.55
30	−11.31	0.237	1.50	3.43	98.25	0.84	17.83	1.17	8.82	1.34	4.32
30	−14.03	0.221	1.50	16.06	88.91	0.81	17.49	1.14	9.23	1.30	5.10
35	0.00	0.246	1.85	0.33	92.91	1.10	10.99	1.54	0.00	1.76	0.00
35	−5.71	0.213	1.65	0.69	97.94	1.00	12.37	1.40	2.48	1.60	0.00
35	−11.31	0.182	1.50	10.14	96.81	0.90	13.08	1.26	5.20	1.44	1.26
35	−14.03	0.166	1.50	30.38	81.43	0.87	12.88	1.22	5.74	1.39	2.17
40	0.00	0.196	1.87	0.64	94.63	1.20	6.86	1.69	0.00	1.93	0.00
40	−5.71	0.166	1.64	2.44	99.92	1.08	8.55	1.51	0.00	1.73	0.00
40	−11.31	0.136	1.50	20.00	94.26	0.96	9.56	1.35	2.73	1.54	0.00
40	−14.03	0.122	1.50	51.97	70.39	0.92	9.63	1.28	3.57	1.47	0.55

B. Physical Dimensions for a Technoeconomic Wall Section

ϕ Degree	α Degree	Top Width, mm	Bottom Width, mm	Horizontal Angle θ Degree	Depth of Shear Key D_{sk}, mm $\mu = 0.50$	$\mu = 0.70$	$\mu = 0.80$	Volume of Wall, m³
25	0.00	300	1795	90.00	900	500	100	3.666
25	−5.71	300	1475	84.29	900	600	400	3.106
25	−11.31	300	1170	78.69	900	700	500	2.573
25	−14.03	300	1060	75.97	900	700	600	2.380
30	0.00	300	1620	90.00	700	200	0	3.360
30	−5.71	300	1290	84.29	700	400	100	2.783
30	−11.31	300	985	78.69	700	500	300	2.249
30	−14.03	300	875	75.97	700	500	300	2.056
35	0.00	300	1455	90.00	400	0	0	3.071

(Continued)

TABLE 6.31 (Continued)

H = 3.5 m, $\gamma_{concrete}$ = 24.52 kN/m³, γ_{sat} = 15.70 kN/m³ and ι = 0°

ϕ Degree	α Degree	Top Width, mm	Bottom Width, mm	Horizontal Angle θ Degree	Depth of Shear Key D_{sk}, mm			Volume of Wall, m³
					$\mu = 0.50$	$\mu = 0.70$	$\mu = 0.80$	
35	−5.71	300	1115	84.29	500	200	0	2.476
35	−11.31	300	820	78.69	500	300	100	1.960
35	−14.03	300	710	75.97	500	300	200	1.768
40	0.00	300	1305	90.00	200	0	0	2.809
40	−5.71	300	960	84.29	300	0	0	2.205
40	−11.31	300	665	78.69	300	200	0	1.689
40	−14.03	300	550	75.97	400	300	100	1.488

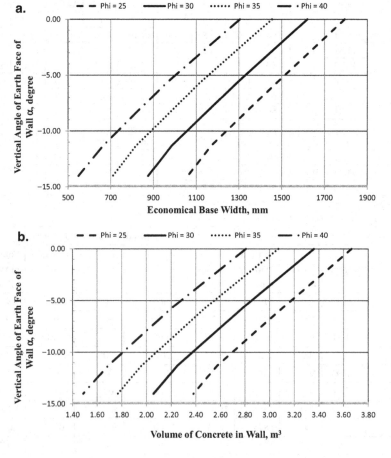

FIGURE 6.32 Design Charts for a Concrete Wall H = 3.5 m, $\gamma_{concrete}$ = 24.52 kN/m³, γ_{sat} = 15.70 kN/m³ and ι = 0°: (a) Economical Base Width for a Wall, (b) Volume of a Concrete Wall

TABLE 6.32

H = 3.5 m, $\gamma_{concrete}$ = 24.52 kN/m³, γ_{sat} = 17.27 kN/m³ and ι = 0°

A. Results for Stability Analysis

ϕ Degree	α Degree	K_{ah}	F_{OVT}	Pressure at Heel, kN/m²	Toe, kN/m²	$\mu = 0.50$ F_{SLD}	Net Force, kN	$\mu = 0.70$ F_{SLD}	Net Force, kN	$\mu = 0.80$ F_{SLD}	Net Force, kN
25	0.00	0.373	1.83	0.24	88.96	0.88	26.33	1.24	11.22	1.41	3.66
25	−5.71	0.339	1.68	0.38	91.61	0.82	26.74	1.15	13.79	1.31	7.32
25	−11.31	0.305	1.53	0.24	95.81	0.75	26.90	1.05	16.07	1.20	10.65
25	−14.03	0.288	1.50	6.09	92.13	0.73	26.41	1.02	16.39	1.17	11.38
30	0.00	0.304	1.84	0.33	90.35	0.97	19.11	1.35	5.25	1.55	0.00
30	−5.71	0.270	1.66	0.21	94.19	0.89	19.78	1.25	8.20	1.43	2.40
30	−11.31	0.237	1.50	1.28	98.78	0.81	20.14	1.13	10.69	1.30	5.97
30	−14.03	0.221	1.50	12.98	89.95	0.79	19.56	1.11	10.87	1.26	6.52
35	0.00	0.246	1.84	0.06	92.36	1.06	13.34	1.48	0.67	1.69	0.00
35	−5.71	0.213	1.65	0.02	97.34	0.96	14.34	1.35	4.01	1.54	0.00
35	−11.31	0.182	1.50	6.77	98.20	0.87	14.76	1.22	6.51	1.40	2.38
35	−14.03	0.166	1.50	25.14	84.05	0.85	14.34	1.19	6.83	1.36	3.08
40	0.00	0.196	1.85	0.13	94.27	1.15	8.67	1.62	0.00	1.85	0.00
40	−5.71	0.166	1.64	1.26	99.65	1.05	10.00	1.46	0.79	1.67	0.00
40	−11.31	0.136	1.50	16.22	95.34	0.94	10.68	1.32	3.51	1.50	0.00
40	−14.03	0.122	1.50	44.65	73.80	0.90	10.54	1.27	4.15	1.45	0.95

B. Physical Dimensions for a Technoeconomic Wall Section

ϕ Degree	α Degree	Top Width, mm	Bottom Width, mm	Horizontal Angle θ Degree	Depth of Shear Key D_{sk}, mm $\mu = 0.50$	$\mu = 0.70$	$\mu = 0.80$	Volume of Wall, m³
25	0.00	300	1885	90.00	900	600	300	3.824
25	−5.71	300	1565	84.29	1000	700	500	3.264
25	−11.31	300	1255	78.69	1000	700	600	2.721
25	−14.03	300	1135	75.97	900	700	600	2.511
30	0.00	300	1700	90.00	600	200	0	3.500
30	−5.71	300	1365	84.29	700	400	200	2.914
30	−11.31	300	1050	78.69	700	500	400	2.363
30	−14.03	300	940	75.97	700	500	400	2.170
35	0.00	300	1525	90.00	400	0	0	3.194
35	−5.71	300	1180	84.29	500	200	0	2.590
35	−11.31	300	875	78.69	500	300	200	2.056

(Continued)

TABLE 6.32 (Continued)

H = 3.5 m, $\gamma_{concrete}$ = 24.52 kN/m³, γ_{sat} = 17.27 kN/m³ and ι = 0°

ϕ Degree	α Degree	Top Width, mm	Bottom Width, mm	Horizontal Angle θ Degree	Depth of Shear Key D_{sk}, mm			Volume of Wall, m³
					μ = 0.50	μ = 0.70	μ = 0.80	
35	−14.03	300	765	75.97	500	300	200	1.864
40	0.00	300	1365	90.00	200	0	0	2.914
40	−5.71	300	1015	84.29	300	100	0	2.301
40	−11.31	300	715	78.69	300	200	0	1.776
40	−14.03	300	600	75.97	400	300	200	1.575

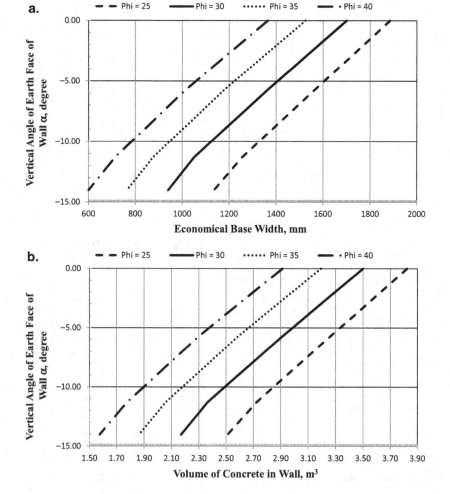

FIGURE 6.33 Design Charts for a Concrete Wall H = 3.5 m, $\gamma_{concrete}$ = 24.52 kN/m³, γ_{sat} = 17.27 kN/m³ and ι = 0°: (a) Economical Base Width for a Wall, (b) Volume of a Concrete Wall

TABLE 6.33

H = 3.5 m, $\gamma_{concrete}$ = 24.52 kN/m³, γ_{sat} = 15.70 kN/m³ and ι = ½ϕ

A. Results for Stability Analysis

ϕ Degree	α Degree	K_{ah}	F_{OVT}	Pressure at Heel, kN/m²	Toe, kN/m²	μ = 0.50 F_{SLD}	Net Force, kN	μ = 0.70 F_{SLD}	Net Force, kN	μ = 0.80 F_{SLD}	Net Force, kN
25	0.00	0.450	1.83	0.24	88.34	0.85	29.95	1.19	14.23	1.36	6.36
25	−5.71	0.407	1.68	0.26	90.93	0.80	29.72	1.12	16.24	1.28	9.49
25	−11.31	0.364	1.54	0.64	94.26	0.74	29.29	1.03	17.94	1.18	12.27
25	−14.03	0.344	1.50	4.06	93.03	0.71	28.74	1.00	18.31	1.14	13.10
30	0.00	0.373	1.84	0.28	89.60	0.93	22.51	1.30	8.01	1.48	0.75
30	−5.71	0.329	1.67	0.33	92.96	0.86	22.42	1.21	10.26	1.38	4.18
30	−11.31	0.286	1.51	0.58	97.95	0.79	22.24	1.11	12.32	1.27	7.36
30	−14.03	0.266	1.50	10.06	91.23	0.77	21.48	1.08	12.37	1.23	7.82
35	0.00	0.305	1.84	0.22	91.22	1.01	16.22	1.41	2.90	1.61	0.00
35	−5.71	0.262	1.65	0.01	95.97	0.93	16.48	1.31	5.60	1.49	0.17
35	−11.31	0.220	1.50	4.38	98.70	0.85	16.41	1.19	7.75	1.37	3.41
35	−14.03	0.201	1.50	20.96	85.98	0.83	15.72	1.17	7.83	1.34	3.89
40	0.00	0.245	1.86	0.58	92.66	1.11	10.89	1.55	0.00	1.77	0.00
40	−5.71	0.204	1.64	0.92	98.35	1.01	11.59	1.42	1.90	1.62	0.00
40	−11.31	0.165	1.50	12.42	96.78	0.92	11.81	1.29	4.30	1.47	0.54
40	−14.03	0.147	1.50	37.88	77.58	0.89	11.42	1.25	4.72	1.43	1.37

B. Physical Dimensions for a Technoeconomic Wall Section

ϕ Degree	α Degree	Top Width, mm	Bottom Width, mm	Horizontal Angle θ Degree	Depth of Shear Key D_{sk}, mm μ = 0.50	μ = 0.70	μ = 0.80	Volume of Wall, m³
25	0.00	300	1975	90.00	1100	700	500	3.981
25	−5.71	300	1645	84.29	1100	800	600	3.404
25	−11.31	300	1330	78.69	1000	800	700	2.853
25	−14.03	300	1195	75.97	1000	800	700	2.616
30	0.00	300	1795	90.00	700	400	0	3.666
30	−5.71	300	1450	84.29	800	500	300	3.063
30	−11.31	300	1120	78.69	800	600	400	2.485
30	−14.03	300	1000	75.97	800	600	400	2.275
35	0.00	300	1620	90.00	500	100	0	3.360
35	−5.71	300	1260	84.29	600	300	0	2.730
35	−11.31	300	935	78.69	600	400	200	2.161

(Continued)

TABLE 6.33 (Continued)

H = 3.5 m, $\gamma_{concrete}$ = 24.52 kN/m³, γ_{sat} = 15.70 kN/m³ and ι = ½ϕ

ϕ Degree	α Degree	Top Width, mm	Bottom Width, mm	Horizontal Angle θ Degree	Depth of Shear Key D_{sk}, mm			Volume of Wall, m³
					μ = 0.50	μ = 0.70	μ = 0.80	
35	−14.03	300	820	75.97	600	400	200	1.960
40	0.00	300	1455	90.00	300	0	0	3.071
40	−5.71	300	1085	84.29	400	100	0	2.424
40	−11.31	300	765	78.69	400	200	0	1.864
40	−14.03	300	645	75.97	400	200	100	1.654

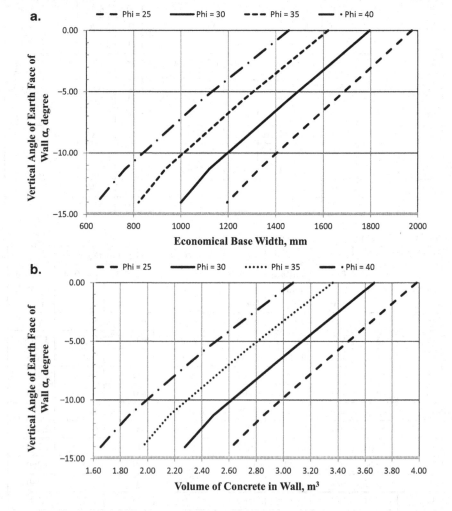

FIGURE 6.34 Design Charts for a Concrete Wall H = 3.5 m, $\gamma_{concrete}$ = 24.52 kN/m³, γ_{sat} = 15.70 kN/m³ and ι = ½ϕ: (a) Economical Base Width for a Wall, (b) Volume of a Concrete Wall

TABLE 6.34

H = 3.5 m, $\gamma_{concrete}$ = 24.52 kN/m³, γ_{sat} = 17.27 kN/m³ and ι = ½ϕ

A. Results for Stability Analysis

ϕ Degree	α Degree	K_{ah}	F_{OVT}	Pressure at Heel, kN/m²	Toe, kN/m²	μ = 0.50 F_{SLD}	Net Force, kN	μ = 0.70 F_{SLD}	Net Force, kN	μ = 0.80 F_{SLD}	Net Force, kN
25	0.00	0.450	1.83	0.21	87.74	0.81	34.74	1.14	18.33	1.30	10.13
25	−5.71	0.407	1.69	0.09	90.25	0.76	33.97	1.07	19.84	1.22	12.78
25	−11.31	0.364	1.55	0.33	93.34	0.71	33.01	1.00	21.05	1.14	15.07
25	−14.03	0.344	1.50	2.12	93.62	0.69	32.34	0.96	21.36	1.10	15.87
30	0.00	0.373	1.83	0.21	88.99	0.88	26.35	1.24	11.24	1.41	3.68
30	−5.71	0.329	1.68	0.15	92.17	0.83	25.72	1.16	12.98	1.33	6.61
30	−11.31	0.286	1.52	0.34	96.66	0.77	25.01	1.07	14.54	1.23	9.31
30	−14.03	0.266	1.50	7.39	92.22	0.75	24.12	1.05	14.54	1.20	9.75
35	0.00	0.305	1.84	0.08	90.61	0.96	19.24	1.35	5.38	1.54	0.00
35	−5.71	0.262	1.67	0.50	94.25	0.90	18.88	1.26	7.47	1.44	1.76
35	−11.31	0.220	1.50	2.65	98.64	0.83	18.48	1.16	9.38	1.32	4.82
35	−14.03	0.201	1.50	16.82	87.98	0.81	17.59	1.14	9.30	1.30	5.15
40	0.00	0.245	1.85	0.31	92.11	1.06	13.23	1.48	0.56	1.69	0.00
40	−5.71	0.204	1.65	0.50	97.43	0.98	13.42	1.37	3.29	1.57	0.00
40	−11.31	0.165	1.50	9.48	97.46	0.90	13.25	1.26	5.37	1.44	1.43
40	−14.03	0.147	1.50	31.02	81.57	0.87	12.69	1.22	5.66	1.39	2.14

B. Physical Dimensions for a Technoeconomic Wall Section

ϕ Degree	α Degree	Top Width, mm	Bottom Width, mm	Horizontal Angle θ Degree	Depth of Shear Key D_{sk}, mm μ = 0.50	μ = 0.70	μ = 0.80	Volume of Wall, m³
25	0.00	300	2075	90.00	1000	700	500	4.156
25	−5.71	300	1740	84.29	1100	800	600	3.570
25	−11.31	300	1420	78.69	1100	800	700	3.010
25	−14.03	300	1275	75.97	1100	800	700	2.756
30	0.00	300	1885	90.00	700	400	200	3.824
30	−5.71	300	1535	84.29	800	600	400	3.211
30	−11.31	300	1200	78.69	800	600	500	2.625
30	−14.03	300	1070	75.97	800	600	500	2.398
35	0.00	300	1700	90.00	500	200	0	3.500
35	−5.71	300	1340	84.29	600	300	100	2.870
35	−11.31	300	1000	78.69	600	400	300	2.275

(Continued)

TABLE 6.34 (Continued)

$H = 3.5$ m, $\gamma_{concrete} = 24.52$ kN/m³, $\gamma_{sat} = 17.27$ kN/m³ and $\iota = \frac{1}{2}\phi$

ϕ Degree	α Degree	Top Width, mm	Bottom Width, mm	Horizontal Angle θ Degree	Depth of Shear Key D_{sk}, mm			Volume of Wall, m³
					$\mu = 0.50$	$\mu = 0.70$	$\mu = 0.80$	
35	−14.03	300	880	75.97	600	400	300	2.065
40	0.00	300	1525	90.00	300	0	0	3.194
40	−5.71	300	1150	84.29	400	200	0	2.538
40	−11.31	300	820	78.69	400	200	100	1.960
40	−14.03	300	695	75.97	400	200	100	1.741

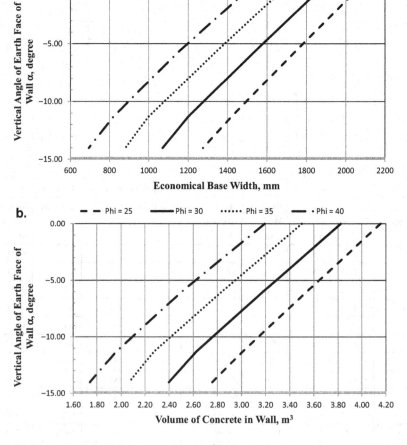

FIGURE 6.35 Design Charts for a Concrete Wall $H = 3.5$ m, $\gamma_{concrete} = 24.52$ kN/m³, $\gamma_{sat} = 17.27$ kN/m³ and $\iota = \frac{1}{2}\phi$: (a) Economical Base Width for a Wall, (b) Volume of a Concrete Wall

TABLE 6.35

$H = 3.5$ m, $\gamma_{concrete} = 24.52$ kN/m³, $\gamma_{sat} = 15.70$ kN/m³ and $\iota = \phi$

A. Results for Stability Analysis

ϕ Degree	α Degree	K_{ah}	F_{OVT}	Pressure at Heel, kN/m²	Toe, kN/m²	$\mu = 0.50$ F_{SLD}	Net Force, kN	$\mu = 0.70$ F_{SLD}	Net Force, kN	$\mu = 0.80$ F_{SLD}	Net Force, kN
25	0.00	0.821	1.82	0.37	84.74	0.64	68.30	0.90	47.71	1.03	37.42
25	−5.71	0.746	1.71	0.21	86.35	0.62	64.15	0.87	45.97	1.00	36.89
25	−11.31	0.675	1.61	0.14	88.30	0.60	60.23	0.83	44.36	0.95	36.43
25	−14.03	0.641	1.56	0.53	89.04	0.58	58.29	0.81	43.51	0.93	36.12
30	0.00	0.750	1.82	0.23	85.34	0.67	60.71	0.94	40.97	1.08	31.11
30	−5.71	0.666	1.71	0.40	86.90	0.65	55.63	0.92	38.45	1.05	29.85
30	−11.31	0.587	1.59	0.30	89.35	0.63	51.07	0.88	36.36	1.00	29.01
30	−14.03	0.549	1.54	0.36	90.79	0.61	48.93	0.86	35.41	0.98	28.65
35	0.00	0.671	1.82	0.06	86.11	0.71	52.42	0.99	33.67	1.13	24.30
35	−5.71	0.580	1.70	0.07	88.22	0.69	46.88	0.97	30.85	1.11	22.83
35	−11.31	0.496	1.58	0.21	91.08	0.67	41.98	0.93	28.56	1.07	21.85
35	−14.03	0.456	1.51	0.42	92.88	0.65	39.75	0.91	27.58	1.04	21.50
40	0.00	0.587	1.82	0.12	86.81	0.75	43.69	1.06	26.03	1.21	17.19
40	−5.71	0.492	1.70	0.39	89.14	0.74	37.99	1.04	23.18	1.18	15.77
40	−11.31	0.406	1.55	0.25	93.30	0.71	33.24	1.00	21.21	1.14	15.20
40	−14.03	0.366	1.50	3.32	92.82	0.70	30.88	0.98	20.07	1.12	14.67

B. Physical Dimensions for a Technoeconomic Wall Section

ϕ Degree	α Degree	Top Width, mm	Bottom Width, mm	Horizontal Angle θ Degree	Depth of Shear Key D_{sk}, mm $\mu = 0.50$	$\mu = 0.70$	$\mu = 0.80$	Volume of Wall, m³
25	0.00	300	2690	90.00	1500	1300	1100	5.233
25	−5.71	300	2335	84.29	1500	1200	1100	4.611
25	−11.31	300	1995	78.69	1500	1300	1200	4.016
25	−14.03	300	1835	75.97	1500	1300	1200	3.736
30	0.00	300	2565	90.00	1100	900	700	5.014
30	−5.71	300	2190	84.29	1200	900	800	4.358
30	−11.31	300	1825	78.69	1100	900	800	3.719
30	−14.03	300	1650	75.97	1200	1000	900	3.413
35	0.00	300	2420	90.00	800	600	500	4.760
35	−5.71	300	2020	84.29	900	700	600	4.060
35	−11.31	300	1635	78.69	800	700	600	3.386

(Continued)

TABLE 6.35 (Continued)

H = 3.5 m, $\gamma_{concrete}$ = 24.52 kN/m³, γ_{sat} = 15.70 kN/m³ and $\iota = \phi$

ϕ Degree	α Degree	Top Width, mm	Bottom Width, mm	Horizontal Angle θ Degree	Depth of Shear Key D_{sk}, mm			Volume of Wall, m³
					$\mu = 0.50$	$\mu = 0.70$	$\mu = 0.80$	
35	−14.03	300	1450	75.97	800	600	600	3.063
40	0.00	300	2260	90.00	600	400	300	4.480
40	−5.71	300	1840	84.29	500	400	300	3.745
40	−11.31	300	1430	78.69	600	400	300	3.028
40	−14.03	300	1250	75.97	600	400	300	2.713

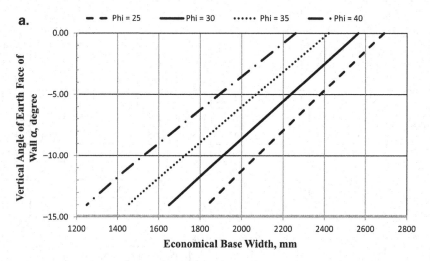

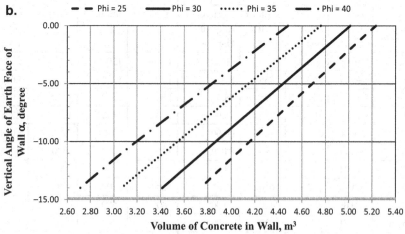

FIGURE 6.36 Design Charts for a Concrete Wall H = 3.5 m, $\gamma_{concrete}$ = 24.52 kN/m³, γ_{sat} = 15.70 kN/m³ and $\iota = \phi$: (a) Economical Base Width for a Wall, (b) Volume of a Concrete Wall

TABLE 6.36

H = 3.5 m, $\gamma_{concrete}$ = 24.52 kN/m³, γ_{sat} = 17.27 kN/m³ and $\iota = \phi$

A. Results for Stability Analysis

ϕ Degree	Y Degree	K_{ah}	F_{OVT}	Pressure at		$\mu = 0.50$		$\mu = 0.70$		$\mu = 0.80$	
				Heel, kN/m²	Toe, kN/m²	F_{SLD}	Net Force, kN	F_{SLD}	Net Force, kN	F_{SLD}	Net Force, kN
25	0.00	0.821	1.82	0.23	84.42	0.61	77.84	0.86	56.34	0.98	45.59
25	−5.71	0.746	1.71	0.08	85.90	0.59	72.69	0.83	53.64	0.95	44.11
25	−11.31	0.675	1.61	0.08	87.61	0.57	67.83	0.80	51.12	0.91	42.76
25	−14.03	0.641	1.57	0.26	88.45	0.56	65.49	0.78	49.89	0.90	42.10
30	0.00	0.750	1.82	0.17	84.92	0.64	69.30	0.90	48.68	1.02	38.38
30	−5.71	0.666	1.71	0.05	86.63	0.62	63.19	0.87	45.18	1.00	36.18
30	−11.31	0.587	1.60	0.08	88.73	0.60	57.57	0.84	42.08	0.97	34.33
30	−14.03	0.549	1.55	0.56	89.57	0.59	54.88	0.83	40.58	0.95	33.43
35	0.00	0.671	1.82	0.15	85.50	0.67	59.97	0.94	40.37	1.08	30.57
35	−5.71	0.580	1.70	0.00	87.60	0.66	53.29	0.93	36.47	1.06	28.06
35	−11.31	0.496	1.59	0.49	89.80	0.64	47.25	0.90	33.09	1.03	26.00
35	−14.03	0.456	1.53	0.68	91.42	0.63	44.54	0.88	31.66	1.01	25.23
40	0.00	0.587	1.82	0.10	86.28	0.72	50.19	1.01	31.74	1.15	22.52
40	−5.71	0.492	1.70	0.24	88.54	0.71	43.30	0.99	27.77	1.13	20.01
40	−11.31	0.406	1.57	0.64	91.73	0.69	37.39	0.96	24.68	1.10	18.33
40	−14.03	0.366	1.50	1.04	93.86	0.67	34.79	0.94	23.44	1.08	17.77

B. Physical Dimensions for a Technoeconomic Wall Section

ϕ Degree	α Degree	Top Width, mm	Bottom Width, mm	Horizontal Angle θ Degree	Depth of Shear Key D_{sk}, mm			Volume of Wall, m³
					$\mu = 0.50$	$\mu = 0.70$	$\mu = 0.80$	
25	0.00	300	2825	90.00	1600	1300	1200	5.469
25	−5.71	300	2465	84.29	1500	1300	1100	4.839
25	−11.31	300	2120	78.69	1400	1200	1100	4.235
25	−14.03	300	1955	75.97	1500	1300	1200	3.946
30	0.00	300	2695	90.00	1200	900	800	5.241
30	−5.71	300	2310	84.29	1200	1000	900	4.568
30	−11.31	300	1940	78.69	1100	900	800	3.920
30	−14.03	300	1765	75.97	1100	900	800	3.614
35	0.00	300	2545	90.00	900	700	500	4.979
35	−5.71	300	2135	84.29	800	600	500	4.261
35	−11.31	300	1745	78.69	800	700	600	3.579

(Continued)

TABLE 6.36 (Continued)

H = 3.5 m, $\gamma_{concrete}$ = 24.52 kN/m³, γ_{sat} = 17.27 kN/m³ and ι = ϕ

					Depth of Shear Key D_{sk}, mm			
ϕ Degree	α Degree	Top Width, mm	Bottom Width, mm	Horizontal Angle θ Degree	μ = 0.50	μ = 0.70	μ = 0.80	Volume of Wall, m³
35	−14.03	300	1555	75.97	800	700	600	3.246
40	0.00	300	2375	90.00	600	400	300	4.681
40	−5.71	300	1945	84.29	500	400	300	3.929
40	−11.31	300	1530	78.69	600	500	400	3.203
40	−14.03	300	1330	75.97	600	400	400	2.853

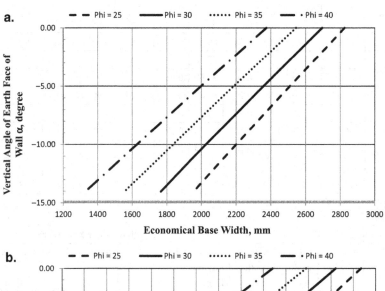

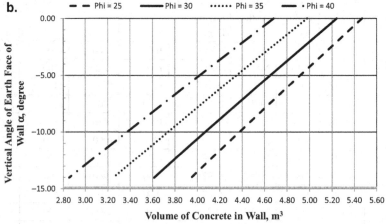

FIGURE 6.37 Design Charts for a Concrete Wall H = 3.5 m, $\gamma_{concrete}$ = 24.52 kN/m³, γ_{sat} = 17.27 kN/m³ and ι = ϕ: (a) Economical Base Width for a Wall, (b) Volume of a Concrete Wall

6.8 CHARTS FOR THE DESIGN OF A 4.0-M-HIGH CONCRETE BREAST WALL

The charts/tables developed for a 4.0-m-high concrete breast wall for the previously discussed six combinations are presented in Tables 6.37 to 6.42 and Figures 6.38 to 6.43.

TABLE 6.37
$H = 4.0$ m, $\gamma_{concrete} = 24.52$ kN/m^3, $\gamma_{sat} = 15.70$ kN/m^3 and $\iota = 0°$

A. Results for Stability Analysis

ϕ Degree	α Degree	K_{ah}	F_{OVT}	Pressure at Heel, kN/m^2	Pressure at Toe, kN/m^2	$\mu = 0.50$ F_{SLD}	$\mu = 0.50$ Net Force, kN	$\mu = 0.70$ F_{SLD}	$\mu = 0.70$ Net Force, kN	$\mu = 0.80$ F_{SLD}	$\mu = 0.80$ Net Force, kN
25	0.00	0.373	1.83	0.63	99.91	0.91	29.97	1.28	11.25	1.46	1.90
25	−5.71	0.339	1.71	3.52	99.77	0.85	30.53	1.19	14.41	1.37	6.36
25	−11.31	0.305	1.58	7.82	99.38	0.79	30.81	1.10	17.22	1.26	10.43
25	−14.03	0.288	1.51	10.20	99.67	0.75	30.93	1.05	18.58	1.20	12.41
30	0.00	0.304	1.87	2.41	99.53	1.00	21.31	1.41	4.03	1.61	0.00
30	−5.71	0.270	1.71	5.58	99.98	0.93	22.28	1.30	7.76	1.49	0.50
30	−11.31	0.237	1.57	11.30	99.76	0.85	22.98	1.18	11.10	1.35	5.16
30	−14.03	0.221	1.50	16.52	98.38	0.80	23.20	1.12	12.56	1.28	7.24
35	0.00	0.246	1.90	4.06	99.48	1.10	14.39	1.54	0.00	1.76	0.00
35	−5.71	0.213	1.73	8.90	99.30	1.01	15.73	1.42	2.64	1.62	0.00
35	−11.31	0.182	1.55	17.04	99.01	0.90	17.01	1.27	6.68	1.45	1.51
35	−14.03	0.166	1.50	30.41	90.95	0.86	17.21	1.20	8.05	1.37	3.46
40	0.00	0.196	1.93	5.73	99.64	1.21	8.87	1.69	0.00	1.93	0.00
40	−5.71	0.166	1.73	11.78	99.81	1.10	10.79	1.53	0.00	1.75	0.00
40	−11.31	0.136	1.53	24.51	98.61	0.95	12.66	1.34	3.81	1.53	0.00
40	−14.03	0.122	1.50	51.94	78.98	0.90	12.91	1.26	5.14	1.44	1.25

B. Physical Dimensions for a Technoeconomic Wall Section

ϕ Degree	α Degree	Top Width, mm	Bottom Width, mm	Horizontal Angle θ Degree	Depth of Shear Key D_{sk}, mm $\mu = 0.50$	Depth of Shear Key D_{sk}, mm $\mu = 0.70$	Depth of Shear Key D_{sk}, mm $\mu = 0.80$	Volume of Wall, m^3
25	0.00	300	2070	90.00	1000	500	100	4.740
25	−5.71	300	1735	84.29	1100	700	500	4.070
25	−11.31	300	1410	78.69	1100	800	600	3.420
25	−14.03	300	1250	75.97	1100	800	700	3.100
30	0.00	300	1885	90.00	700	200	0	4.370
30	−5.71	300	1530	84.29	800	400	100	3.660
30	−11.31	300	1190	78.69	800	500	400	2.980
30	−14.03	300	1030	75.97	800	600	400	2.660
35	0.00	300	1710	90.00	400	0	0	4.020
35	−5.71	300	1345	84.29	600	200	0	3.290
35	−11.31	300	990	78.69	600	300	100	2.580

(Continued)

TABLE 6.37 (Continued)

$H = 4.0$ m, $\gamma_{concrete} = 24.52$ kN/m³, $\gamma_{sat} = 15.70$ kN/m³ and $\iota = 0°$

ϕ Degree	α Degree	Top Width, mm	Bottom Width, mm	Horizontal Angle θ Degree	Depth of Shear Key D_{sk}, mm			Volume of Wall, m³
					$\mu = 0.50$	$\mu = 0.70$	$\mu = 0.80$	
35	−14.03	300	840	75.97	600	400	200	2.280
40	0.00	300	1545	90.00	200	0	0	3.690
40	−5.71	300	1165	84.29	400	0	0	2.930
40	−11.31	300	800	78.69	400	200	0	2.200
40	−14.03	300	660	75.97	400	200	100	1.920

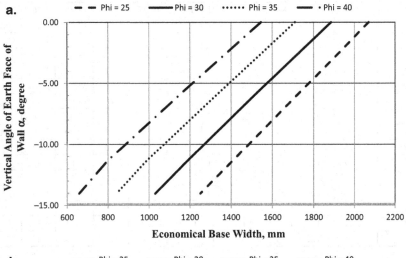

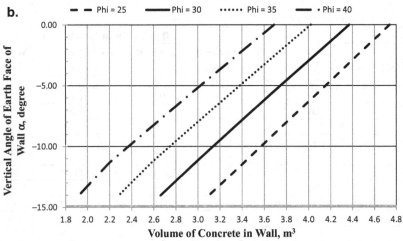

FIGURE 6.38 Design Charts for a Concrete Wall $H = 4.0$ m, $\gamma_{concrete} = 24.52$ kN/m³, $\gamma_{sat} = 15.70$ kN/m³ and $\iota = 0°$: (a) Economical Base Width for a Wall, (b) Volume of a Concrete Wall

TABLE 6.38

$H = 4.0$ m, $\gamma_{concrete} = 24.52$ kN/m³, $\gamma_{sat} = 17.27$ kN/m³ and $\iota = 0°$

A. Results for Stability Analysis

				Pressure at		$\mu = 0.50$		$\mu = 0.70$		$\mu = 0.80$	
ϕ Degree	α Degree	K_{ah}	F_{OVT}	Heel, kN/m²	Toe, kN/m²	F_{SLD}	Net Force, kN	F_{SLD}	Net Force, kN	F_{SLD}	Net Force, kN
25	0.00	0.373	1.83	0.26	99.63	0.87	35.04	1.22	15.55	1.40	5.80
25	−5.71	0.339	1.70	2.69	99.72	0.82	35.06	1.15	18.21	1.31	9.79
25	−11.31	0.305	1.58	6.34	99.61	0.76	34.81	1.06	20.52	1.22	13.38
25	−14.03	0.288	1.52	8.79	99.50	0.73	34.61	1.02	21.56	1.16	15.04
30	0.00	0.304	1.85	1.53	99.74	0.96	25.37	1.34	7.44	1.53	0.00
30	−5.71	0.270	1.71	4.69	99.86	0.89	25.71	1.25	10.53	1.43	2.94
30	−11.31	0.237	1.57	9.62	99.88	0.82	25.89	1.15	13.39	1.31	7.14
30	−14.03	0.221	1.50	13.59	99.37	0.78	25.90	1.09	14.68	1.25	9.07
35	0.00	0.246	1.88	3.03	99.78	1.05	17.55	1.47	1.05	1.68	0.00
35	−5.71	0.213	1.73	7.78	99.27	0.98	18.28	1.37	4.62	1.56	0.00
35	−11.31	0.182	1.56	14.82	99.27	0.88	19.07	1.23	8.19	1.41	2.76
35	−14.03	0.166	1.50	24.45	94.58	0.83	19.18	1.17	9.55	1.34	4.73
40	0.00	0.196	1.92	5.06	99.49	1.16	11.20	1.62	0.00	1.85	0.00
40	−5.71	0.166	1.73	10.34	99.91	1.06	12.65	1.48	0.46	1.70	0.00
40	−11.31	0.136	1.53	21.29	99.26	0.94	14.06	1.31	4.74	1.50	0.08
40	−14.03	0.122	1.50	44.37	83.12	0.89	14.14	1.24	5.95	1.42	1.85

B. Physical Dimensions for a Technoeconomic Wall Section

					Depth of Shear Key D_{sk}, mm			
ϕ Degree	α Degree	Top Width, mm	Bottom Width, mm	Horizontal Angle θ Degree	$\mu = 0.50$	$\mu = 0.70$	$\mu = 0.80$	Volume of Wall, m³
25	0.00	300	2170	90.00	1000	600	300	4.940
25	−5.71	300	1830	84.29	1100	800	600	4.260
25	−11.31	300	1500	78.69	1100	800	700	3.600
25	−14.03	300	1340	75.97	1100	800	700	3.280
30	0.00	300	1970	90.00	700	300	0	4.540
30	−5.71	300	1615	84.29	800	500	200	3.830
30	−11.31	300	1270	78.69	800	600	400	3.140
30	−14.03	300	1105	75.97	800	600	500	2.810
35	0.00	300	1785	90.00	500	0	0	4.170
35	−5.71	300	1420	84.29	500	200	0	3.440
35	−11.31	300	1060	78.69	600	400	200	2.720

(Continued)

TABLE 6.38 (Continued)

H = 4.0 m, $\gamma_{concrete}$ = 24.52 kN/m³, γ_{sat} = 17.27 kN/m³ and ι = 0°

φ Degree	α Degree	Top Width, mm	Bottom Width, mm	Horizontal Angle θ Degree	Depth of Shear Key D_{sk}, mm μ = 0.50	μ = 0.70	μ = 0.80	Volume of Wall, m³
35	−14.03	300	900	75.97	600	400	300	2.400
40	0.00	300	1615	90.00	300	0	0	3.830
40	−5.71	300	1230	84.29	300	0	0	3.060
40	−11.31	300	860	78.69	400	200	0	2.320
40	−14.03	300	715	75.97	400	200	100	2.030

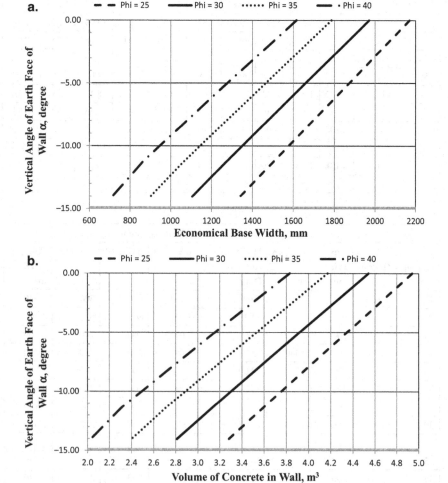

FIGURE 6.39 Design Charts for a Concrete Wall H = 4.0 m, $\gamma_{concrete}$ = 24.52 kN/m³, γ_{sat} = 17.27 kN/m³ and ι = 0°: (a) Economical Base Width for a Wall, (b) Volume of a Concrete Wall

TABLE 6.39

$H = 4.0$ m, $\gamma_{concrete} = 24.52$ kN/m^3, $\gamma_{sat} = 15.70$ kN/m^3 and $\iota = \frac{1}{2}\phi$

A. Results for Stability Analysis

ϕ Degree	α Degree	K_{ah}	F_{OVT}	Pressure at Heel, kN/m^2	Toe, kN/m^2	$\mu = 0.50$ F_{SLD}	Net Force, kN	$\mu = 0.70$ F_{SLD}	Net Force, kN	$\mu = 0.80$ F_{SLD}	Net Force, kN
25	0.00	0.450	1.82	0.00	99.29	0.84	39.82	1.18	19.56	1.34	9.42
25	−5.71	0.407	1.70	1.74	99.96	0.79	39.09	1.11	21.58	1.27	12.82
25	−11.31	0.364	1.59	5.53	99.42	0.74	38.03	1.04	23.13	1.19	15.67
25	−14.03	0.344	1.53	7.65	99.48	0.71	37.53	1.00	23.90	1.14	17.09
30	0.00	0.373	1.83	0.60	99.94	0.91	29.99	1.28	11.28	1.46	1.92
30	−5.71	0.329	1.71	3.65	99.94	0.86	29.30	1.21	13.42	1.38	5.48
30	−11.31	0.286	1.57	8.30	99.83	0.80	28.63	1.12	15.51	1.28	8.95
30	−14.03	0.266	1.51	11.99	99.28	0.77	28.28	1.07	16.48	1.23	10.57
35	0.00	0.305	1.86	2.13	99.81	1.00	21.47	1.40	4.20	1.60	0.00
35	−5.71	0.262	1.72	6.31	99.57	0.94	21.20	1.32	6.88	1.51	0.00
35	−11.31	0.220	1.57	13.20	99.06	0.86	21.09	1.21	9.64	1.38	3.91
35	−14.03	0.201	1.50	20.96	95.86	0.82	20.93	1.15	10.79	1.31	5.73
40	0.00	0.245	1.89	3.79	99.79	1.10	14.35	1.54	0.00	1.76	0.00
40	−5.71	0.204	1.73	9.30	99.50	1.03	14.72	1.44	1.90	1.64	0.00
40	−11.31	0.165	1.55	19.28	98.86	0.92	15.38	1.29	5.56	1.48	0.64
40	−14.03	0.147	1.50	37.40	87.40	0.88	15.32	1.23	6.73	1.40	2.44

B. Physical Dimensions for a Technoeconomic Wall Section

ϕ Degree	α Degree	Top Width, mm	Bottom Width, mm	Horizontal Angle θ Degree	Depth of Shear Key D_{sk}, mm $\mu = 0.50$	$\mu = 0.70$	$\mu = 0.80$	Volume of Wall, m^3
25	0.00	300	2270	90.00	1100	700	500	5.140
25	−5.71	300	1915	84.29	1200	900	700	4.430
25	−11.31	300	1580	78.69	1200	900	800	3.760
25	−14.03	300	1415	75.97	1200	900	800	3.430
30	0.00	300	2070	90.00	800	400	100	4.740
30	−5.71	300	1705	84.29	800	500	300	4.010
30	−11.31	300	1350	78.69	900	600	500	3.300
30	−14.03	300	1180	75.97	900	700	500	2.960
35	0.00	300	1885	90.00	600	200	0	4.370
35	−5.71	300	1505	84.29	500	200	0	3.610
35	−11.31	300	1135	78.69	600	400	200	2.870

(Continued)

TABLE 6.39 (Continued)

H = 4.0 m, $\gamma_{concrete}$ = 24.52 kN/m³, γ_{sat} = 15.70 kN/m³ and ι = ½ϕ

ϕ Degree	α Degree	Top Width, mm	Bottom Width, mm	Horizontal Angle θ Degree	Depth of Shear Key D_{sk}, mm			Volume of Wall, m³
					μ = 0.50	μ = 0.70	μ = 0.80	
35	−14.03	300	965	75.97	600	400	300	2.530
40	0.00	300	1705	90.00	300	0	0	4.010
40	−5.71	300	1310	84.29	300	0	0	3.220
40	−11.31	300	925	78.69	400	200	100	2.450
40	−14.03	300	765	75.97	400	300	100	2.130

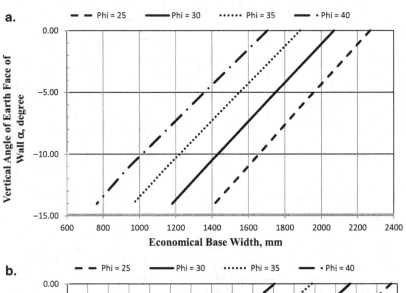

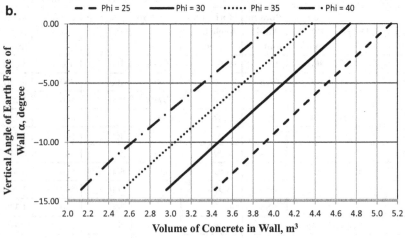

FIGURE 6.40 Design Charts for a Concrete Wall H = 4.0 m, $\gamma_{concrete}$ = 24.52 kN/m³, γ_{sat} = 15.70 kN/m³ and ι = ½ϕ: (a) Economical Base Width for a Wall and (b) Volume of a Concrete Wall

TABLE 6.40

H = 4.0 m, $\gamma_{concrete}$ = 24.52 kN/m³, γ_{sat} = 17.27 kN/m³ and ι = ½ϕ

A. Results for Stability Analysis

ϕ Degree	α Degree	K_{ah}	F_{OVT}	Pressure at Heel, kN/m²	Toe, kN/m²	μ = 0.50 F_{SLD}	Net Force, kN	μ = 0.70 F_{SLD}	Net Force, kN	μ = 0.80 F_{SLD}	Net Force, kN
25	0.00	0.450	1.82	0.01	98.66	0.80	46.05	1.12	24.90	1.28	14.32
25	−5.71	0.407	1.70	1.08	99.81	0.76	44.71	1.06	26.39	1.21	17.22
25	−11.31	0.364	1.59	4.47	99.38	0.72	42.95	1.00	27.27	1.15	19.42
25	−14.03	0.344	1.53	5.96	99.86	0.69	42.15	0.97	27.79	1.10	20.61
30	0.00	0.373	1.83	0.23	99.66	0.87	35.06	1.22	15.57	1.40	5.83
30	−5.71	0.329	1.71	2.98	99.70	0.83	33.65	1.16	17.03	1.33	8.72
30	−11.31	0.286	1.58	7.22	99.54	0.78	32.27	1.09	18.45	1.24	11.54
30	−14.03	0.266	1.52	10.12	99.47	0.74	31.63	1.04	19.17	1.19	12.94
35	0.00	0.305	1.85	1.71	99.52	0.96	25.46	1.34	7.48	1.53	0.00
35	−5.71	0.262	1.71	4.91	99.99	0.90	24.57	1.27	9.62	1.45	2.15
35	−11.31	0.220	1.57	10.93	99.72	0.84	23.79	1.17	11.75	1.34	5.73
35	−14.03	0.201	1.50	17.18	97.58	0.80	23.35	1.12	12.67	1.28	7.33
40	0.00	0.245	1.88	3.31	99.51	1.05	17.41	1.48	0.91	1.69	0.00
40	−5.71	0.204	1.72	7.66	99.99	0.99	17.19	1.39	3.84	1.58	0.00
40	−11.31	0.165	1.55	16.38	99.68	0.90	17.23	1.26	6.90	1.44	1.74
40	−14.03	0.147	1.50	31.63	90.37	0.86	16.93	1.20	7.88	1.37	3.35

B. Physical Dimensions for a Technoeconomic Wall Section

ϕ Degree	α Degree	Top Width, mm	Bottom Width, mm	Horizontal Angle θ Degree	Depth of Shear Key D_{sk}, mm μ = 0.50	μ = 0.70	μ = 0.80	Volume of Wall, m³
25	0.00	300	2385	90.00	1200	800	600	5.370
25	−5.71	300	2020	84.29	1100	800	700	4.640
25	−11.31	300	1680	78.69	1200	1000	800	3.960
25	−14.03	300	1510	75.97	1200	1000	800	3.620
30	0.00	300	2170	90.00	800	500	300	4.940
30	−5.71	300	1800	84.29	800	500	300	4.200
30	−11.31	300	1440	78.69	900	700	500	3.480
30	−14.03	300	1265	75.97	900	700	600	3.130
35	0.00	300	1975	90.00	600	200	0	4.550
35	−5.71	300	1585	84.29	600	300	100	3.770
35	−11.31	300	1210	78.69	700	400	300	3.020

(Continued)

TABLE 6.40 (Continued)

$H = 4.0$ m, $\gamma_{concrete} = 24.52$ kN/m³, $\gamma_{sat} = 17.27$ kN/m³ and $\iota = \frac{1}{2}\phi$

ϕ Degree	α Degree	Top Width, mm	Bottom Width, mm	Horizontal Angle θ Degree	Depth of Shear Key D_{sk}, mm			Volume of Wall, m³
					$\mu = 0.50$	$\mu = 0.70$	$\mu = 0.80$	
35	−14.03	300	1035	75.97	600	500	300	2.670
40	0.00	300	1785	90.00	400	0	0	4.170
40	−5.71	300	1380	84.29	400	100	0	3.360
40	−11.31	300	990	78.69	500	300	100	2.580
40	−14.03	300	825	75.97	400	300	200	2.250

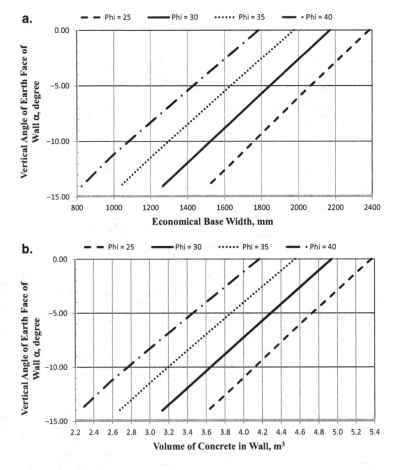

FIGURE 6.41 Design Charts for a Concrete Wall $H = 4.0$ m, $\gamma_{concrete} = 24.52$ kN/m³, $\gamma_{sat} = 17.27$ kN/m³ and $\iota = \frac{1}{2}\phi$: (a) Economical Base Width for a Wall, (b) Volume of a Concrete Wall

TABLE 6.41

$H = 4.0$ m, $\gamma_{concrete} = 24.52$ kN/m^3, $\gamma_{sat} = 15.70$ kN/m^3 and $\iota = \phi$

A. Results for Stability Analysis

ϕ Degree	α Degree	K_{ah}	F_{OVT}	Pressure at Heel, kN/m^2	Toe, kN/m^2	$\mu = 0.50$ F_{SLD}	Net Force, kN	$\mu = 0.70$ F_{SLD}	Net Force, kN	$\mu = 0.80$ F_{SLD}	Net Force, kN
25	0.00	0.821	1.81	0.27	95.58	0.64	89.85	0.89	63.22	1.02	49.90
25	−5.71	0.746	1.71	0.17	97.10	0.62	84.31	0.86	60.79	0.99	49.03
25	−11.31	0.675	1.61	0.44	98.63	0.59	79.03	0.83	58.45	0.95	48.16
25	−14.03	0.641	1.56	0.41	99.76	0.58	76.54	0.81	57.40	0.92	47.83
30	0.00	0.750	1.82	0.30	96.00	0.67	79.88	0.93	54.33	1.07	41.56
30	−5.71	0.666	1.70	0.04	97.97	0.65	73.28	0.91	51.07	1.03	39.97
30	−11.31	0.587	1.60	0.55	99.69	0.62	67.08	0.87	48.02	1.00	38.49
30	−14.03	0.549	1.55	2.16	99.42	0.61	63.99	0.86	46.37	0.98	37.55
35	0.00	0.671	1.82	0.19	96.71	0.70	69.04	0.98	44.78	1.12	32.65
35	−5.71	0.580	1.70	0.02	98.94	0.68	61.76	0.96	41.03	1.10	30.67
35	−11.31	0.496	1.59	1.90	99.83	0.66	54.97	0.93	37.50	1.06	28.76
35	−14.03	0.456	1.54	3.68	99.86	0.65	51.83	0.91	35.92	1.04	27.96
40	0.00	0.587	1.82	0.14	97.51	0.75	57.68	1.04	34.85	1.19	23.44
40	−5.71	0.492	1.69	0.46	99.71	0.73	50.13	1.03	30.99	1.17	21.42
40	−11.31	0.406	1.59	4.15	99.59	0.71	43.28	1.00	27.52	1.14	19.63
40	−14.03	0.366	1.53	7.08	99.13	0.70	40.27	0.98	26.14	1.12	19.07

B. Physical Dimensions for a Technoeconomic Wall Section

ϕ Degree	α Degree	Top Width, mm	Bottom Width, mm	Horizontal Angle θ Degree	Depth of Shear Key D_{sk}, mm $\mu = 0.50$	$\mu = 0.70$	$\mu = 0.80$	Volume of Wall, m^3
25	0.00	300	3090	90.00	1700	1400	1200	6.780
25	−5.71	300	2690	84.29	1700	1400	1300	5.980
25	−11.31	300	2310	78.69	1700	1400	1300	5.220
25	−14.03	300	2125	75.97	1600	1400	1300	4.850
30	0.00	300	2950	90.00	1300	1100	900	6.500
30	−5.71	300	2520	84.29	1300	1000	900	5.640
30	−11.31	300	2115	78.69	1300	1100	900	4.830
30	−14.03	300	1930	75.97	1300	1100	900	4.460
35	0.00	300	2785	90.00	1000	800	600	6.170
35	−5.71	300	2330	84.29	900	700	600	5.260
35	−11.31	300	1910	78.69	1000	800	700	4.420

(Continued)

TABLE 6.41 (Continued)

H = 4.0 m, $\gamma_{concrete}$ = 24.52 kN/m³, γ_{sat} = 15.70 kN/m³ and $\iota = \phi$

ϕ Degree	α Degree	Top Width, mm	Bottom Width, mm	Horizontal Angle θ Degree	Depth of Shear Key D_{sk}, mm			Volume of Wall, m³
					$\mu = 0.50$	$\mu = 0.70$	$\mu = 0.80$	
35	−14.03	300	1710	75.97	900	800	700	4.020
40	0.00	300	2600	90.00	600	400	300	5.800
40	−5.71	300	2125	84.29	600	500	300	4.850
40	−11.31	300	1690	78.69	700	500	400	3.980
40	−14.03	300	1480	75.97	700	500	400	3.560

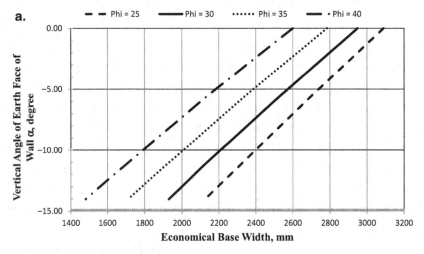

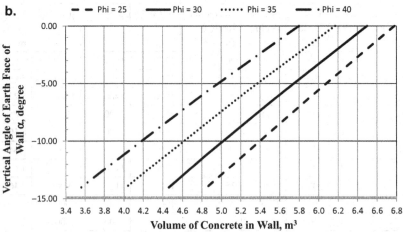

FIGURE 6.42 Design Charts for a Concrete Wall H = 4.0 m, $\gamma_{concrete}$ = 24.52 kN/m³, γ_{sat} = 15.70 kN/m³ and $\iota = \phi$: (a) Economical Base Width for a Wall, (b) Volume of a Concrete Wall

TABLE 6.42

$H = 4.0$ m, $\gamma_{concrete} = 24.52$ kN/m³, $\gamma_{sat} = 17.27$ kN/m³ and $\iota = \phi$

A. *Results for Stability Analysis*

				Pressure at		$\mu = 0.50$		$\mu = 0.70$		$\mu = 0.80$	
ϕ Degree	α Degree	K_{ah}	F_{OVT}	Heel, kN/ m²	Toe, kN/ m²	F_{SLD}	Net Force, kN	F_{SLD}	Net Force, kN	F_{SLD}	Net Force, kN
25	0.00	0.821	1.81	0.14	95.25	0.61	102.29	0.85	74.46	0.97	60.54
25	−5.71	0.746	1.71	0.12	96.57	0.59	95.45	0.82	70.76	0.94	58.41
25	−11.31	0.675	1.61	0.11	98.23	0.57	89.01	0.79	67.35	0.91	56.52
25	−14.03	0.641	1.57	0.43	98.89	0.56	85.88	0.78	65.65	0.89	55.54
30	0.00	0.750	1.81	0.00	95.84	0.63	91.17	0.89	64.50	1.01	51.17
30	−5.71	0.666	1.71	0.26	97.11	0.62	82.99	0.87	59.66	0.99	47.99
30	−11.31	0.587	1.60	0.17	99.27	0.60	75.59	0.84	55.52	0.96	45.49
30	−14.03	0.549	1.56	0.98	99.70	0.59	71.99	0.82	53.43	0.94	44.16
35	0.00	0.671	1.81	0.08	96.31	0.67	78.95	0.94	53.60	1.07	40.92
35	−5.71	0.580	1.70	0.23	98.04	0.66	70.06	0.92	48.28	1.05	37.39
35	−11.31	0.496	1.59	0.93	99.93	0.64	62.07	0.89	43.71	1.02	34.52
35	−14.03	0.456	1.55	3.08	99.38	0.63	58.19	0.88	41.38	1.01	32.97
40	0.00	0.587	1.81	0.00	97.11	0.71	66.19	0.99	42.35	1.14	30.44
40	−5.71	0.492	1.70	0.28	99.16	0.70	57.06	0.98	36.99	1.12	26.96
40	−11.31	0.406	1.59	3.13	99.60	0.69	48.88	0.96	32.30	1.10	24.01
40	−14.03	0.366	1.54	5.75	99.20	0.68	45.22	0.95	30.31	1.08	22.86

B. *Physical Dimensions for a Technoeconomic Wall Section*

					Depth of Shear Key D_{sk}, mm			
ϕ Degree	α Degree	Top Width, mm	Bottom Width, mm	Horizontal Angle θ Degree	$\mu = 0.50$	$\mu = 0.70$	$\mu = 0.80$	Volume of Wall, m³
25	0.00	300	3245	90.00	1700	1400	1300	7.090
25	−5.71	300	2840	84.29	1700	1500	1300	6.280
25	−11.31	300	2450	78.69	1700	1400	1300	5.500
25	−14.03	300	2265	75.97	1600	1400	1300	5.130
30	0.00	300	3095	90.00	1400	1100	1000	6.790
30	−5.71	300	2665	84.29	1300	1100	900	5.930
30	−11.31	300	2245	78.69	1300	1100	1000	5.090
30	−14.03	300	2050	75.97	1300	1100	1000	4.700
35	0.00	300	2925	90.00	900	700	600	6.450
35	−5.71	300	2465	84.29	1000	800	600	5.530
35	−11.31	300	2025	78.69	1000	800	700	4.650

(Continued)

TABLE 6.42 (Continued)

$H = 4.0$ m, $\gamma_{concrete} = 24.52$ kN/m³, $\gamma_{sat} = 17.27$ kN/m³ and $\iota = \phi$

ϕ Degree	α Degree	Top Width, mm	Bottom Width, mm	Horizontal Angle θ Degree	Depth of Shear Key D_{sk}, mm			Volume of Wall, m³
					$\mu = 0.50$	$\mu = 0.70$	$\mu = 0.80$	
35	-14.03	300	1825	75.97	1000	800	700	4.250
40	0.00	300	2730	90.00	600	400	300	6.060
40	-5.71	300	2245	84.29	700	500	400	5.090
40	-11.31	300	1795	78.69	700	500	400	4.190
40	-14.03	300	1580	75.97	700	500	400	3.760

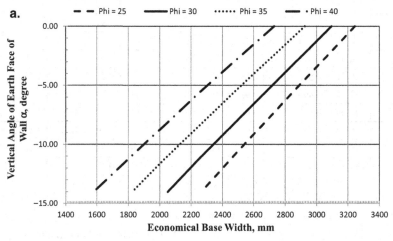

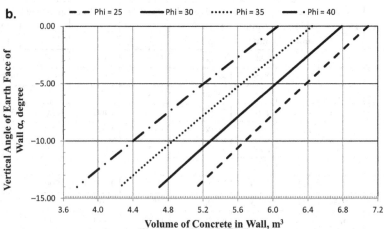

FIGURE 6.43 Design Charts for a Concrete Wall $H = 4.0$ m, $\gamma_{concrete} = 24.52$ kN/m³, $\gamma_{sat} = 17.27$ kN/m³ and $\iota = \phi$: (a) Economical Base Width for a Wall, (b) Volume of a Concrete Wall

6.9 CHARTS FOR THE DESIGN OF A 4.5-M-HIGH CONCRETE BREAST WALL

The charts/tables developed for a 4.5-m-high concrete breast wall for previously discussed six combinations are presented in Tables 6.43 to 6.48 and Figures 6.44 to 6.49.

TABLE 6.43

$H = 4.5$ m, $\gamma_{concrete} = 24.52$ kN/m^3, $\gamma_{sat} = 15.70$ kN/m^3 and $\iota = 0°$

A. Results for Stability Analysis

ϕ Degree	α Degree	K_{ah}	F_{OVT}	Pressure at Heel, kN/m^2	Toe, kN/m^2	$\mu = 0.50$ F_{SLD}	Net Force, kN	$\mu = 0.70$ F_{SLD}	Net Force, kN	$\mu = 0.80$ F_{SLD}	Net Force, kN
25	0.00	0.373	1.97	10.67	99.88	0.95	35.88	1.32	11.38	1.51	0.00
25	−5.71	0.339	1.82	13.43	99.76	0.88	37.20	1.23	16.24	1.40	5.76
25	−11.31	0.305	1.66	17.33	99.61	0.80	38.10	1.12	20.54	1.28	11.76
25	−14.03	0.288	1.59	20.27	99.15	0.76	38.36	1.07	22.41	1.22	14.44
30	0.00	0.304	2.00	12.07	99.83	1.04	25.29	1.45	2.76	1.66	0.00
30	−5.71	0.270	1.82	15.86	99.45	0.95	26.97	1.34	8.10	1.53	0.00
30	−11.31	0.237	1.64	21.43	99.08	0.86	28.35	1.21	13.02	1.38	5.35
30	−14.03	0.221	1.55	25.04	99.22	0.81	28.96	1.14	15.33	1.30	8.51
35	0.00	0.246	2.02	13.56	99.87	1.13	16.82	1.58	0.00	1.81	0.00
35	−5.71	0.213	1.83	18.51	99.37	1.03	19.05	1.45	2.15	1.65	0.00
35	−11.31	0.182	1.62	26.14	99.16	0.92	21.12	1.28	7.88	1.47	1.26
35	−14.03	0.166	1.51	31.85	99.22	0.85	22.06	1.19	10.57	1.36	4.83
40	0.00	0.196	2.06	15.37	99.79	1.24	10.04	1.73	0.00	1.98	0.00
40	−5.71	0.166	1.83	21.83	99.19	1.12	12.97	1.56	0.00	1.79	0.00
40	−11.31	0.136	1.58	32.58	99.39	0.96	15.86	1.34	4.59	1.54	0.00
40	−14.03	0.122	1.50	52.00	88.12	0.89	16.66	1.24	6.96	1.42	2.11

B. Physical Dimensions for a Technoeconomic Wall Section

ϕ Degree	α Degree	Top Width, mm	Bottom Width, mm	Horizontal Angle θ Degree	Depth of Shear Key D_{sk}, mm $\mu = 0.50$	$\mu = 0.70$	$\mu = 0.80$	Volume of Wall, m^3
25	0.00	300	2465	90.00	1100	500	0	6.221
25	−5.71	300	2060	84.29	1100	700	300	5.310
25	−11.31	300	1670	78.69	1200	900	600	4.433
25	−14.03	300	1485	75.97	1200	900	700	4.016
30	0.00	300	2240	90.00	700	100	0	5.715
30	−5.71	300	1820	84.29	800	400	0	4.770
30	−11.31	300	1415	78.69	900	600	400	3.859
30	−14.03	300	1220	75.97	900	600	500	3.420
35	0.00	300	2030	90.00	500	0	0	5.243
35	−5.71	300	1595	84.29	500	100	0	4.264
35	−11.31	300	1175	78.69	600	400	100	3.319

(Continued)

TABLE 6.43 (Continued)

H = 4.5 m, $\gamma_{concrete}$ = 24.52 kN/m³, γ_{sat} = 15.70 kN/m³ and ι = 0°

ϕ Degree	α Degree	Top Width, mm	Bottom Width, mm	Horizontal Angle θ Degree	Depth of Shear Key D_{sk}, mm			Volume of Wall, m³
					μ = 0.50	μ = 0.70	μ = 0.80	
35	−14.03	300	975	75.97	700	400	300	2.869
40	0.00	300	1835	90.00	200	0	0	4.804
40	−5.71	300	1385	84.29	300	0	0	3.791
40	−11.31	300	950	78.69	500	200	0	2.813
40	−14.03	300	770	75.97	500	300	100	2.408

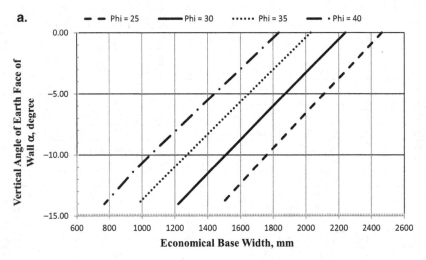

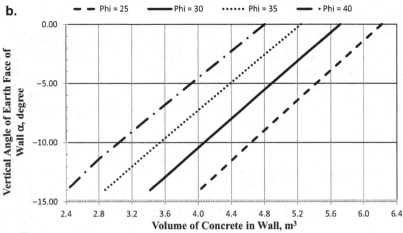

FIGURE 6.44 Design Charts for a Concrete Wall H = 4.5 m, $\gamma_{concrete}$ = 24.52 kN/m³, γ_{sat} = 15.70 kN/m³ and ι = 0°: (a) Economical Base Width for a Wall, (b) Volume of a Concrete Wall

TABLE 6.44

$H = 4.5$ m, $\gamma_{concrete} = 24.52$ kN/m^3, $\gamma_{sat} = 17.27$ kN/m^3 and $\iota = 0°$

A. Results for Stability Analysis

ϕ Degree	α Degree	K_{ah}	F_{OVT}	Pressure at Heel, kN/m^2	Toe, kN/m^2	$\mu = 0.50$ F_{SLD}	Net Force, kN	$\mu = 0.70$ F_{SLD}	Net Force, kN	$\mu = 0.80$ F_{SLD}	Net Force, kN
25	0.00	0.373	1.96	10.00	99.95	0.90	42.24	1.26	16.74	1.44	3.99
25	−5.71	0.339	1.82	12.82	99.52	0.84	42.77	1.18	20.80	1.35	9.82
25	−11.31	0.305	1.67	16.38	99.34	0.78	42.97	1.09	24.45	1.24	15.19
25	−14.03	0.288	1.60	18.73	99.21	0.74	42.93	1.04	26.07	1.19	17.64
30	0.00	0.304	1.99	11.49	99.74	0.99	30.23	1.39	6.78	1.58	0.00
30	−5.71	0.270	1.82	14.85	99.51	0.92	31.22	1.29	11.48	1.47	1.61
30	−11.31	0.237	1.65	19.87	99.17	0.84	31.93	1.17	15.76	1.34	7.68
30	−14.03	0.221	1.56	22.67	99.72	0.79	32.25	1.11	17.84	1.27	10.63
35	0.00	0.246	2.01	13.04	99.66	1.08	20.59	1.52	0.00	1.74	0.00
35	−5.71	0.213	1.82	16.91	99.92	1.00	22.27	1.40	4.62	1.59	0.00
35	−11.31	0.182	1.62	23.55	99.95	0.89	23.70	1.25	9.76	1.43	2.79
35	−14.03	0.166	1.52	29.24	99.26	0.83	24.28	1.17	12.09	1.34	5.99
40	0.00	0.196	2.05	14.86	99.51	1.19	12.86	1.67	0.00	1.90	0.00
40	−5.71	0.166	1.83	20.09	99.70	1.08	15.28	1.51	0.00	1.73	0.00
40	−11.31	0.136	1.59	30.78	98.64	0.95	17.47	1.32	5.55	1.51	0.00
40	−14.03	0.122	1.50	44.11	92.89	0.87	18.27	1.22	8.05	1.40	2.93

B. Physical Dimensions for a Technoeconomic Wall Section

ϕ Degree	α Degree	Top Width, mm	Bottom Width, mm	Horizontal Angle θ Degree	Depth of Shear Key D_{sk}, mm $\mu = 0.50$	$\mu = 0.70$	$\mu = 0.80$	Volume of Wall, m^3
25	0.00	300	2580	90.00	1100	600	200	6.480
25	−5.71	300	2175	84.29	1100	700	500	5.569
25	−11.31	300	1780	78.69	1200	900	700	4.680
25	−14.03	300	1590	75.97	1200	900	800	4.253
30	0.00	300	2345	90.00	800	300	0	5.951
30	−5.71	300	1920	84.29	800	400	100	4.995
30	−11.31	300	1510	78.69	900	600	400	4.073
30	−14.03	300	1310	75.97	900	700	500	3.623
35	0.00	300	2125	90.00	400	0	0	5.456
35	−5.71	300	1680	84.29	500	200	0	4.455
35	−11.31	300	1255	78.69	700	400	200	3.499

(Continued)

TABLE 6.44 (Continued)

$H = 4.5$ m, $\gamma_{concrete} = 24.52$ kN/m³, $\gamma_{sat} = 17.27$ kN/m³ and $\iota = 0°$

φ Degree	α Degree	Top Width, mm	Bottom Width, mm	Horizontal Angle θ Degree	Depth of Shear Key D_{sk}, mm			Volume of Wall, m³
					$\mu = 0.50$	$\mu = 0.70$	$\mu = 0.80$	
35	−14.03	300	1055	75.97	700	500	300	3.049
40	0.00	300	1920	90.00	200	0	0	4.995
40	−5.71	300	1460	84.29	300	0	0	3.960
40	−11.31	300	1025	78.69	500	200	0	2.981
40	−14.03	300	830	75.97	500	300	200	2.543

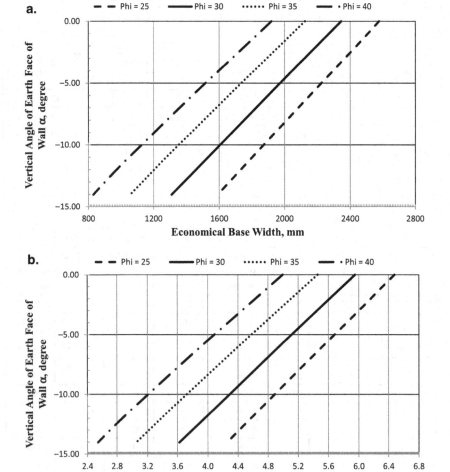

FIGURE 6.45 Design Charts for a Concrete Wall $H = 4.5$ m, $\gamma_{concrete} = 24.52$ kN/m³, $\gamma_{sat} = 17.27$ kN/m³ and $\iota = 0°$: (a) Economical Base Width for a Wall, (b) Volume of a Concrete Wall

TABLE 6.45

H = 4.5 m, $\gamma_{concrete}$ = 24.52 kN/m³, γ_{sat} = 15.70 kN/m³ and ι = ½ϕ

A. Results for Stability Analysis

				Pressure at		μ = 0.50		μ = 0.70		μ = 0.80	
ϕ Degree	α Degree	K_{ah}	F_{OVT}	Heel, kN/ m²	Toe, kN/ m²	F_{SLD}	Net Force, kN	F_{SLD}	Net Force, kN	F_{SLD}	Net Force, kN
25	0.00	0.450	1.95	9.47	99.94	0.87	48.24	1.22	21.74	1.39	8.48
25	−5.71	0.407	1.81	11.77	99.90	0.82	47.77	1.14	24.93	1.31	13.51
25	−11.31	0.364	1.67	15.01	99.81	0.76	47.05	1.06	27.74	1.22	18.09
25	−14.03	0.344	1.60	17.11	99.78	0.73	46.62	1.02	29.01	1.17	20.21
30	0.00	0.373	1.97	10.63	99.92	0.95	35.91	1.32	11.41	1.51	0.00
30	−5.71	0.329	1.82	13.62	99.85	0.89	35.68	1.24	15.03	1.42	4.70
30	−11.31	0.286	1.65	17.89	99.92	0.82	35.39	1.14	18.45	1.31	9.97
30	−14.03	0.266	1.58	21.54	99.23	0.78	35.14	1.09	19.94	1.25	12.34
35	0.00	0.305	2.00	12.20	99.66	1.03	25.39	1.45	2.81	1.66	0.00
35	−5.71	0.262	1.82	15.97	99.69	0.96	25.76	1.35	7.20	1.54	0.00
35	−11.31	0.220	1.63	22.38	99.36	0.88	26.14	1.23	11.42	1.40	4.06
35	−14.03	0.201	1.54	27.20	99.09	0.83	26.34	1.16	13.46	1.32	7.01
40	0.00	0.245	2.03	13.83	99.60	1.13	16.65	1.59	0.00	1.82	0.00
40	−5.71	0.204	1.83	19.09	99.32	1.05	17.78	1.47	1.23	1.68	0.00
40	−11.31	0.165	1.60	27.40	99.96	0.93	19.21	1.30	6.67	1.49	0.40
40	−14.03	0.147	1.50	38.55	95.77	0.87	19.67	1.21	8.93	1.39	3.55

B. Physical Dimensions for a Technoeconomic Wall Section

					Depth of Shear Key D_{sk}, mm			
ϕ Degree	α Degree	Top Width, mm	Bottom Width, mm	Horizontal Angle θ Degree	μ = 0.50	μ = 0.70	μ = 0.80	Volume of Wall, m³
25	0.00	300	2695	90.00	1300	800	400	6.739
25	−5.71	300	2275	84.29	1300	900	600	5.794
25	−11.31	300	1870	78.69	1300	1000	800	4.883
25	−14.03	300	1675	75.97	1300	1000	900	4.444
30	0.00	300	2465	90.00	900	400	0	6.221
30	−5.71	300	2025	84.29	900	500	200	5.231
30	−11.31	300	1600	78.69	1000	700	500	4.275
30	−14.03	300	1400	75.97	1000	700	600	3.825
35	0.00	300	2245	90.00	500	0	0	5.726
35	−5.71	300	1785	84.29	600	300	0	4.691
35	−11.31	300	1345	78.69	700	500	300	3.701

(Continued)

TABLE 6.45 (Continued)

$H = 4.5$ m, $\gamma_{concrete} = 24.52$ kN/m³, $\gamma_{sat} = 15.70$ kN/m³ and $\iota = \frac{1}{2}\phi$

ϕ Degree	α Degree	Top Width, mm	Bottom Width, mm	Horizontal Angle θ Degree	Depth of Shear Key D_{sk}, mm			Volume of Wall, m³
					$\mu = 0.50$	$\mu = 0.70$	$\mu = 0.80$	
35	−14.03	300	1135	75.97	700	500	400	3.229
40	0.00	300	2030	90.00	300	0	0	5.243
40	−5.71	300	1555	84.29	400	0	0	4.174
40	−11.31	300	1095	78.69	500	300	0	3.139
40	−14.03	300	890	75.97	500	300	200	2.678

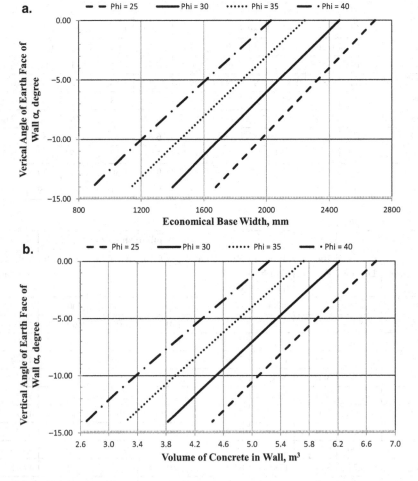

FIGURE 6.46 Design Charts for a Concrete Wall $H = 4.5$ m, $\gamma_{concrete} = 24.52$ kN/m³, $\gamma_{sat} = 15.70$ kN/m³ and $\iota = \frac{1}{2}\phi$: (a) Economical Base Width for a Wall, (b) Volume of a Concrete Wall

TABLE 6.46

$H = 4.5$ m, $\gamma_{concrete} = 24.52$ kN/m³, $\gamma_{sat} = 17.27$ kN/m³ and $\iota = \frac{1}{2}\phi$

A. Results for Stability Analysis

				Pressure at		$\mu = 0.50$		$\mu = 0.70$		$\mu = 0.80$	
ϕ Degree	α Degree	K_{ah}	F_{OVT}	Heel, kN/ m²	Toe, kN/ m²	F_{SLD}	Net Force, kN	F_{SLD}	Net Force, kN	F_{SLD}	Net Force, kN
25	0.00	0.450	1.94	9.05	99.79	0.83	56.12	1.16	28.48	1.32	14.66
25	−5.71	0.407	1.81	11.15	99.77	0.78	54.73	1.10	30.80	1.25	18.83
25	−11.31	0.364	1.68	14.20	99.55	0.73	53.11	1.03	32.76	1.17	22.58
25	−14.03	0.344	1.61	15.89	99.73	0.71	52.29	0.99	33.68	1.13	24.38
30	0.00	0.373	1.96	9.96	99.99	0.90	42.28	1.26	16.77	1.44	4.02
30	−5.71	0.329	1.81	12.68	99.94	0.85	41.12	1.19	19.50	1.36	8.69
30	−11.31	0.286	1.66	16.69	99.85	0.79	39.92	1.11	22.06	1.27	13.13
30	−14.03	0.266	1.59	19.74	99.45	0.76	39.29	1.06	23.21	1.21	15.18
35	0.00	0.305	1.99	11.61	99.60	0.99	30.35	1.38	6.86	1.58	0.00
35	−5.71	0.262	1.82	15.15	99.54	0.93	29.80	1.30	10.37	1.49	0.65
35	−11.31	0.220	1.64	20.46	99.73	0.85	29.42	1.19	13.92	1.36	6.16
35	−14.03	0.201	1.55	24.42	99.84	0.81	29.27	1.13	15.64	1.29	8.83
40	0.00	0.245	2.01	12.88	99.86	1.09	20.52	1.52	0.00	1.74	0.00
40	−5.71	0.204	1.82	17.71	99.60	1.01	20.78	1.42	3.48	1.62	0.00
40	−11.31	0.165	1.61	25.47	99.83	0.91	21.40	1.28	8.17	1.46	1.55
40	−14.03	0.147	1.50	32.11	99.68	0.85	21.77	1.19	10.45	1.36	4.79

B. Physical Dimensions for a Technoeconomic Wall Section

					Depth of Shear Key D_{sk}, mm			
ϕ Degree	α Degree	Top Width, mm	Bottom Width, mm	Horizontal Angle θ Degree	$\mu = 0.50$	$\mu = 0.70$	$\mu = 0.80$	Volume of Wall, m³
25	0.00	300	2825	90.00	1300	900	600	7.031
25	−5.71	300	2400	84.29	1300	900	700	6.075
25	−11.31	300	1990	78.69	1400	1100	900	5.153
25	−14.03	300	1790	75.97	1400	1100	900	4.703
30	0.00	300	2580	90.00	800	400	100	6.480
30	−5.71	300	2135	84.29	900	600	300	5.479
30	−11.31	300	1705	78.69	900	600	500	4.511
30	−14.03	300	1500	75.97	1000	800	600	4.050
35	0.00	300	2350	90.00	500	100	0	5.963
35	−5.71	300	1885	84.29	600	300	0	4.916
35	−11.31	300	1435	78.69	600	400	200	3.904

(Continued)

TABLE 6.46 (Continued)

H = 4.5 m, $\gamma_{concrete}$ = 24.52 kN/m³, γ_{sat} = 17.27 kN/m³ and $\iota = \frac{1}{2}\phi$

ϕ Degree	α Degree	Top Width, mm	Bottom Width, mm	Horizontal Angle θ Degree	Depth of Shear Key D_{sk}, mm			Volume of Wall, m³
					$\mu = 0.50$	$\mu = 0.70$	$\mu = 0.80$	
35	−14.03	300	1220	75.97	700	500	400	3.420
40	0.00	300	2120	90.00	300	0	0	5.445
40	−5.71	300	1640	84.29	400	100	0	4.365
40	−11.31	300	1175	78.69	500	300	100	3.319
40	−14.03	300	955	75.97	500	300	200	2.824

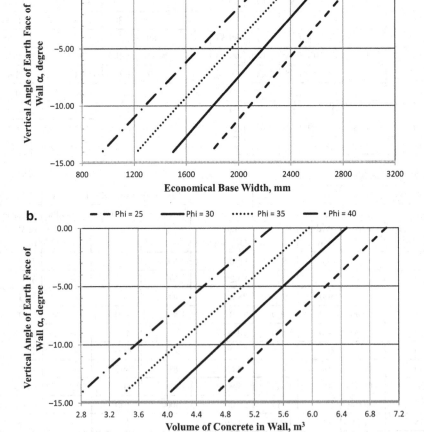

FIGURE 6.47 Design Charts for Concrete Wall H = 4.5 m, $\gamma_{concrete}$ = 24.52 kN/m³, γ_{sat} = 17.27 kN/m³ and $\iota = \frac{1}{2}\phi$: (a) Economical Base Width for a Wall, (b) Volume of a Concrete Wall

TABLE 6.47

$H = 4.5$ m, $\gamma_{concrete} = 24.52$ kN/m^3, $\gamma_{sat} = 15.70$ kN/m^3 and $\iota = \phi$

A. Results for Stability Analysis

ϕ Degree	α Degree	K_{ah}	F_{OVT}	Pressure at Heel, kN/m^2	Toe, kN/m^2	$\mu = 0.50$ F_{SLD}	Net Force, kN	$\mu = 0.70$ F_{SLD}	Net Force, kN	$\mu = 0.80$ F_{SLD}	Net Force, kN
25	0.00	0.821	1.90	6.52	99.78	0.65	111.83	0.91	77.38	1.04	60.15
25	−5.71	0.746	1.80	7.64	99.97	0.63	104.83	0.88	74.30	1.01	59.04
25	−11.31	0.675	1.70	9.40	99.87	0.61	98.19	0.85	71.42	0.97	58.04
25	−14.03	0.641	1.65	10.73	99.53	0.60	94.98	0.83	69.99	0.95	57.50
30	0.00	0.750	1.91	7.00	99.72	0.68	99.16	0.96	66.06	1.09	49.51
30	−5.71	0.666	1.80	8.48	99.78	0.67	90.73	0.93	61.82	1.06	47.36
30	−11.31	0.587	1.69	10.37	99.97	0.64	83.09	0.90	58.24	1.03	45.81
30	−14.03	0.549	1.64	12.05	99.56	0.63	79.44	0.88	56.51	1.01	45.04
35	0.00	0.671	1.91	7.32	99.94	0.72	85.43	1.01	53.95	1.15	38.21
35	−5.71	0.580	1.81	9.48	99.64	0.71	76.07	0.99	48.99	1.13	35.45
35	−11.31	0.496	1.69	11.97	99.77	0.68	67.96	0.96	45.21	1.09	33.83
35	−14.03	0.456	1.62	13.52	99.95	0.67	64.32	0.94	43.66	1.07	33.33
40	0.00	0.587	1.92	7.99	99.95	0.77	70.98	1.07	41.28	1.23	26.43
40	−5.71	0.492	1.81	10.55	99.68	0.76	61.45	1.06	36.42	1.21	23.91
40	−11.31	0.406	1.68	13.68	100.00	0.73	53.55	1.03	33.11	1.17	22.89
40	−14.03	0.366	1.61	16.35	99.69	0.72	50.05	1.00	31.79	1.14	22.66

B. Physical Dimensions for a Technoeconomic Wall Section

ϕ Degree	α Degree	Top Width, mm	Bottom Width, mm	Horizontal Angle θ Degree	Depth of Shear Key D_{sk}, mm $\mu = 0.50$	$\mu = 0.70$	$\mu = 0.80$	Volume of Wall, m^3
25	0.00	300	3605	90.00	1900	1500	1300	8.786
25	−5.71	300	3155	84.29	1800	1500	1300	7.774
25	−11.31	300	2725	78.69	1900	1600	1400	6.806
25	−14.03	300	2520	75.97	1800	1500	1400	6.345
30	0.00	300	3450	90.00	1400	1100	900	8.438
30	−5.71	300	2970	84.29	1400	1100	1000	7.358
30	−11.31	300	2505	78.69	1400	1100	900	6.311
30	−14.03	300	2285	75.97	1400	1200	1000	5.816
35	0.00	300	3265	90.00	1000	800	600	8.021
35	−5.71	300	2760	84.29	1100	800	700	6.885
35	−11.31	300	2265	78.69	1000	800	600	5.771

(Continued)

TABLE 6.47 (Continued)

$H = 4.5$ m, $\gamma_{concrete} = 24.52$ kN/m³, $\gamma_{sat} = 15.70$ kN/m³ and $\iota = \phi$

| ϕ Degree | α Degree | Top Width, mm | Bottom Width, mm | Horizontal Angle θ Degree | Depth of Shear Key D_{sk}, mm | | | Volume of Wall, m³ |
					$\mu = 0.50$	$\mu = 0.70$	$\mu = 0.80$	
35	−14.03	300	2025	75.97	1100	900	700	5.231
40	0.00	300	3060	90.00	600	400	200	7.560
40	−5.71	300	2525	84.29	600	400	300	6.356
40	−11.31	300	2000	78.69	700	500	400	5.175
40	−14.03	300	1750	75.97	700	600	500	4.613

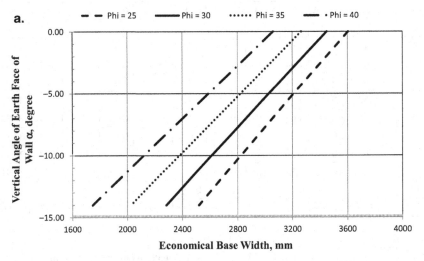

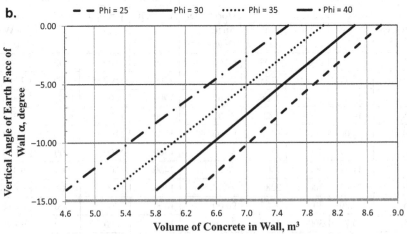

FIGURE 6.48 Design Charts for a Concrete Wall $H = 4.5$ m, $\gamma_{concrete} = 24.52$ kN/m³, $\gamma_{sat} = 15.70$ kN/m³ and $\iota = \phi$: (a) Economical Base Width for a Wall, (b) Volume of a Concrete Wall

TABLE 6.48

$H = 4.5$ m, $\gamma_{concrete} = 24.52$ kN/m³, $\gamma_{sat} = 17.27$ kN/m³ and $\iota = \phi$

A. Results for Stability Analysis

ϕ Degree	α Degree	K_{ah}	F_{OVT}	Pressure at Heel, kN/m²	Toe, kN/m²	$\mu = 0.50$ F_{SLD}	Net Force, kN	$\mu = 0.70$ F_{SLD}	Net Force, kN	$\mu = 0.80$ F_{SLD}	Net Force, kN
25	0.00	0.821	1.89	6.11	99.77	0.62	127.57	0.87	91.59	0.99	73.59
25	−5.71	0.746	1.80	7.25	99.83	0.60	118.89	0.84	86.88	0.97	70.88
25	−11.31	0.675	1.70	8.69	99.91	0.58	110.78	0.82	82.61	0.93	68.52
25	−14.03	0.641	1.66	9.82	99.68	0.57	106.85	0.80	80.52	0.91	67.35
30	0.00	0.750	1.90	6.46	99.82	0.65	113.41	0.91	78.87	1.04	61.60
30	−5.71	0.666	1.80	8.01	99.68	0.64	103.09	0.89	72.78	1.02	57.62
30	−11.31	0.587	1.70	9.67	99.92	0.62	93.78	0.86	67.62	0.99	54.54
30	−14.03	0.549	1.65	11.19	99.56	0.61	89.34	0.85	65.15	0.97	53.05
35	0.00	0.671	1.91	7.00	99.79	0.68	97.92	0.96	65.03	1.10	48.59
35	−5.71	0.580	1.80	8.66	99.86	0.67	86.72	0.94	58.37	1.08	44.20
35	−11.31	0.496	1.69	11.36	99.53	0.66	76.72	0.92	52.74	1.05	40.75
35	−14.03	0.456	1.63	12.85	99.60	0.65	72.24	0.90	50.41	1.03	39.49
40	0.00	0.587	1.92	7.68	99.75	0.73	81.67	1.02	50.67	1.17	35.16
40	−5.71	0.492	1.81	9.79	99.78	0.72	70.22	1.01	44.02	1.16	30.91
40	−11.31	0.406	1.68	12.77	99.93	0.71	60.49	0.99	38.96	1.13	28.19
40	−14.03	0.366	1.62	15.36	99.46	0.69	56.15	0.97	36.84	1.11	27.19

B. Physical Dimensions for a Technoeconomic Wall Section

ϕ Degree	α Degree	Top Width, mm	Bottom Width, mm	Horizontal Angle θ Degree	Depth of Shear Key D_{sk}, mm $\mu = 0.50$	$\mu = 0.70$	$\mu = 0.80$	Volume of Wall, m³
25	0.00	300	3780	90.00	1900	1600	1400	9.180
25	−5.71	300	3325	84.29	1900	1600	1400	8.156
25	−11.31	300	2885	78.69	1900	1600	1500	7.166
25	−14.03	300	2675	75.97	1900	1600	1400	6.694
30	0.00	300	3615	90.00	1400	1100	1000	8.809
30	−5.71	300	3130	84.29	1500	1200	1000	7.718
30	−11.31	300	2655	78.69	1400	1100	1000	6.649
30	−14.03	300	2430	75.97	1400	1200	1100	6.143
35	0.00	300	3425	90.00	1100	800	700	8.381
35	−5.71	300	2905	84.29	1000	800	600	7.211
35	−11.31	300	2405	78.69	1000	800	700	6.086

(Continued)

TABLE 6.48 (Continued)

H = 4.5 m, $\gamma_{concrete}$ = 24.52 kN/m³, γ_{sat} = 17.27 kN/m³ and ι = φ

φ Degree	α Degree	Top Width, mm	Bottom Width, mm	Horizontal Angle θ Degree	Depth of Shear Key D_{sk}, mm			Volume of Wall, m³
					μ = 0.50	μ = 0.70	μ = 0.80	
35	−14.03	300	2160	75.97	1000	800	700	5.535
40	0.00	300	3210	90.00	600	400	300	7.898
40	−5.71	300	2660	84.29	700	500	300	6.660
40	−11.31	300	2125	78.69	700	500	400	5.456
40	−14.03	300	1870	75.97	700	500	400	4.883

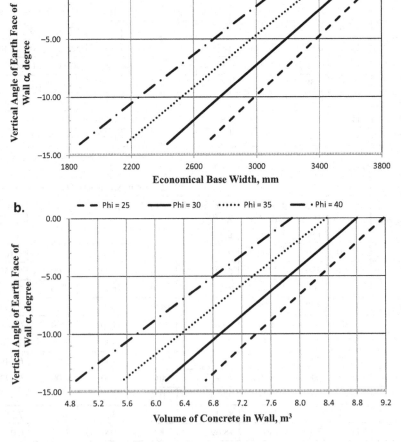

FIGURE 6.49 Design Charts for Concrete Wall H = 4.5 m, $\gamma_{concrete}$ = 24.52 kN/m³, γ_{sat} = 17.27 kN/m³ and ι = φ: (a) Economical Base Width for a Wall, (b) Volume of a Concrete Wall

6.10 CHARTS FOR THE DESIGN OF A 5.0-M-HIGH CONCRETE BREAST WALL

The charts/tables developed for a 5.0-m-high concrete breast wall for the previously discussed six combinations are presented in Tables 6.49 to 6.54 and Figures 6.50 to 6.55.

TABLE 6.49

$H = 5.0$ m, $\gamma_{concrete} = 24.52$ kN/m³, $\gamma_{sat} = 15.70$ kN/m³ and $\iota = 0°$

A. Results for Stability Analysis

ϕ Degree	α Degree	K_{ah}	F_{OVT}	Pressure at Heel, kN/m²	Toe, kN/m²	$\mu = 0.50$ F_{SLD}	Net Force, kN	$\mu = 0.70$ F_{SLD}	Net Force, kN	$\mu = 0.80$ F_{SLD}	Net Force, kN
25	0.00	0.373	2.11	20.95	99.71	0.98	41.67	1.37	10.38	1.57	0.00
25	−5.71	0.339	1.92	23.34	99.90	0.90	44.15	1.26	17.56	1.44	4.27
25	−11.31	0.305	1.73	26.88	99.99	0.82	45.91	1.15	23.78	1.31	12.71
25	−14.03	0.288	1.65	29.74	99.49	0.78	46.45	1.09	26.41	1.25	16.39
30	0.00	0.304	2.13	22.23	99.72	1.07	28.97	1.50	0.24	1.71	0.00
30	−5.71	0.270	1.92	25.72	99.54	0.98	31.84	1.37	7.97	1.57	0.00
30	−11.31	0.237	1.71	30.94	99.33	0.88	34.16	1.23	14.89	1.40	5.26
30	−14.03	0.221	1.61	34.47	99.35	0.82	35.11	1.15	18.02	1.32	9.48
35	0.00	0.246	2.16	23.77	99.62	1.17	18.78	1.63	0.00	1.87	0.00
35	−5.71	0.213	1.92	28.00	99.74	1.06	22.43	1.48	1.12	1.69	0.00
35	−11.31	0.182	1.67	34.99	99.87	0.93	25.55	1.30	9.00	1.48	0.72
35	−14.03	0.166	1.56	41.43	98.75	0.86	26.78	1.20	12.41	1.37	5.23
40	0.00	0.196	2.19	25.44	99.62	1.27	10.73	1.78	0.00	2.04	0.00
40	−5.71	0.166	1.91	31.17	99.59	1.14	15.20	1.59	0.00	1.82	0.00
40	−11.31	0.136	1.63	42.07	98.99	0.97	19.18	1.36	5.10	1.55	0.00
40	−14.03	0.122	1.50	52.15	97.56	0.88	20.90	1.23	9.05	1.41	3.13

B. Physical Dimensions for a Technoeconomic Wall Section

ϕ Degree	α Degree	Top Width, mm	Bottom Width, mm	Horizontal Angle θ Degree	Depth of Shear Key D_{sk}, mm $\mu = 0.50$	$\mu = 0.70$	$\mu = 0.80$	Volume of Wall, m³
25	0.00	300	2885	90.00	1200	500	0	7.963
25	−5.71	300	2400	84.29	1200	700	300	6.750
25	−11.31	300	1940	78.69	1300	900	700	5.600
25	−14.03	300	1725	75.97	1300	1000	800	5.063
30	0.00	300	2620	90.00	700	0	0	7.300
30	−5.71	300	2120	84.29	800	300	0	6.050
30	−11.31	300	1645	78.69	1000	600	400	4.863
30	−14.03	300	1420	75.97	1000	700	500	4.300
35	0.00	300	2375	90.00	400	0	0	6.688
35	−5.71	300	1855	84.29	600	0	0	5.388
35	−11.31	300	1365	78.69	700	400	100	4.163

(Continued)

TABLE 6.49 (Continued)

H = 5.0 m, $\gamma_{concrete}$ = 24.52 kN/m³, γ_{sat} = 15.70 kN/m³ and $\iota = 0°$

ϕ Degree	α Degree	Top Width, mm	Bottom Width, mm	Horizontal Angle θ Degree	Depth of Shear Key D_{sk}, mm			Volume of Wall, m³
					$\mu = 0.50$	$\mu = 0.70$	$\mu = 0.80$	
35	−14.03	300	1140	75.97	700	500	300	3.600
40	0.00	300	2145	90.00	200	0	0	6.113
40	−5.71	300	1610	84.29	300	0	0	4.775
40	−11.31	300	1110	78.69	500	200	0	3.525
40	−14.03	300	880	75.97	500	300	200	2.950

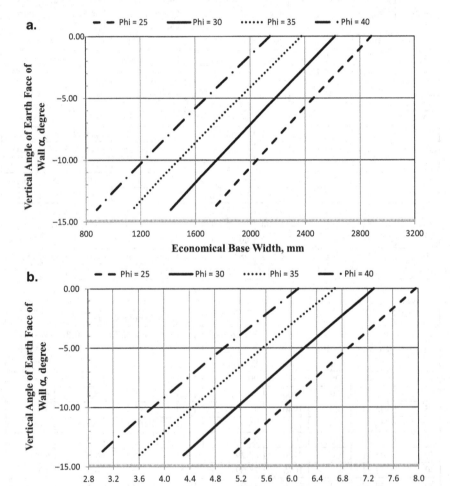

FIGURE 6.50 Design Charts for a Concrete Wall H = 5.0 m, $\gamma_{concrete}$ = 24.52 kN/m³, γ_{sat} = 15.70 kN/m³ and $\iota = 0°$: (a) Economical Base Width for a Wall, (b) Volume of a Concrete Wall

TABLE 6.50
H = 5.0 m, $\gamma_{concrete}$ = 24.52 kN/m³, γ_{sat} = 17.27 kN/m³ and ι = 0°

A. Results for Stability Analysis

ϕ Degree	α Degree	K_{ah}	F_{OVT}	Pressure at Heel, kN/m²	Toe, kN/m²	F_{SLD} (μ = 0.50)	Net Force, kN	F_{SLD} (μ = 0.70)	Net Force, kN	F_{SLD} (μ = 0.80)	Net Force, kN
25	0.00	0.373	2.10	20.31	99.79	0.93	49.35	1.31	16.75	1.49	0.44
25	−5.71	0.339	1.92	22.82	99.60	0.87	50.85	1.21	22.95	1.39	9.00
25	−11.31	0.305	1.75	26.26	99.42	0.80	51.73	1.11	28.34	1.27	16.64
25	−14.03	0.288	1.66	28.06	99.79	0.76	52.02	1.06	30.81	1.21	20.21
30	0.00	0.304	2.12	21.46	99.87	1.02	34.98	1.43	5.09	1.63	0.00
30	−5.71	0.270	1.92	24.58	99.79	0.94	37.00	1.32	12.01	1.51	0.00
30	−11.31	0.237	1.72	29.42	99.45	0.85	38.46	1.20	18.13	1.37	7.96
30	−14.03	0.221	1.62	32.39	99.64	0.81	39.05	1.13	20.95	1.29	11.90
35	0.00	0.246	2.14	22.79	99.95	1.12	23.46	1.56	0.00	1.79	0.00
35	−5.71	0.213	1.92	27.08	99.60	1.02	26.15	1.43	3.83	1.63	0.00
35	−11.31	0.182	1.69	33.84	99.19	0.91	28.46	1.27	10.94	1.45	2.18
35	−14.03	0.166	1.57	38.46	99.36	0.85	29.49	1.18	14.25	1.35	6.63
40	0.00	0.196	2.17	24.57	99.77	1.22	14.20	1.71	0.00	1.96	0.00
40	−5.71	0.166	1.91	29.81	99.74	1.10	17.89	1.54	0.00	1.76	0.00
40	−11.31	0.136	1.64	38.92	99.90	0.95	21.25	1.34	6.40	1.53	0.00
40	−14.03	0.122	1.51	47.32	99.11	0.87	22.68	1.22	10.11	1.39	3.82

B. Physical Dimensions for a Technoeconomic Wall Section

ϕ Degree	α Degree	Top Width, mm	Bottom Width, mm	Horizontal Angle θ Degree	Depth of Shear Key D_{sk}, mm (μ = 0.50)	(μ = 0.70)	(μ = 0.80)	Volume of Wall, m³
25	0.00	300	3020	90.00	1100	500	0	8.300
25	−5.71	300	2535	84.29	1200	800	400	7.088
25	−11.31	300	2070	78.69	1200	900	600	5.925
25	−14.03	300	1845	75.97	1300	1000	800	5.363
30	0.00	300	2740	90.00	700	100	0	7.600
30	−5.71	300	2235	84.29	900	400	0	6.338
30	−11.31	300	1755	78.69	900	600	300	5.138
30	−14.03	300	1525	75.97	1000	700	500	4.563
35	0.00	300	2480	90.00	400	0	0	6.950
35	−5.71	300	1960	84.29	600	100	0	5.650
35	−11.31	300	1465	78.69	600	300	100	4.413

(Continued)

TABLE 6.50 (Continued)

H = 5.0 m, $\gamma_{concrete}$ = 24.52 kN/m³, γ_{sat} = 17.27 kN/m³ and ι = 0°

ϕ Degree	α Degree	Top Width, mm	Bottom Width, mm	Horizontal Angle θ Degree	Depth of Shear Key D_{sk}, mm			Volume of Wall, m³
					μ = 0.50	μ = 0.70	μ = 0.80	
35	−14.03	300	1230	75.97	700	500	300	3.825
40	0.00	300	2240	90.00	200	0	0	6.350
40	−5.71	300	1700	84.29	400	0	0	5.000
40	−11.31	300	1190	78.69	500	300	0	3.725
40	−14.03	300	955	75.97	500	300	200	3.138

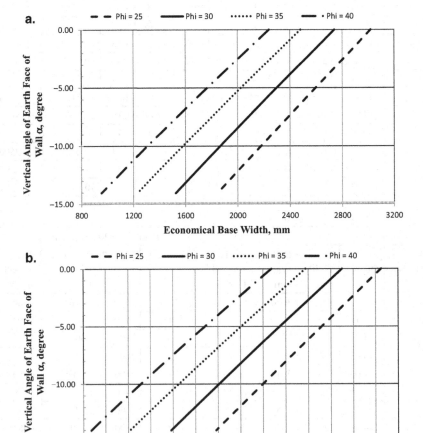

FIGURE 6.51 Design Charts for a Concrete Wall H = 5.0 m, $\gamma_{concrete}$ = 24.52 kN/m³, γ_{sat} = 17.27 kN/m³ and ι = 0°: (a) Economical Base Width for a Wall, (b) Volume of a Concrete Wall

TABLE 6.51

$H = 5.0$ m, $\gamma_{concrete} = 24.52$ kN/m^3, $\gamma_{sat} = 15.70$ kN/m^3 and $\iota = \frac{1}{2}\phi$

A. Results for Stability Analysis

				Pressure at		$\mu = 0.50$		$\mu = 0.70$		$\mu = 0.80$	
ϕ Degree	α Degree	K_{ah}	F_{OVT}	Heel, kN/m^2	Toe, kN/m^2	F_{SLD}	Net Force, kN	F_{SLD}	Net Force, kN	F_{SLD}	Net Force, kN
25	0.00	0.450	2.09	19.80	99.77	0.90	56.59	1.26	22.67	1.44	5.71
25	−5.71	0.407	1.92	22.04	99.72	0.84	56.81	1.18	27.74	1.35	13.21
25	−11.31	0.364	1.75	24.96	99.87	0.78	56.65	1.09	32.24	1.24	20.03
25	−14.03	0.344	1.67	27.36	99.41	0.74	56.31	1.04	34.09	1.19	22.98
30	0.00	0.373	2.11	20.91	99.75	0.98	41.71	1.37	10.41	1.57	0.00
30	−5.71	0.329	1.92	23.59	99.90	0.91	42.30	1.28	16.10	1.46	3.00
30	−11.31	0.286	1.73	28.22	99.41	0.84	42.50	1.17	21.10	1.34	10.40
30	−14.03	0.266	1.63	30.65	99.89	0.79	42.62	1.11	23.54	1.27	14.01
35	0.00	0.305	2.13	21.95	100.00	1.07	29.22	1.49	0.49	1.71	0.00
35	−5.71	0.262	1.92	25.91	99.69	0.99	30.38	1.38	6.89	1.58	0.00
35	−11.31	0.220	1.70	32.00	99.42	0.89	31.48	1.25	12.99	1.43	3.74
35	−14.03	0.201	1.59	36.09	99.69	0.84	32.01	1.17	15.90	1.34	7.84
40	0.00	0.245	2.16	23.66	99.77	1.17	18.70	1.63	0.00	1.87	0.00
40	−5.71	0.204	1.92	28.76	99.48	1.07	20.88	1.50	0.01	1.71	0.00
40	−11.31	0.165	1.66	37.63	99.02	0.94	23.10	1.32	7.37	1.51	0.00
40	−14.03	0.147	1.53	44.29	99.33	0.87	24.21	1.22	10.91	1.39	4.26

B. Physical Dimensions for a Technoeconomic Wall Section

					Depth of Shear Key D_{sk}, mm			
ϕ Degree	α Degree	Top Width, mm	Bottom Width, mm	Horizontal Angle θ Degree	$\mu = 0.50$	$\mu = 0.70$	$\mu = 0.80$	Volume of Wall, m^3
25	0.00	300	3155	90.00	1300	700	200	8.638
25	−5.71	300	2655	84.29	1400	900	600	7.388
25	−11.31	300	2175	78.69	1400	1000	800	6.188
25	−14.03	300	1950	75.97	1500	1100	900	5.625
30	0.00	300	2885	90.00	900	300	0	7.963
30	−5.71	300	2360	84.29	1000	600	200	6.650
30	−11.31	300	1865	78.69	1000	700	400	5.413
30	−14.03	300	1625	75.97	1100	800	600	4.813
35	0.00	300	2620	90.00	600	0	0	7.300
35	−5.71	300	2080	84.29	700	200	0	5.950
35	−11.31	300	1565	78.69	700	400	100	4.663

(Continued)

TABLE 6.51 (Continued)

$H = 5.0$ m, $\gamma_{concrete} = 24.52$ kN/m³, $\gamma_{sat} = 15.70$ kN/m³ and $\iota = \frac{1}{2}\phi$

ϕ Degree	α Degree	Top Width, mm	Bottom Width, mm	Horizontal Angle θ Degree	Depth of Shear Key D_{sk}, mm			Volume of Wall, m³
					$\mu = 0.50$	$\mu = 0.70$	$\mu = 0.80$	
35	−14.03	300	1320	75.97	800	600	400	4.050
40	0.00	300	2370	90.00	300	0	0	6.675
40	−5.71	300	1810	84.29	400	0	0	5.275
40	−11.31	300	1280	78.69	500	200	0	3.950
40	−14.03	300	1030	75.97	600	400	200	3.325

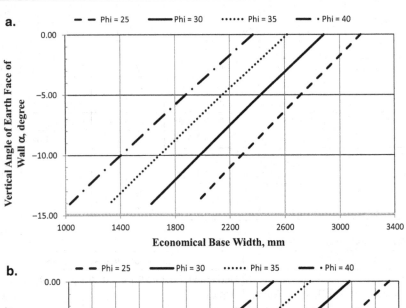

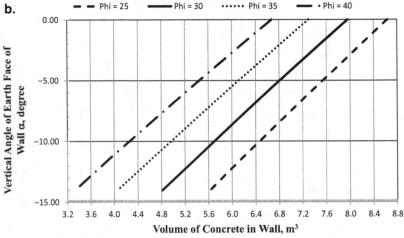

FIGURE 6.52 Design Charts for a Concrete Wall $H = 5.0$ m, $\gamma_{concrete} = 24.52$ kN/m³, $\gamma_{sat} = 15.70$ kN/m³ and $\iota = \frac{1}{2}\phi$: (a) Economical Base Width for a Wall, (b) Volume of a Concrete Wall

TABLE 6.52

H = 5.0 m, $\gamma_{concrete}$ = 24.52 kN/m³, γ_{sat} = 17.27 kN/m³ and ι = ½ϕ

A. Results for Stability Analysis

				Pressure at		$\mu = 0.50$		$\mu = 0.70$		$\mu = 0.80$	
ϕ Degree	α Degree	K_{ah}	F_{OVT}	Heel, kN/ m²	Toe, kN/ m²	F_{SLD}	Net Force, kN	F_{SLD}	Net Force, kN	F_{SLD}	Net Force, kN
25	0.00	0.450	2.08	19.27	99.78	0.86	66.18	1.20	30.80	1.37	13.12
25	−5.71	0.407	1.92	21.36	99.69	0.81	65.26	1.13	34.78	1.29	19.55
25	−11.31	0.364	1.76	24.22	99.58	0.75	63.97	1.05	38.20	1.20	25.31
25	−14.03	0.344	1.68	25.75	99.85	0.72	63.26	1.01	39.78	1.16	28.03
30	0.00	0.373	2.10	20.27	99.82	0.93	49.39	1.31	16.78	1.49	0.48
30	−5.71	0.329	1.92	22.81	99.87	0.88	48.82	1.23	21.36	1.40	7.63
30	−11.31	0.286	1.74	26.79	99.65	0.81	48.01	1.13	25.44	1.30	14.16
30	−14.03	0.266	1.65	29.63	99.36	0.77	47.52	1.08	27.28	1.24	17.17
35	0.00	0.305	2.12	21.52	99.79	1.02	35.14	1.43	5.20	1.63	0.00
35	−5.71	0.262	1.92	24.94	99.73	0.95	35.27	1.33	10.67	1.53	0.00
35	−11.31	0.220	1.71	30.25	99.69	0.87	35.41	1.22	15.90	1.39	6.15
35	−14.03	0.201	1.60	33.84	99.98	0.82	35.50	1.15	18.41	1.31	9.87
40	0.00	0.245	2.15	23.07	99.68	1.12	23.23	1.57	0.00	1.79	0.00
40	−5.71	0.204	1.92	27.51	99.66	1.04	24.44	1.45	2.60	1.66	0.00
40	−11.31	0.165	1.68	35.26	99.50	0.93	25.78	1.30	9.18	1.48	0.88
40	−14.03	0.147	1.55	41.10	99.80	0.86	26.49	1.20	12.36	1.37	5.30

B. Physical Dimensions for a Technoeconomic Wall Section

					Depth of Shear Key D_{sk}, mm			
ϕ Degree	α Degree	Top Width, mm	Bottom Width, mm	Horizontal Angle θ Degree	$\mu = 0.50$	$\mu = 0.70$	$\mu = 0.80$	Volume of Wall, m³
25	0.00	300	3305	90.00	1300	800	500	9.013
25	−5.71	300	2800	84.29	1400	1000	700	7.750
25	−11.31	300	2315	78.69	1400	1000	800	6.538
25	−14.03	300	2080	75.97	1400	1100	900	5.950
30	0.00	300	3020	90.00	900	400	0	8.300
30	−5.71	300	2490	84.29	900	500	200	6.975
30	−11.31	300	1985	78.69	1000	700	500	5.713
30	−14.03	300	1745	75.97	1000	700	500	5.113
35	0.00	300	2745	90.00	600	100	0	7.613
35	−5.71	300	2195	84.29	600	200	0	6.238
35	−11.31	300	1670	78.69	700	400	200	4.925

(Continued)

TABLE 6.52 (Continued)

H = 5.0 m, $\gamma_{concrete}$ = 24.52 kN/m³, γ_{sat} = 17.27 kN/m³ and ι = ½ϕ

ϕ Degree	α Degree	Top Width, mm	Bottom Width, mm	Horizontal Angle θ Degree	Depth of Shear Key D_{sk}, mm			Volume of Wall, m³
					μ = 0.50	μ = 0.70	μ = 0.80	
35	−14.03	300	1420	75.97	700	500	300	4.300
40	0.00	300	2480	90.00	200	0	0	6.950
40	−5.71	300	1910	84.29	300	0	0	5.525
40	−11.31	300	1370	78.69	500	200	0	4.175
40	−14.03	300	1115	75.97	600	400	200	3.538

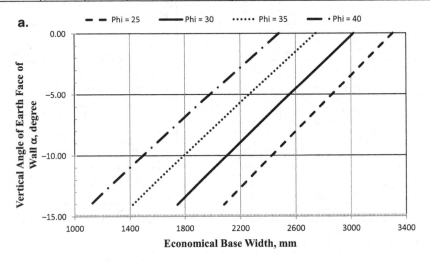

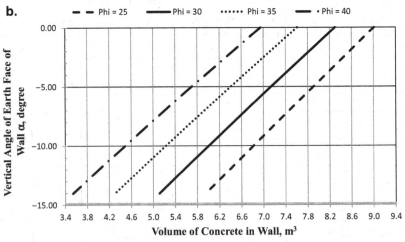

FIGURE 6.53 Design Charts for a Concrete Wall H = 5.0 m, $\gamma_{concrete}$ = 24.52 kN/m³, γ_{sat} = 17.27 kN/m³ and ι = ½ϕ: (a) Economical Base Width for a Wall, (b) Volume of a Concrete Wall

TABLE 6.53

$H = 5.0$ m, $\gamma_{concrete} = 24.52$ kN/m³, $\gamma_{sat} = 15.70$ kN/m³ and $\iota = \phi$

A. Results for Stability Analysis

				Pressure at		$\mu = 0.50$		$\mu = 0.70$		$\mu = 0.80$	
ϕ Degree	α Degree	K_{ah}	F_{OVT}	Heel, kN/ m²	Toe, kN/ m²	F_{SLD}	Net Force, kN	F_{SLD}	Net Force, kN	F_{SLD}	Net Force, kN
25	0.00	0.821	2.04	16.66	99.98	0.68	133.90	0.95	89.70	1.09	67.60
25	−5.71	0.746	1.92	18.07	99.82	0.65	125.99	0.92	86.93	1.05	67.40
25	−11.31	0.675	1.80	19.82	99.68	0.63	118.57	0.88	84.46	1.00	67.41
25	−14.03	0.641	1.74	20.50	99.99	0.61	115.14	0.86	83.45	0.98	67.61
30	0.00	0.750	2.04	17.19	99.84	0.71	118.45	0.99	75.99	1.13	54.76
30	−5.71	0.666	1.92	18.68	99.84	0.69	108.92	0.96	72.00	1.10	53.54
30	−11.31	0.587	1.79	20.90	99.61	0.66	100.18	0.93	68.54	1.06	52.72
30	−14.03	0.549	1.73	22.05	99.71	0.65	96.17	0.90	67.11	1.03	52.57
35	0.00	0.671	2.05	17.57	99.97	0.75	101.73	1.05	61.36	1.20	41.18
35	−5.71	0.580	1.92	19.36	99.99	0.73	91.23	1.02	56.73	1.17	39.48
35	−11.31	0.496	1.78	22.04	99.83	0.70	81.95	0.98	53.08	1.12	38.64
35	−14.03	0.456	1.70	23.68	99.85	0.69	77.78	0.96	51.63	1.10	38.55
40	0.00	0.587	2.06	18.22	99.96	0.80	84.18	1.11	46.14	1.27	27.12
40	−5.71	0.492	1.92	20.70	99.70	0.78	73.39	1.09	41.52	1.25	25.58
40	−11.31	0.406	1.76	24.04	99.66	0.75	64.41	1.05	38.50	1.20	25.54
40	−14.03	0.366	1.68	26.37	99.62	0.73	60.51	1.02	37.46	1.17	25.93

B. Physical Dimensions for a Technoeconomic Wall Section

					Depth of Shear Key D_{sk}, mm			
ϕ Degree	α Degree	Top Width, mm	Bottom Width, mm	Horizontal Angle θ Degree	$\mu = 0.50$	$\mu = 0.70$	$\mu = 0.80$	Volume of Wall, m³
25	0.00	300	4215	90.00	2000	1600	1300	11.288
25	−5.71	300	3685	84.29	2000	1600	1400	9.963
25	−11.31	300	3175	78.69	2000	1600	1400	8.688
25	−14.03	300	2925	75.97	2000	1700	1500	8.063
30	0.00	300	4035	90.00	1600	1200	1000	10.838
30	−5.71	300	3465	84.29	1500	1100	900	9.413
30	−11.31	300	2920	78.69	1500	1200	1000	8.050
30	−14.03	300	2655	75.97	1500	1200	1000	7.388
35	0.00	300	3820	90.00	1100	700	500	10.300
35	−5.71	300	3215	84.29	1100	800	600	8.788
35	−11.31	300	2635	78.69	1100	900	700	7.338

(Continued)

TABLE 6.53 (Continued)

$H = 5.0$ m, $\gamma_{concrete} = 24.52$ kN/m³, $\gamma_{sat} = 15.70$ kN/m³ and $\iota = \phi$

| ϕ Degree | α Degree | Top Width, mm | Bottom Width, mm | Horizontal Angle θ Degree | Depth of Shear Key D_{sk}, mm | | | Volume of Wall, m³ |
					$\mu = 0.50$	$\mu = 0.70$	$\mu = 0.80$	
35	−14.03	300	2355	75.97	1100	800	700	6.638
40	0.00	300	3580	90.00	700	400	200	9.700
40	−5.71	300	2945	84.29	700	500	300	8.113
40	−11.31	300	2330	78.69	800	500	400	6.575
40	−14.03	300	2035	75.97	700	500	400	5.838

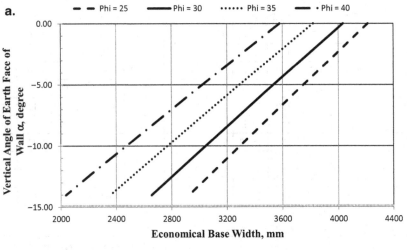

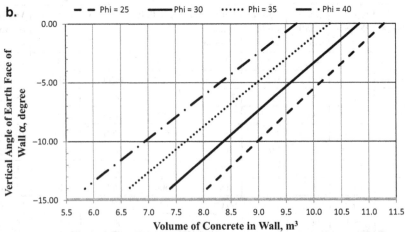

FIGURE 6.54 Design Charts for a Concrete Wall $H = 5.0$ m, $\gamma_{concrete} = 24.52$ kN/m³, $\gamma_{sat} = 15.70$ kN/m³ and $\iota = \phi$: (a) Economical Base Width for a Wall, (b) Volume of a Concrete Wall

TABLE 6.54

$H = 5.0$ m, $\gamma_{concrete} = 24.52$ kN/m³, $\gamma_{sat} = 17.27$ kN/m³ and $\iota = \phi$

A. Results for Stability Analysis

ϕ Degree	α Degree	K_{ah}	F_{OVT}	Pressure at Heel, kN/m²	Toe, kN/m²	$\mu = 0.50$ F_{SLD}	Net Force, kN	$\mu = 0.70$ F_{SLD}	Net Force, kN	$\mu = 0.80$ F_{SLD}	Net Force, kN
25	0.00	0.821	2.03	16.27	99.96	0.64	153.08	0.90	106.88	1.03	83.79
25	−5.71	0.746	1.92	17.48	99.91	0.63	143.20	0.88	102.25	1.00	81.77
25	−11.31	0.675	1.80	19.00	99.86	0.60	133.93	0.84	98.03	0.96	80.07
25	−14.03	0.641	1.75	19.97	99.77	0.59	129.49	0.82	96.01	0.94	79.27
30	0.00	0.750	2.04	16.75	99.86	0.67	135.76	0.94	91.41	1.08	69.24
30	−5.71	0.666	1.92	18.11	99.87	0.66	124.01	0.92	85.29	1.05	65.93
30	−11.31	0.587	1.80	19.84	99.98	0.64	113.30	0.89	80.01	1.02	63.37
30	−14.03	0.549	1.74	21.33	99.60	0.62	108.19	0.87	77.47	1.00	62.11
35	0.00	0.671	2.04	17.15	99.94	0.71	116.96	1.00	74.80	1.14	53.72
35	−5.71	0.580	1.92	18.91	99.86	0.70	104.04	0.98	67.85	1.12	49.75
35	−11.31	0.496	1.79	21.20	99.87	0.68	92.65	0.95	62.23	1.08	47.01
35	−14.03	0.456	1.72	22.86	99.71	0.66	87.44	0.93	59.78	1.06	45.95
40	0.00	0.587	2.06	17.90	99.81	0.76	97.18	1.06	57.44	1.21	37.57
40	−5.71	0.492	1.92	20.11	99.65	0.75	83.98	1.05	50.55	1.20	33.84
40	−11.31	0.406	1.77	23.10	99.68	0.73	72.83	1.02	45.51	1.16	31.85
40	−14.03	0.366	1.69	24.93	99.94	0.71	68.00	0.99	43.64	1.13	31.46

B. Physical Dimensions for a Technoeconomic Wall Section

ϕ Degree	α Degree	Top Width, mm	Bottom Width, mm	Horizontal Angle θ Degree	Depth of Shear Key D_{sk}, mm $\mu = 0.50$	$\mu = 0.70$	$\mu = 0.80$	Volume of Wall, m³
25	0.00	300	4420	90.00	2000	1700	1400	11.800
25	−5.71	300	3880	84.29	2100	1700	1500	10.450
25	−11.31	300	3360	78.69	2000	1700	1500	9.150
25	−14.03	300	3110	75.97	2000	1600	1500	8.525
30	0.00	300	4230	90.00	1500	1200	1000	11.325
30	−5.71	300	3650	84.29	1500	1200	1000	9.875
30	−11.31	300	3090	78.69	1500	1300	1100	8.475
30	−14.03	300	2825	75.97	1500	1200	1100	7.813
35	0.00	300	4005	90.00	1100	800	600	10.763
35	−5.71	300	3390	84.29	1100	800	700	9.225
35	−11.31	300	2795	78.69	1100	900	700	7.738

(Continued)

TABLE 6.54 (Continued)

$H = 5.0$ m, $\gamma_{concrete} = 24.52$ kN/m³, $\gamma_{sat} = 17.27$ kN/m³ and $\iota = \phi$

ϕ Degree	α Degree	Top Width, mm	Bottom Width, mm	Horizontal Angle θ Degree	Depth of Shear Key D_{sk}, mm			Volume of Wall, m³
					$\mu = 0.50$	$\mu = 0.70$	$\mu = 0.80$	
35	−14.03	300	2510	75.97	1100	900	700	7.025
40	0.00	300	3755	90.00	600	400	200	10.138
40	−5.71	300	3105	84.29	700	400	300	8.513
40	−11.31	300	2475	78.69	700	500	300	6.938
40	−14.03	300	2170	75.97	700	500	400	6.175

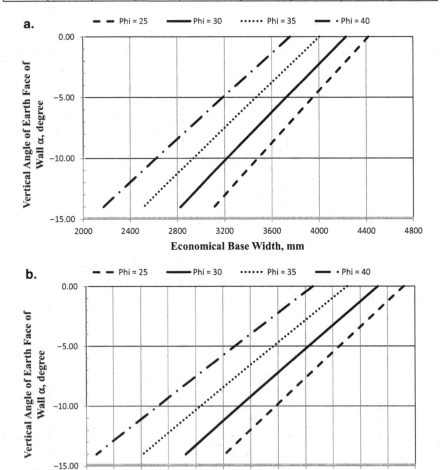

FIGURE 6.55 Design Charts for a Concrete Wall $H = 5.0$ m, $\gamma_{concrete} = 24.52$ kN/m³, $\gamma_{sat} = 17.27$ kN/m³ and $\iota = \phi$: (a) Economical Base Width for a Wall, (b) Volume of a Concrete Wall

7 Detailing of Breast Walls

7.1 RANDOM RUBBLE MASONRY WALLS

Walls should be made in random rubble (R. R.) masonry consisting of hammer-dressed hard stones brought to course approximately every 600 mm. Masonry courses must be normal to face batter, and the back of the wall can be left rough. Masonry work should proceed at a uniform level.

The least dimension of stone should be 200 mm × 150 mm × 100 mm. Approximately half the stones should tail into the wall by twice their height. Stones must break the joint by half the height of the course.

7.1.1 THROUGH-STONES/BONDING ELEMENT

Through-stones should be provided at every 1.0-m interval both horizontally and vertically throughout the length and height of the wall. In the absence of through-stones, reinforced concrete blocks of 100 mm × 100 mm can be used. The through-stones or bonding element should be staggered both horizontally and vertically.

7.1.2 CEMENT MORTAR

A cement mortar ratio of 1:5 (cement:sand) or richer should be provided.

7.1.3 COPING

The coping should consist of large stones, laid and pointed in cement mortar or plain cement concrete 50 mm to 75 mm thick. The top of the coping should be weather-sloped toward the valley side. Coping preferably should be with stones on edge so that these are not readily dislodged.

7.2 DRAINAGE

Provision should be made to prevent water accumulation behind the retaining wall. Accumulated water causes increased pressure, seepage and, in areas subject to frost action, an expansive force of considerable magnitude near the top of the wall. Drainage is an important aspect of retaining wall construction, and the build-up of water behind the wall is the principal cause of breast wall failure. Good piped drainage that disposes water away from behind the wall to limit this build-up of pressure is essential, except for walls that retain free-draining sand.

DOI: 10.1201/9781003162995-7

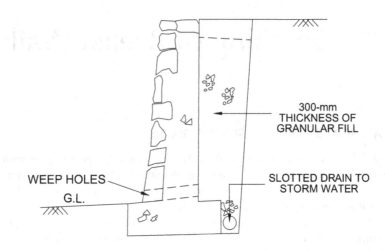

FIGURE 7.1   Typical Vertical Drainage Arrangement behind a Masonry/Concrete Wall

The drainage system should consist of granular free-draining gravel placed imme-
diately behind the wall for a minimum width of 300 mm and for the full height of
the wall. The typical drainage arrangement behind a wall is depicted in Figure 7.1.

7.2.1 SPECIFYING FILTERS BY PERMEABILITY

Recall that the specification for granular material for the vertical drainage layer to be
used in conjunction with granular fill for a retaining wall is based on the particle-size
distribution. The aim is to provide a filter with greater permeability than the fill. It
has been suggested recently that instead of specifying the grading of the filter, its
permeability can be specified instead. This, of course, implies measuring the permeability of the fill and knowing how much more permeable the filter is desired than
the fill. If the specification by permeability was to replace the specification by grading completely, for filters, it is then necessary to specify the two values of permeability to cover the filtration criterion as well as the permeability criterion. Alternatively,
a mixed system can be used in which the permeability criterion of the filter is specified
by permeability, and the filtration criterion is specified by grading.

7.3 WEEP HOLES

Weep holes provide an opening that allows the drainage of any moisture that may come
from the back of the wall through penetration, capillary action or leakage. From an engineering point of view, weep holes relieve hydrostatic pressure on walls. This reduces the
structural design demand of the water and earth pressure by reducing thickness. With
this, weep holes also reduce the buoyancy and uplift on the structure, making it possible
to construct a lighter structure without any uplift-related stability issues.

Weep holes should be provided in cement stone masonry/concrete walls at a spacing
of about 1.0 m to 1.5 m centre to centre vertically and horizontally. Square-shaped

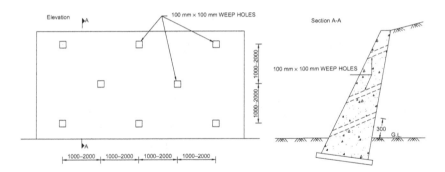

FIGURE 7.2 Elevation and Section A-A of a Wall with Weep Holes

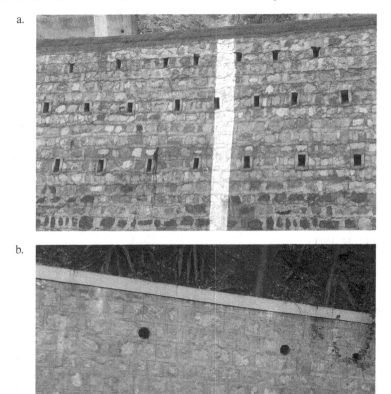

FIGURE 7.3 R. R. Masonry Breast Wall with Weep Holes: (a) Rectangular Weep Holes (b) PVC Pipes as Weep Holes

holes 75 mm to 100 mm or 100 mm to 150 mm PVC (flexible) pipes should be embedded at 10° down from the horizontal towards the valley side to effectively drain water from the ground. The lowest weep hole is generally kept 300 mm above ground level.

7.3.1 DRAINAGE PIPE

A perforated or slotted drainage pipe, protected with a filter cloth to prevent it from being clogged by silt, must be installed behind the base of the wall to collect and remove water, which is then drained to a storm-water system.

7.3.2 CATCH WATER DRAIN

A catch water drain should be provided at the toe of the breast wall to collect water from weep holes and surface run-off from the slope.

7.4 DIP OF THE FOUNDATION AT THE BASE

The base of the breast wall should be placed on firm, undisturbed soil. In areas where freezing is expected, the base of the wall should be placed below the frost line. If the soil under the wall consists of soft or silty clay, it is usually advisable to place 100 mm to 150 mm of well-compacted sand or gravel under the base of the wall. It has also been observed that the dip of the foundation base towards the hillside increases the factor of safety against sliding significantly, and therefore, it is recommended to provide a dip of 6:1 (horizontal:vertical) in the base of the wall (MWHS, 2010).

7.5 WALL JOINTS

7.5.1 CONSTRUCTION JOINTS IN CONCRETE WALLS

Vertical or horizontal joints are placed between two successive pours of concrete. To increase the shear resistance at the joints, keys may be used as shown in Figure 7.4.

7.5.2 CONTRACTION JOINT

A separation gap of 50 mm to 100 mm should be provided as vertical joints placed in the wall (from the top of the base to the top of the wall), which will allow the concrete to shrink without noticeable harm (see Figure 7.5). Also, gaps should be provided at bends and junctions.

7.5.3 EXPANSION JOINT

These vertical joints are provided in large retaining walls to allow for the expansion of the concrete due to temperature changes, and usually extend from the top to the bottom of the wall (see Figure 7.6). These joints may be filled with flexible joint fillers.

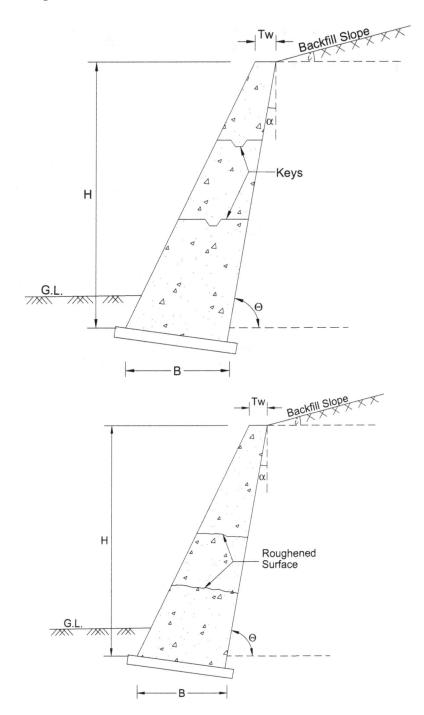

FIGURE 7.4   Construction Joints in Concrete Wall

FIGURE 7.5   Contraction Joints in a Wall

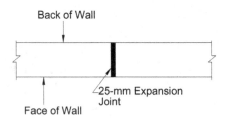

FIGURE 7.6   Expansion Joints in a Concrete Wall

7.6   TYPICAL DRAWINGS

Typical detailed drawings for R. R. masonry and concrete breast walls are depicted in Figures 7.7 and 7.8.

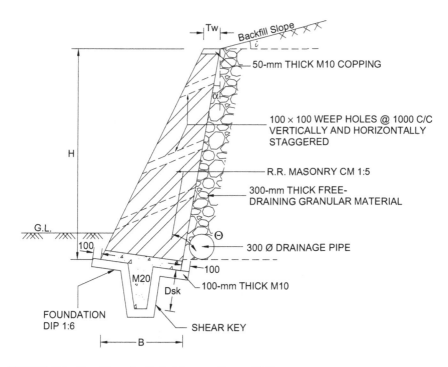

FIGURE 7.7 Detailing of a R. R. Masonry Breast Wall

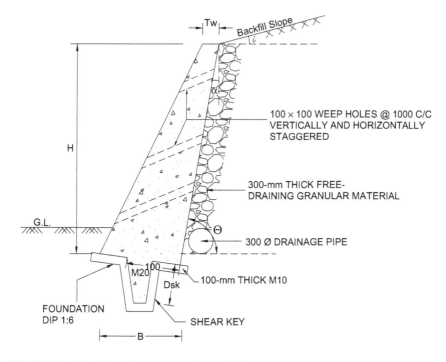

FIGURE 7.8 Detailing of a Concrete Breast Wall

8 Application of Design Tables and Charts

8.1 APPLICATION OF THE DESIGN CHARTS

8.1.1 Masonry Breast Walls

Combinations of following design parameters were made for carrying out the design of masonry breast walls, and in all, 54 design charts were prepared and presented in Chapter 5 in Figures 5.2 to 5.55.

Height of Wall	Parameters Considered
1.0 m	$\phi = 25°, 30°, 35°, 40°$
1.5 m	$\delta = \frac{2}{3}\phi$ or 22.50°, whichever is less
2.0 m	$\iota = 0°, \frac{1}{2}\phi, \phi$
2.5 m	$\alpha = 0°, -5.71°, -11.31°, -14.03°$
3.0 m	$\gamma_{masonry} = 22.56$ kN/m^3
3.5 m	$\gamma_{sat} = 15.70$ kN/m^3 and 17.27 kN/m^3
4.0 m	
4.5 m	
5.0 m	

The charts give the economical base width and the volume of masonry wall per m running length obtained for a combination of various parameters.

8.1.2 Concrete Breast Walls

Combinations of following design parameters were made for carrying out the design of concrete breast walls, and in all, 54 design charts were prepared and presented in Chapter 6 in Figures 6.2 to 6.55:

Height of Wall	Parameters Considered
1.0 m	$\phi = 25°, 30°, 35°, 40°$
1.5 m	$\delta = \frac{1}{3}\phi$
2.0 m	$\iota = 0°, \frac{1}{2}\phi, \phi$
2.5 m	$\alpha = 0°, -5.71°, -11.31°, -14.03°$
3.0 m	$\gamma_{concrete} = 24.52$ kN/m^3
3.5 m	$\gamma_{sat} = 15.70$ kN/m^3 and 17.27 kN/m^3
4.0 m	
4.5 m	
5.0 m	

DOI: 10.1201/9781003162995-8

The charts give the economical base width and the volume of concrete wall per m running length obtained for a combination of various parameters.

8.1.3 Utility of Charts

When the designer is not sure about fixing the vertical angle of face of wall α for economic design, the charts presented in Figures 5.2 to 5.55 and Figures 6.2 to 6.55 can be referred depending on the masonry/concrete wall; the angle of the surcharge, ι; and the saturated density of soil, γ_{sat}. The designer can directly read the data from the charts and can compare the base width requirement and volume of masonry/concrete. For detailed results and exact dimensions refer to the respective design tables.

8.2 APPLICATION OF THE DESIGN TABLES

8.2.1 Masonry Breast Walls

Combinations of following design parameters were made to carry out the design of masonry breast walls, and in all, 54 design tables were prepared and presented in Chapter 5 in Tables 5.1 to 5.54:

Height of Wall	Parameters Considered
1.0 m	$\phi = 25°, 30°, 35°, 40°$
1.5 m	$\delta = \frac{2}{3}\phi$ or 22.50°, whichever is less
2.0 m	$\iota = 0°, \frac{1}{2}\phi, \phi$
2.5 m	$\alpha = 0°, -5.71°, -11.31°, -14.03°$
3.0 m	$\gamma_{masonry} = 22.56$ kN/m³
3.5 m	$\gamma_{sat} = 15.70$ kN/m³ and 17.27 kN/m³
4.0 m	
4.5 m	
5.0 m	

The tables give detailed results of stability analysis like the factor of safety against overturning, base pressures at the toe and heel, the factor of safety against sliding for different values of the coefficient of friction between the masonry and the soil/rock mass and the net horizontal force required for design of the shear key.

The tables also give the top and bottom widths of wall, the volume of masonry in wall and the depth of the shear key required for different values of the coefficient of friction between the masonry and the soil/rock mass.

8.2.2 Concrete Breast Walls

Combinations of the following design parameters were made to carry out the design of concrete breast walls, and in all, 54 design tables were prepared and presented in Chapter 6 in Tables 6.1 to 6.54:

Height of Wall	Parameters Considered
1.0 m	$\phi = 25°, 30°, 35°, 40°$
1.5 m	$\delta = \frac{1}{3}\phi$
2.0 m	$\iota = 0°, \frac{1}{2}\phi, \phi$
2.5 m	$\alpha = 0°, -5.71°, -11.31°, -14.03°$
3.0 m	$\gamma_{concrete} = 24.52$ kN/m³
3.5 m	$\gamma_{sat} = 15.70$ kN/m³ and 17.27 kN/m³
4.0 m	
4.5 m	
5.0 m	

The tables give detailed results for the stability analysis, such as the factor of safety against overturning, base pressures at the toe and heel, the factor of safety against sliding for different values of the coefficient of friction between the concrete and the soil/rock mass and the net horizontal force required for designing the shear key.

The tables also give the top and bottom widths of the wall, the volume of concrete in the wall and the depth of the shear key required for different values of the coefficient of friction between the concrete and the soil/rock mass.

8.2.3 Application of the Tables

The following steps should be followed to carry out the design of masonry/concrete breast walls:

Step 1   For a given value of the angle of repose of the soil, ϕ; the angle of friction, δ; and the angle of the surcharge, ι, and provided vertical angle to the earth face of the wall, α, read the value of the active earth pressure coefficient, K_a, from Table 2.1.

Step 2   For a given value of the angle of repose of the soil, ϕ, and the angle of friction, δ, read the value of the passive earth pressure coefficient, K_p, from Table 2.2.

Step 3   For a given value of the height of the wall, H; the angle of repose of the soil, ϕ; the angle of the surcharge, ι; and the unit weight of saturated soil, γ_{sat}, and provided vertical angle to the earth face of the wall, α, read the value of the horizontal component of the active earth pressure coefficient, K_{ah}; the factor of safety against overturning; the maximum base pressure at the toe, P_{Toe}, when the pressure at the heel is nearly zero; base pressures at the heel, P_{Heel}; the factor of safety against sliding for different values of the coefficient of friction between the masonry/concrete and the soil/rock mass and net horizontal force required for design of shear key from results of stability analysis of the respective table.

Step 4   For a given value of the height of wall, H; the angle of repose of the soil, ϕ; the angle of the surcharge, ι; and the unit weight of saturated soil, γ_{sat}, and provided vertical angle to the earth face of the wall, α, read the top and bottom widths of the wall; the volume of masonry/concrete in wall, V_{Wall}; and the depth of the shear key, D_{sk}, required for the different values of the coefficient of friction between the masonry/concrete and the soil/rock mass from physical dimensions for technoeconomic wall section of the respective table already selected in Step 3.

Step 5   Design the shear key as explained in Section 8.3.

Step 6   On the basis of the design data obtained from Step 4, the drawing of the masonry/concrete wall can be prepared as per details shown in Figure 7.7 (masonry wall) or Figure 7.8 (concrete wall).

8.3 DESIGN OF THE SHEAR KEY

The passive resistance to the shear key has already been explained in Section 3.2.3. The key is also subjected to the pressure of the soil from behind, but it is safer to assume the bending moment due to the passive resistance only, since the back pressure of the soil reduces the bending moment exerted on the shear key. It has also been assumed in the design that there is no soil cover over the shear key, that is height h_1 is zero, to calculate the passive resistance. The width of the shear key at the bottom has been assumed as 150 mm, and the location of the shear key has been assumed at the centre of the base. However, the shear key is best positioned near the heel to get the maximum advantage. The depth of the shear key has already been calculated and given in the design tables for different combinations. Thus, the moment at the base of the wall should be

$$M_{skey} = 1.5 K_{ph} \gamma_{sub} \frac{D_{sk}^3}{3} \text{ or}$$
$$= 0.5 K_{ph} \gamma_{sub} D_{sk}^3$$

(8.1)

Limiting Moment of Resistance $M_R = \text{Const} * b * d^2$,

where Const depends on characteristic strength of concrete, f_{ck}; and the yield stress of steel, f_s, is equal to 2.755 for M_{20} concrete and $F_e 415$ steel; the width/thickness of the shear key at the base will be equal to

$$W_{sk} = \sqrt{\frac{M_{skey}}{2.755}}.$$

(8.2)

Thus, provide the width, W_{sk}, at the top of the shear key, that is at the base of the wall.

The main reinforcement and distribution reinforcement in the shear key can be easily calculated by using formulae of the limit state design method and may be

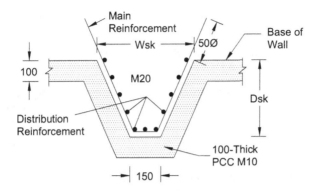

FIGURE 8.1   Reinforcement Details for the Shear Key

provided accordingly. Typical reinforcement details for the shear key are shown in Figure 8.1.

8.4 DESIGN EXAMPLES

To illustrate the use of the tables, the following examples have been chosen.

8.4.1 EXAMPLE 1

Design a masonry breast wall for the following parameters:

1. Height of the wall: H = 3.5 m
2. Angle of repose of the soil: $\phi = 30°$
3. Angle of the surcharge: $\iota = 15°$
4. Unit weight of masonry: $\gamma_{masonry} = 22.56$ kN/m³
5. Unit weight of saturated soil: $\gamma_{sat} = 17.27$ kN/m³
6. Safe bearing capacity of the soil = 100.00 kN/m²
7. Coefficient of the friction between the masonry and the soil/rock mass: $\mu = 0.80$

Let the earth face of the breast wall make an angle with the vertical, $\alpha = -5.71°$ or Slope 1.0(H):10.0(V).

Solution:

Step 1   For $\phi = 30°$, $\delta = 20°$, $\iota = 15°$ and $\alpha = -5.71°$:
           Read from Table 2.1 the value of the active earth pressure coefficient: $K_a = 0.319$.

Step 2   For $\phi = 30°$ and $\delta = 20°$:
           Read from Table 2.2 the value of the passive earth pressure coefficient: $K_p = 6.112$.

Step 3   For H = 3.5 m, $\gamma_{masonry}$ = 22.56 kN/m³, γ_{sat} = 17.27 kN/m³, ϕ = 30°, ι = 15°
          and α = −5.71°:
          Read the detailed results of stability analysis from Table 5.34 to get the
          value of following parameters:

		Pressure at		μ = 0.80	
K_{ah}	F_{OVT}	Heel, kN/m²	Toe, kN/m²	FF_{SLD}	Net Force, kN
0.309	1.67	0.15	83.65	1.30	7.39

Step 4   For H = 3.5 m, $\gamma_{masonry}$ = 22.56 kN/m³, γ_{sat} = 17.27 kN/m³, ϕ = 30°, ι = 15°
          and α = −5.71°:
          Read the top and bottom widths of wall, the volume of the masonry in the
          wall and the depth of the shear key from Table 5.34:

ϕ Degree	α Degree	Top Width, mm	Bottom Width, mm	Horizontal Angle θ Degree	Depth of Shear Key D_{sk} mm	Volume of Wall, m³
30	−5.71	300	1565	84.29	300	3.264

Step 5   Since the factor of safety against sliding is less than 1.5, the depth of the
          shear key has been indicated in the Table 5.34 as 400 mm.

          Design of Shear Key
          • K_{ph} = $K_p \cdot Cos(\delta)$ = 6.112. cos(20) = 5.743
          • γ_{sub} = 7.46 kN/m³
          • Using Equation 8.1, M_{skey} = 1.34 kN-m.
          • Using M_{20} concrete and F_e 415 steel, the width of key = 34 mm,
            using Equation 8.2.
          • Therefore, provide 150 mm width throughout the depth of 300 mm.
          • Provide the main reinforcement as 12ϕ @ 300 mm c/c and the
            distribution reinforcement as 12ϕ @ 450 mm c/c.

8.4.2 EXAMPLE 2

Design a concrete breast wall for the following parameters:

1. Height of the wall: H = 4.0 m
2. Angle of repose of the soil: ϕ = 35°
3. Angle of the surcharge: ι = 0°
4. Unit weight of the concrete: $\gamma_{concrete}$ = 24.52 kN/m³
5. Unit weight of saturated soil: γ_{sat} = 15.70 kN/m³
6. Safe bearing capacity of the soil = 100.00 kN/m²
7. Coefficient of friction between masonry and soil/rock mass: μ = 0.50

Let the earth face of the breast wall make an angle with the vertical face, α = −11.31°
or Slope 1.0(H):5.0(V).

Solution:

Step 1   For $\phi = 35°$, $\delta = 11.67°$, $\iota = 0°$ and $\alpha = -11.31°$:
Read from Table 2.1, the value of the active earth pressure coefficient:
$K_a = 0.182$.

Step 2   For $\phi = 35°$ and $\delta = 11.67°$:
Read from Table 2.2, the value of the passive earth pressure coefficient:
$K_p = 5.686$.

Step 3   For $H = 4.0$ m, $\gamma_{concrete} = 24.52$ kN/m³, $\gamma_{sat} = 15.70$ kN/m³, $\phi = 35°$, $\iota = 0°$ and $\alpha = -11.31°$:
Read detailed results of stability analysis from Table 6.37 to get the values for the following parameters:

		Pressure at		$\mu = 0.50$	
K_{ah}	F_{OVT}	Heel, kN/m²	Toe, kN/m²	F_{SLD}	Net Force, kN
0.182	1.55	17.04	99.01	0.90	17.01

Step 4   For $H = 4.0$ m, $\gamma_{concrete} = 24.52$ kN/m³, $\gamma_{sat} = 15.70$ kN/m³, $\phi = 35°$, $\iota = 0°$ and $\alpha = -11.31°$:
Read the top and bottom widths of wall, the volume of concrete in the wall and the depth of the shear key from Table 6.37:

ϕ Degree	α Degree	Top Width, mm	Bottom Width, mm	Horizontal Angle θ Degree	Depth of Shear Key D_{sk}, mm	Volume of Wall, m³
35	−11.31	300	990	78.69	600	2.580

Step 5   Since the factor of safety against sliding is less than 1.5, the depth of the shear key is indicated in the Table 6.37 as 600 mm.

Design of Shear Key
- $K_{ph} = K_p \cdot \cos(\delta) = 5.686 \cdot \cos(11.67) = 5.568$
- $\gamma_{sub} = 5.89$ kN/m³
- Using Equation 8.1, $M_{skey} = 9.44$ kN-m.
- Using M_{20} concrete and F_e 415 steel, the width of the key = 59 mm, using Equation 8.2.
- Therefore, provide 150 mm width throughout the depth of 600 mm.
- Provide the main reinforcement as 12ϕ @ 300 mm c/c and the distribution reinforcement as 12ϕ @ 450 mm c/c.

8.5 APPLICATION OF THE TABLES FOR DIFFERENT DESIGN PARAMETERS

8.5.1 TOP WIDTH OF BREAST WALLS

The top width of the wall has been assumed to be 200 mm, but if, due to other requirements, more top width is needed, then locally extend the top of the wall

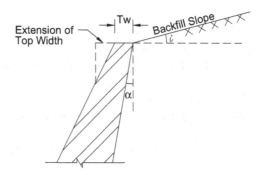

FIGURE 8.2 Extension of the Top Width of the Wall

without disturbing the vertical angle, α, which the earth face of the wall makes with the vertical, as shown in Figure 8.2.

8.5.2 ANGLE OF REPOSE OF THE SOIL, ϕ

Standard design charts and tables have been developed for the angle of repose of the soil, ϕ, as 25°, 30°, 35° and 40°. But in the field, actual values for the angle of repose of the soil, ϕ, may differ from what has been assumed in developing the charts and tables.

In this context, it can be said that if the actual value of the angle of repose of the soil, ϕ, is just greater than 25°, 30°, 35° or 40°, the value of the angle of repose of the soil, ϕ, may be adopted as 25°, 30°, 35° or 40° for the design of masonry/concrete breast walls.

However, if the actual value of the angle of repose of the soils, ϕ, is between the two values adopted for developing the design charts/tables, the linear interpolation technique can be used for computing the base width and other parameters.

8.5.3 UNIT WEIGHT OF MATERIALS

The unit weight of masonry, $\gamma_{masonry}$, has been adopted as 22.56 kN/m³ (2.30 t/m³) for developing the standard design charts and tables. But in the field, actual values may differ slightly from what has been assumed in developing the charts and tables.

In this context, it can be said that if the actual value of the unit weight of masonry, $\gamma_{masonry}$, is more than 22.56 kN/m³ (2.30 t/m³), the base width indicated in the design tables/charts may be adopted (refer to Section 4.3.2).

However, if the actual value of the unit weight of masonry, $\gamma_{masonry}$, is less than 22.56 kN/m³ (2.30 t/m³), the base width indicated in the design tables/charts may be increased by 5% for every 0.1 t/m³ (1.0 kN/m³; refer to Section 4.3.2).

Similar rules should be followed in the case of concrete breast walls.

8.5.4 UNIT WEIGHT OF SOIL

Two values of unit weight of saturated soil, γ_{sat} = 15.70 kN/m³ (1.60 t/m³) and 17.27 kN/m³ (1.76 t/m³) have been adopted for developing the standard design charts and tables. But in the field, actual values may differ slightly from what has been assumed in developing the charts and tables.

In this context, it can be said that if the actual value of unit weight of saturated soil, γ_{sat}, is less than 15.70 kN/m³ (1.60 t/m³), the base width indicated in the design tables/charts may be adopted (refer to Section 4.3.1).

However, if the actual value of unit weight of saturated soil, γ_{sat}, is more than 17.27 kN/m³ (1.76 t/m³), the base width indicated in the design tables/charts may be increased on the basis of results described in Section 4.3.1.

8.5.5 ANGLE OF THE SURCHARGE, ι

Three values for the angle of the surcharge, $\iota = 0°$, ½ϕ, ϕ have been considered for developing standard design charts and tables. But in the field, actual values may differ from what has been assumed in developing the charts and tables.

From Section 4.3.3, it is clearly seen that the base width increases linearly with the increase in the angle of the surcharge from 0° to the angle of repose, ϕ. In this context, it can be said that if the actual value of the angle of the surcharge, ι, is slightly less than ½ϕ or ϕ, the base width indicated in the design tables/charts for ½ϕ or ϕ may be adopted.

However, if the actual value of the angle of the surcharge, ι, is between these values, linear interpolation technique should be used for estimating the base width and volume of masonry/concrete in a wall, and a higher depth of the shear key should be adopted for the defined value of the coefficient of friction between the masonry/concrete and the soil/rock mass.

8.5.6 HEIGHT OF WALLS

The standard design charts and tables have been developed for different height of walls ranging from 1.0 m to 5.0 m at a 0.50 m interval. But in the field, the actual height of a breast wall may differ from the adopted values.

In this context, it can be said that if the actual height of the breast wall is just less than the height assumed for developing the table/charts, the higher value of height may be adopted for the design of the breast masonry/concrete wall.

However, if the actual value of the height of the breast wall is between the two values adopted for developing the design charts/tables, the linear interpolation technique can be used for computing base width and other parameters.

8.5.7 EXAMPLE 3

Design a masonry breast wall for the following parameters:

1. Height of the wall, H = 3.0 m,
2. Angle of repose of the soil, $\phi = 30°$
3. Angle of the surcharge, $\iota = 20°$
4. Unit weight of the masonry, $\gamma_{masonry} = 22.56$ kN/m³

5. Unit weight of the saturated soil, γ_{sat} = 17.27 kN/m³
6. Coefficient of friction between the masonry and the soil/rock mass, μ = 0.80

Let the earth face of the breast wall make an angle with the vertical face, α = −5.71°
or Slope 1.0(H):10.0(V).

Solution:

Step 1   Since the angle of the surcharge does not match with the tables/charts
              presented in Chapter 5, the values for the angle of the surcharge equal
              to ½ϕ and ϕ are obtained from the Table 5.28 and Table 5.30, and then
              the interpolation technique has to be applied to get the design values.

Step 2   For H = 3.0 m, $\gamma_{masonry}$ = 22.56 kN/m³, γ_{sat} = 17.27 kN/m³, ϕ = 30°,
              ι = 15° and α = −5.71°:
              Read the top and bottom widths of the wall, the volume of masonry in
              the wall and the depth of the shear key from Table 5.28:

ϕ Degree	α Degree	Top Width, mm	Bottom Width, mm	Horizontal Angle θ Degree	Depth of Shear Key D_{sk}, mm	Volume of Wall, m³
30	−5.71	300	1320	84.29	300	2.430

Step 3   For H = 3.0 m, $\gamma_{masonry}$ = 22.56 kN/m³, γ_{sat} = 17.27 kN/m³, ϕ = 30°,
              ι = 30° and α = −5.71°:
              Read the top and bottom widths of wall, the volume of masonry in wall
              and the depth of the shear key from Table 5.30:

ϕ Degree	α Degree	Top Width, mm	Bottom Width, mm	Horizontal Angle θ degree	Depth of Shear Key D_{sk}, mm	Volume of Wall, m³
30	−5.71	300	2070	84.29	600	3.555

Step 4   Interpolated Values
              Base width = 1320 + (2070 − 1320) *5/15
              Therefore, the base width for ι = 20° = 1570 mm
              Volume of wall = 2.430 + (3.555 − 2.430) *5/15
              Therefore, the volume of the wall = 2.805 m³

Step 5   Since the factor of safety against sliding is less than 1.5, provide the
              maximum depth of the shear key as 600 mm.

              The shear key can be designed by reading the analysis data from
              Table 5.30.

Bibliography

Chalisgaonkar, Rajendra (1987), Computer Aided Design of Retaining Walls, *Indian Concrete Journal*, Bombay, Vol. 61, No. 12, December.

Chalisgaonkar, Rajendra (1988), Inclined Retaining Walls, *Indian Concrete Journal*, Bombay, Vol. 62, No. 8, August.

Chalisgaonkar, Rajendra (1993), Computer Based Parametric Study for Cantilever Type Wall, *Bridge and Structural Engineer Journal*, Indian National Group of International Association for Bridge and Structural Engineering (IABSE), New Delhi, Vol. XXXIII, No. 2, June.

Chalisgaonkar, Rajendra (2018), Influence of Design Parameters on Earth Pressures behind Retaining Walls, *Journal of Indian Highways*, Indian Road Congress, New Delhi, Vol. 46, No. 11, November.

Chalisgaonkar, Rajendra (2019), Charts for Techno Economic Design of Masonry Breast Walls, *Water and Energy International*, Central Board of Irrigation and Power, New Delhi, Vol. 62/RNI, No. 3, ISSN: 0974-4711, June.

Chalisgaonkar, Rajendra (2020), Revisiting Design of Gravity Retaining Walls, *Journal of Indian Highways*, Indian Road Congress, New Delhi, Vol. 48, No.4, April.

Coulomb, C. A. (1776), Essai sur une application des regles des maximis et minimis a quelques problems de statique relatifis a Parchitecture, *Memoires de Mathematique de l'Academic Royale De Science*, Vol. 7, Paris.

Coyle, H. M., et al. (1972), *Field Measurements of Lateral Earth Pressures on a Cantilever Retaining Wall*, TTI Research Report 169–2, College Station, TX.

Donkada, Shravya, and Devdas Menon (2012), Optimal Design of Reinforced Concrete Retaining Walls, *Indian Concrete Journal*, Bombay, Vol. 86, No. 4, April.

Huntington, Whitney Clark (1957), *Earth Pressures and Retaining Walls*, John Wiley & Sons, Inc., London.

IRC:6 (2017), Standard Specifications and Code of Practice for Road Bridges Section: II Loads and Stresses IRC:6, Indian Road Congress, New Delhi.

IRC:78 (2014), Standard Specifications and Code of Practice for Road Bridges Section: VII Foundations and Substructure IRC:78, Indian Road Congress, New Delhi.

IS-875 (2008), Indian Standard Code of Practice for Design Loads (Other than Earthquake) for Buildings and Structures Part—Dead Loads–Unit Weights of Building Materials and Stored Materials, IS-875: Part 1, Bureau of Indian Standards, New Delhi.

IS:1893 (1984, Reaffirmed 1998), Indian Standard Code of Practice: Criteria for Earthquake Resistant Design of Structures, IS:1893, Bureau of Indian Standards, New Delhi.

IS:1893 (2003), Indian Railway Standard Code of Practice for the Design of Sub Structure and Foundation of Bridges, Research Designs and Standards Organisation, Lucknow.

IS:1893 (2014), Indian Standard Code of Practice: Criteria for Earthquake Resistant Design of Structures Part 3—Bridges and Retaining Walls, IS:1893: Part 3, Bureau of Indian Standards, New Delhi.

IS:14458 (1997), Indian Standard Code of Practice: Retaining Wall for Hilly Area—Guidelines IS:14458: Part 2, Bureau of Indian Standards, New Delhi.

IS:14458 (1998), Indian Standard Code of Practice: Retaining Wall for Hill Area—Guidelines, Part 1 Selection of Type of Wall, IS:14458 Part 1, Bureau of Indian Standards, New Delhi.

James, R. G., and P. L. Bransby (1970), Experimental and Theoretical Investigations of a Passive Earth Pressure Problem, *Geotechnique*, Vol. 20, No. 1, March.

Jumikis, A. R. (1964), *Mechanics of Soils*, D. Van Nostrand Company Inc, Princeton, NJ.

Mackey, R. D., and D. P. Kirk (1967), At Rest, Active and Passive Earth Pressures, *Proc. Southeastern Asian Regional Conf. on Soil Engineering*, Bangkok.

MWHS (2010), Design and Construction of Stone Masonry Retaining Walls—A Quick Guide, The Department of Engineering Services, Ministry of Works & Human Settlement (MWHS), Thimphu, Bhutan.

Padhye, R. D., and P. B. Ullagaddi (2011), Analysis of Retaining Wall with Pressure Relief Shelf by Coulomb's Method, *Proceedings of Indian Geotechnical Conference*, December 15–17, 2011, Kochi (Paper No. K—106).

Peck, R. B., and H. O. Ireland (1961), Full-Scale Lateral Load Test of a Retaining Wall Foundation, *5th ICSMFE*, Vol. 2.

Pillai, S. Unnikrishna, and Devdas Menon (2005), *Reinforced Concrete Design*, Tata McGraw Hill Publishing Company Limited, New Delhi.

Rankine, W. J. M. (1857), On the Stability of Loose Earth, *Philosophical Transactions of Royal Society*, London, Vol. 147.

Rehnman, S. E., and B. B. Broms (1972), Lateral Pressures on Basement Wall: Results from Full-Scale Tests, *Proc. 5th European Conf. SMFE*, Vol. 1.

Rowe, P. W., and K. Peaker (1965), Passive Earth Pressure Measurements, *Geotechnique*, Vol. 15, No. 1, March.

Terzaghi, Karl (1934), Large Retaining Wall Tests, *Engineering-News Record*, February.

Terzaghi, Karl (1941), General Wedge Theory of Earth Pressure, *Transactions of the American Society of Civil Engineers*, Vol. 106.

Index

Printed in the United States
by Baker & Taylor Publisher Services